MATHEMATICAL METHODS IN KINETIC THEORY

Second Edition

MATHEMATICAL METHODS IN KINETIC THEORY

Second Edition

CARLO CERCIGNANI

Politecnico di Milano
Milan, Italy

Springer Science+Business Media, LLC

Library of Congress Cataloging-in-Publication Data

Cercignani, Carlo.
Mathematical methods in kinetic theory / Carlo Cercignani. -- 2nd ed.
p. cm.
Includes bibliographical references.
ISBN 0-306-43460-1
1. Gases, Kinetic theory of. 2. Boundary value problems.
I. Title.
QC175.C46 1990
531'.113--dc20 90-35539
CIP

ISBN 978-1-4899-7293-4 ISBN 978-1-4899-7291-0 (eBook)
DOI 10.1007/978-1-4899-7291-0

Originally published by Plenum Press, New York 1990
Softcover reprint of the hardcover 1st edition 1990

PREFACE TO THE SECOND EDITION

The motivations that suggested the publication of this book in 1969 are still largely valid. In this revised edition I have updated the literature on the subjects that were covered in the original version; I have also added an appendix to Chapter VI, explaining the technique for solving singular integral equations and systems of such equations. In addition I have tried to give an introduction to the important developments concerning the purely mathematical theory (existence and uniqueness theorems). This part of kinetic theory has reached a mature stage in the last few years and is relevant for both the physical foundations of the subject (validity of the Boltzmann equation) and the application of kinetic theory to rarefied gas dynamics (numerical and simulation methods).

Milan, Italy

CARLO CERCIGNANI

PREFACE TO THE FIRST EDITION

When the density of a gas becomes sufficiently low that the mean free path is no longer negligibly small compared to a characteristic dimension of the flow geometry, the results obtained in continuum fluid dynamics require corrections which become more and more important as the degree of rarefaction increases. When this rarefaction becomes sufficiently great, continuum dynamics must be replaced by the kinetic theory of gases, and the Navier–Stokes equations by the Boltzmann equation. The latter is a rather complicated nonlinear integrodifferential equation whose solution for practical problems appears feasible only through suitable approximate mathematical techniques.

Flows of gases of arbitrary rarefaction have become problems of practical interest to aerodynamicists in the last twenty years and, consequently, solving the Boltzmann equation is no longer an academic problem. On the other hand, the mathematical nature of this equation is such that the classic methods of mathematical physics require substantial modifications in order to be applied with success to the kinetic theory of gases. Hence a special treatment of the mathematical methods used in kinetic theory seems in order.

The aim of this book is to present the mathematical theory of rarefied gases in systematic and modern fashion. It is therefore fundamentally different from the classic work by Chapman and Cowling on *The Mathematical Theory of Non-Uniform Gases*. The latter appeared as a definitive treatise on the subject when the theory of the Boltzmann equation was considered to be coextensive with the problem of computing transport coefficients in order to complete the conservation equations of continuum mechanics. It also differs somewhat from the book by M. N. Kogan on *Rarefied Gas Dynamics*,* which has been recently translated into English and is a complementary reading to the present work. On the one hand, Kogan's book deals more with results than with methods, and on the other, it considers a more general subject, including such topics as polyatomic gases and mixtures.

The present book is devoted to the boundary value problems arising in connection with the Boltzmann equation. The latter is usually taken in its

* Plenum Press, New York, 1969.

linearized version, the only one for which a systematic theory exists at present. The fundamental concepts are of course developed in full generality, i.e., without restriction to the linearized theory.

Considerable emphasis has been given to the use of model equations, which constitute, in our opinion, not only a very effective tool for computing special solutions, but also a rigorous method of solution, provided each model is considered as an element of an unending hierarchy approximating the Boltzmann equation as closely as we wish.

An effort has also been made to give proper place to a discussion of the boundary conditions to be matched with the Boltzmann equation. The importance of the boundary conditions lies in the fact that they describe the effect of the interaction of the gas molecules with the molecules of a solid body in contact with the gas. This interaction is responsible for the drag exerted upon a body moving in a gas and for the heat transfer between the gas and the neighboring solid boundaries.

To achieve even the limited goals set forth above, it has been necessary to impose severe restrictions on the discussions of certain methods. The most obvious victims of this attitude are the various moment and difference methods, as well as the kinetic theory of free and nearly-free molecular flows, which are only briefly described in Chapter VIII. The decision to curtail the discussion of these topics was taken more light-heartedly once the author became aware of the translation of Kogan's book, which gives a more detailed treatment of these subjects.

The book is intended for a mature audience with little or no prior knowledge of the subject, but reaches the heart of the matter very quickly. The degree of mathematical sophistication is not unusually high, although recourse to the techniques of functional analysis is had, especially in Chapter VI (Sections 2 to 5). However, these sections can be skipped by readers interested in applications rather than the mathematical foundations of the subject. In the spirit of not requiring too high a standard of mathematical knowledge from the reader, a brief summary of certain concepts from the theory of generalized functions is given in Sections 2 and 3 of Chapter I.

In conclusion, although the choice of some parts of the material reflects the tastes and the inclinations of the author, it is hoped that this book can constitute a useful reference for people engaged in research on rarefied gas dynamics, and perhaps a textbook for an advanced course in applied mathematics or aerodynamics. People interested in different fields where similar mathematical problems arise, such as neutron transport in reactor theory, electron transport in solid state theory, phonon transport in liquid helium, and radiative transport, may also find this book interesting reading.

Milan, Italy

CARLO CERCIGNANI

CONTENTS

Chapter I

BASIC PRINCIPLES

1. Introduction

The kinetic theory of gases is a part of statistical mechanics, i.e., of the statistical theory of the dynamics of mechanical systems formed by a great number of particles, such as the number of molecules contained in a lump of matter of macroscopic dimensions. The aim of statistical mechanics is to explain the macroscopic behavior of matter in terms of the mechanical behavior of the constituent molecules, i.e., in terms of motions and interactions of a large number of particles. We shall assume that classical mechanics can be applied, and, therefore, the molecules are subject to Newton's second law

$$\dot{\boldsymbol{\xi}}_i = \mathbf{X}_i; \qquad \dot{\mathbf{x}}_i = \boldsymbol{\xi}_i \tag{1.1a}$$

or

$$\ddot{\mathbf{x}}_i = \mathbf{X}_i \tag{1.1b}$$

where $\mathbf{x}_i$ is the position vector of the ith particle ($i = 1, \ldots, N$) and $\boldsymbol{\xi}_i$ its velocity vector; both $\mathbf{x}_i$ and $\boldsymbol{\xi}_i$ are functions of the time variable t, and the dots denote, as usual, differentiation with respect to t. Here $\mathbf{X}_i$ is the force acting upon the ith particle divided by the mass of the particle. Such a force will in general be the sum of an external force (e.g., gravity or centrifugal forces) and the force describing the action of the other particles of the system on the ith particle. The expression of such forces must be given as part of the description of the mechanical system.

In order to obtain the actual dynamics of the system, one would have to solve the $6N$ first-order differential equations, Eqs. (1.1a), in the $6N$ unknowns constituting the components of the $2N$ vectors $(\mathbf{x}_i, \boldsymbol{\xi}_i)$ $(i = 1, \ldots, N)$. A prerequisite for this is the knowledge of the $6N$ initial conditions

$$\mathbf{x}_i(0) = \mathbf{x}_i^0; \qquad \dot{\mathbf{x}}_i(0) = \boldsymbol{\xi}_i(0) = \boldsymbol{\xi}_i^0 \tag{1.2}$$

where $\mathbf{x}_i^0$ and $\boldsymbol{\xi}_i^0$ are $6N$ given constants which describe the initial state of the system.

Now, as is frequently pointed out, it is impossible, even with high-speed computers, to solve the above initial value problem for a number of particles of a realistic order of magnitude (say, $N = 10^{20}$). In fact, the above $6N$ differential equations are coupled, the force $\mathbf{X}_i$ being a function of the position vectors of all the particles $\mathbf{x}_j$ ($j = 1, \ldots, N$). However, this would be only a practical restriction, which could, in principle, be overcome by some future progress of the computational art. There are, on the other hand, two more basic reasons which force us to look at the problem from a different point of view, and bring us to the basic idea of statistical mechanics.

The first is that even if we were able to overcome the above computational difficulties, we would have to supply the initial data $\mathbf{x}_i^0$ and $\boldsymbol{\xi}_i^0$, i.e., the positions and velocities of all the molecules at $t = 0$, and obtaining these data appears difficult, even in principle: in fact, it would involve the simultaneous measurement of the positions and velocities of $6N$ particles without appreciably disturbing their state, in particular, without separating one from the influence of the others.

The second reason is that even if somebody (say, Maxwell's demon) supplied us with the initial data and we were clever enough to be able to compute the subsequent evolution of their system, this detailed information would be useless because knowledge of where the single molecules are and of what their velocities are is information which, in this form, does not tell us what we really want to know, e.g., the pressure exerted on a wall by a gas at a given density and temperature.

The basic idea of statistical mechanics is that of averaging over our ignorance (meaning the incapability of macroscopic bodies to detect certain microscopic details of another macroscopic body) and treating our system in a statistical sense. The various averages are what matter, provided they are related to such macroscopic quantities as pressure, temperature, heat flow, stresses, etc.

The first kind of averaging which is present in any treatment based upon statistical mechanics is, as suggested by the above considerations, over our ignorance of initial data. However, other averaging processes are usually required, as, e.g., over the details of interactions between particles. The latter also include the interactions of the molecules of a fluid with the solid boundaries which bound the flow and are also formed of molecules. What is the practical result of this averaging? It is that we shall have to talk about probabilities instead of certainties: i.e., in our description, a given particle will not have a definite position and velocity, but only different probabilities of having different positions and velocities.

The statistical mechanics of bodies in thermal and mechanical equilibrium has been developed in a satisfactory way (at least when only one phase is present), while the statistical mechanics of nonequilibrium phenomena has a firm basis only in the case of gases which are not too dense.

For gases an equation exists, the so-called Boltzmann equation, which contains, in principle, all the behavior of a not-too-dense gas. The mathematical treatment of this equation, especially in the case of small deviations from equilibrium, is substantially the subject of this book. However, in this chapter a short account will be given of the meaning and the position of the Boltzmann equation in the general frame of statistical mechanics.

2. Probabilities and Certainties. Ordinary and Generalized Functions

As mentioned before, probability concepts are of basic importance in the kinetic theory of gases and, more generally, in statistical mechanics. We shall talk about probability in a rather intuitive sense, illustrated by the familiar experience of throwing dice or tossing coins and asking for the probability of getting one of the numbers from one to six or of getting heads or tails.

The probability of getting a certain result is a number between zero and one, which, in a rough sense, can be experimentally interpreted as the frequency of that result in a long series of trials [in the case of tossing a coin $P(H) = P(T) = \frac{1}{2}$, where $P(H)$ and $P(T)$ denote the probabilities of getting heads and tails, respectively]. If the events are mutually exclusive, then the sum of the probabilities of all the possible events must be one, this meaning that one of the events will certainly happen (either heads or tails, in the above example).

However, it is to be stressed that the variables which appear in statistical mechanics usually range through a continuous set of values, instead of being restricted to a discrete set (as the set of two elements, heads and tails, which describe the result of tossing a coin). Therefore, strictly speaking, the probability of obtaining any given value of the continuum of possible values will be, in general, zero; on the other hand, the "sum" of the probabilities must be one. There is nothing strange in this, since it is the exact parallel of the statement that a geometrical point has no length, while a segment, which is a set of points, has a nonzero length. Therefore we have to talk about the probability of obtaining a result which lies in an infinitesimal interval (or, more generally, set), instead of one having a fixed value: this probability will also, in general, be an infinitesimal quantity of the same order as the length of the interval (or measure of the set). In other words, we have to introduce a probability density $P(\mathbf{z})$ [where $\mathbf{z} = (z_1, z_2, \ldots, z_n)$ is a vector summarizing all the continuous real variables required to describe the set], such that $P(\mathbf{z})\, d^n\mathbf{z}$ is the probability that $\mathbf{z}$ lies between $\mathbf{z}$ and $\mathbf{z} + d\mathbf{z}$, with $d^n\mathbf{z}$ denoting the volume of an infinitesimal cell, equal to the product

$dz_1 \cdots dz_n$. In this case the property that the "sum" of probabilities is one becomes

$$\int_Z P(\mathbf{z})\, d\mathbf{z} = 1 \tag{2.1}$$

where Z is the region of the n-dimensional space where $\mathbf{z}$ varies (possibly the whole n-dimensional space) and we omit the superscript n in the volume element, since no confusion arises.

What is the use of a probability density? The answer is simple: a probability density is needed to compute averages; if we know the probability density $P(\mathbf{z})$, we can compute the average value of any given function $\varphi(\mathbf{z})$ of the vector $\mathbf{z}$. As a matter of fact, we can define averages as follows

$$\langle \varphi(\mathbf{z}) \rangle = \overline{\varphi(\mathbf{z})} = \int_Z P(\mathbf{z})\varphi(\mathbf{z})\, d\mathbf{z} \tag{2.2}$$

where brackets or a bar is conventional notation for averaging. In other words, to compute the average value of a function $\varphi(\mathbf{z})$, we integrate it over all values of $\mathbf{z}$, weighting each $d\mathbf{z}$ with the probability density that this value has of being realized. It is clear that this definition is in agreement with our intuition about averages.

Speaking about probability densities is very useful, but at first sight would seem to present a serious inconvenience. It can happen, either because we want to talk about a highly idealized case or because it is actually required to have some detailed information about a mechanical system, that we occasionally need to consider some variable as known with complete certainty. Then, if $\mathbf{z}$ definitely has the value $\mathbf{z}_0$, the probability density for any $\mathbf{z} \neq \mathbf{z}_0$ will obviously be zero. On the other hand, Eq. (2.1) has to be satisfied. Thus for considering the case of certainty the "probability" density cannot be an ordinary function. We have to enlarge our concept of function if we want to include certainty as a particular case of probability.

The required generalization is achieved by means of the so-called "generalized functions" or "distributions." Because of the importance of this concept in the following, we shall give a quick survey of this theory; more details can be found in the literature.

Generalized functions can be defined in many ways, e.g., as ideal limits of sequences of sufficiently regular functions in the same way as real numbers are ideal limits of sequences of rational numbers. Strictly speaking, the generalized functions have no local meaning; i.e., it has no meaning to ask what value is taken by a given generalized function at a given point. However, the important fact is that a precise meaning can be attached to the "scalar product" of a generalized function $T(\mathbf{z})$ with a sufficiently smooth ordinary function $\varphi(\mathbf{z})$ (test function):

$$\langle T, \varphi \rangle = \int_Z T(\mathbf{z})\varphi(\mathbf{z})\, d\mathbf{z} \tag{2.3}$$

To give an idea of the meaning of such integral, we can consider a sequence $\{T_m(z)\}$ of ordinary functions which has $T(z)$ as its ideal limit; then

$$\int_Z T(\mathbf{z})\varphi(\mathbf{z})\, d\mathbf{z} = \lim_{m\to\infty} \int_Z T_m(\mathbf{z})\varphi(\mathbf{z})\, d\mathbf{z} \tag{2.4}$$

for any test function $\varphi(\mathbf{z})$ such that the indicated limit does exist. It is usually assumed that the set of test functions is dense in the set of continuous functions.

The simplest example of a generalized function is given by the so-called Dirac delta function, which is illustrated by the probability density corresponding to the above mentioned case when the n-dimensional vector $\mathbf{z}$ has the value $\mathbf{z}_0$ with certainty. We can define the delta function $\delta(\mathbf{z} - \mathbf{z}_0)$ as follows:

$$\delta(\mathbf{z} - \mathbf{z}_0) = \lim_{m\to\infty} (m/\pi^{1/2})^n \exp[-m^2(\mathbf{z} - \mathbf{z}_0)^2] \tag{2.5}$$

It is clear from this definition that $\delta(\mathbf{z} - \mathbf{z}_0)$ cannot be an ordinary function: as a matter of fact, the indicated limit gives 0 for $\mathbf{z} \neq \mathbf{z}_0$ and $+\infty$ for $\mathbf{z} = \mathbf{z}_0$, and can hardly be thought of as defining a function. However, as we said before, the relation (2.4) is completely meaningful provided $\varphi(\mathbf{z})$ is continuous at $\mathbf{z} = \mathbf{z}_0$ and satisfies mild integrability conditions. It can be shown easily, in fact, that under such assumptions

$$\int_Z \delta(\mathbf{z} - \mathbf{z}_0)\varphi(\mathbf{z})\, d\mathbf{z} = \lim_{m\to\infty} \int (m/\pi^{1/2})^n \exp[-m^2(\mathbf{z} - \mathbf{z}_0)^2]\varphi(\mathbf{z})\, d\mathbf{z} \tag{2.6}$$

exists and is equal to $\varphi(\mathbf{z}_0)$:

$$\int \delta(\mathbf{z} - \mathbf{z}_0)\varphi(\mathbf{z})\, d\mathbf{z} = \varphi(\mathbf{z}_0) \tag{2.7}$$

where we omit indicating explicitly the domain of integration Z. In particular, if $\varphi(\mathbf{z}) = 1$,

$$\int \delta(\mathbf{z} - \mathbf{z}_0)\, d\mathbf{z} = 1 \tag{2.8}$$

and if $\varphi(\mathbf{z}) = \mathbf{z}$,

$$\int \delta(\mathbf{z} - \mathbf{z}_0)\mathbf{z}\, d\mathbf{z} = \mathbf{z}_0 \tag{2.9}$$

Therefore $\delta(\mathbf{z} - \mathbf{z}_0)$ is the probability density which describes a situation where the average of $\mathbf{z}$ is $\mathbf{z}_0$ [see Eqs. (2.1) and (2.2)]; but we can show more; i.e., as was anticipated, the delta function describes a situation where we know with absolute certainty that the variable $\mathbf{z}$ has the value $\mathbf{z}_0$. As a matter of fact, we can inquire about the possibility of getting values different from $\mathbf{z}_0$ by evaluating the average "deviation" of $\mathbf{z}$ from $\mathbf{z}_0$, and this will be shown to be zero when the probability density is $\delta(\mathbf{z} - \mathbf{z}_0)$. In order to do this, we have to introduce a suitable way of measuring the deviation from the average value. To this end, we note that for any probability density if we define $\mathbf{z}_0$ as the average of $\mathbf{z}$,

$$\mathbf{z}_0 = \int \mathbf{z}P(\mathbf{z})\, d\mathbf{z} \tag{2.10}$$

it follows, because of Eq. (2.1), that

$$\int (\mathbf{z} - \mathbf{z}_0)P(\mathbf{z})\, d\mathbf{z} = \int \mathbf{z}P(\mathbf{z})\, d\mathbf{z} - \mathbf{z}_0 \int P(\mathbf{z})\, d\mathbf{z} = \mathbf{z}_0 - \mathbf{z}_0 = 0 \tag{2.11}$$

Therefore the average of $(\mathbf{z} - \mathbf{z}_0)$ does not give us any information about the deviation of $\mathbf{z}$ from $\mathbf{z}_0$. The reason for this is that, because of the very definition of average, departures from $\mathbf{z}_0$ on its left exactly compensate departures on its right.

An average measure of these departures can be obtained by evaluating

$$\int (\mathbf{z} - \mathbf{z}_0)^2 P(\mathbf{z})\, d\mathbf{z} \tag{2.12}$$

i.e., the mean-square deviation. In this case no compensation can occur and we obtain useful information about the spread of $\mathbf{z}$ about $\mathbf{z}_0$. However, as we said above, in the case of the δ-function we obtain zero for the integral (2.12) also, which means that there is no spread, i.e., $\mathbf{z}$ has exactly the value $\mathbf{z}_0$; as a matter of fact we have

$$\int (\mathbf{z} - \mathbf{z}_0)^2\, \delta(\mathbf{z} - \mathbf{z}_0)\, d\mathbf{z} = (\mathbf{z}_0 - \mathbf{z}_0)^2 = 0 \tag{2.13}$$

because of Eq. (2.7).

We notice here that another possible way of measuring the deviation of $\mathbf{z}$ from $\mathbf{z}_0$, once the probability density is given, is that of computing the average of $|\mathbf{z} - \mathbf{z}_0|$ instead of $(\mathbf{z} - \mathbf{z}_0)^2 = |\mathbf{z} - \mathbf{z}_0|^2$. In general, the result will not be the square root of the previous one, i.e., in general, the square of the average of $|\mathbf{z} - \mathbf{z}_0|$ is different from the average of the square of $|\mathbf{z} - \mathbf{z}_0|$. As a matter of fact, because of Schwarz's inequality

$$\left(\int |\mathbf{z} - \mathbf{z}_0| P(\mathbf{z})\, d\mathbf{z}\right)^2 \leq \int |\mathbf{z} - \mathbf{z}_0|^2 P(\mathbf{z})\, d\mathbf{z} \int P(\mathbf{z})\, d\mathbf{z} = \int |\mathbf{z} - \mathbf{z}_0|^2 P(\mathbf{z})\, d\mathbf{z} \tag{2.14}$$

i.e., the square of the average is not larger than the average of the square. The equality sign holds, however, for $P(\mathbf{z}) = \delta(\mathbf{z} - \mathbf{z}_0)$, since in this case both sides are zero.

3. Some Properties of Generalized Functions and Further Examples

As we said before, a generalized function T is defined by a sequence $\{T_m\}$ of ordinary functions. However, the sequence itself is by no means uniquely defined by T; i.e., different sequences can have as ideal limit the same generalized function, in the same way that different sequences of rational numbers can have the same real number as ideal limit. To give an example, the delta function could also be defined by the following sequence:

$$\delta_m(\mathbf{z} - \mathbf{z}_0) = \frac{m^n}{\tau_n} H\left(\frac{1}{m} - |\mathbf{z} - \mathbf{z}_0|\right) \tag{3.1}$$

where τ_n is the volume of the unit sphere in n dimensions, related to the area of the same sphere ω_n by $\tau_n = \omega_n/n$, with ω_n given by [$\Gamma(x)$ denotes the gamma function]

$$\omega_n = 2\pi^{n/2}/\Gamma(n/2) \tag{3.2}$$

while $H(x)$ denotes the Heaviside step function equal to one for $x > 0$ and zero for $x < 0$. In the one-dimensional case Eq. (3.1) becomes

$$\delta_m(z - z_0) = \frac{m}{2} H\left(\frac{1}{m} - |z - z_0|\right) \tag{3.3}$$

which offers a simple visualization of the limit $m \to \infty$. Another sequence converging to the delta function in the one-dimensional case is

$$\delta_m(z - z_0) = m/\pi[1 + (z - z_0)^2 m^2] \tag{3.4}$$

It is easily verified that $\delta_m(z - z_0) \xrightarrow[m \to \infty]{} \delta(z - z_0)$. We also note that although in the above examples m was assumed to go to infinity through integral values, no difference arises if we let $m \to \infty$ through a continuous set of values. More generally, if η is a real parameter, we can say that the limit for

$\eta \to \eta_0$ of the ordinary function $T(\mathbf{z};\eta)$ is the generalized function $T(\mathbf{z})$ if, for any sequence $\{\eta_m\}$ such that $\eta_m \to \eta_0$ as $m \to \infty$, $T(\mathbf{z};\eta_m)$ constitutes a sequence defining $T(\mathbf{z})$. In particular, from the above considerations we deduce that

$$\lim_{\epsilon\to 0+} \frac{\epsilon}{\pi[(z - z_0)^2 + \epsilon^2]} = \delta(z - z_0) \tag{3.5}$$

by simply putting $\epsilon = 1/m$ in Eq. (3.4) and letting m go to $+\infty$ through any sequence of positive real numbers.

Particular cases of generalized functions are the ordinary functions (this justifies the name of the former!); as a matter of fact, if a sequence converges (almost everywhere) in the ordinary sense to an ordinary function, it will be taken as defining that ordinary function as generalized function. In particular, one can consider sequences converging to zero almost everywhere as defining the zero generalized function.

This fact can be used to define in a precise fashion what we mean by equality between generalized functions. Two generalized functions will be taken to be equal if the difference between any sequence defining the first one and any sequence defining the second one defines the zero (generalized) function. Another interesting property of generalized functions is that they can be differentiated as many times as we wish, and the result is a generalized function for which Eq. (2.1) is meaningful, provided the envisaged test functions are sufficiently smooth. The derivative T' of a generalized function T can be thought of as defined by a sequence of ordinary functions consisting of the derivatives of the functions constituting a sequence specifying T (of course, we have to choose a sequence consisting of continuously differentiable functions to specify T).

As an example, we can consider the derivative of a one-dimensional delta function: $\delta'(z - z_0)$ can be defined by the sequence

$$\delta'_m(z - z_0) = -2m^3\pi^{-1/2}(z - z_0)\exp[-m^2(z - z_0)^2] \tag{3.6}$$

which is obtained by differentiating the sequence appearing in Eq. (2.5) for $n = 1$.

The class of test functions suitable for $\delta'(z - z_0)$ is that of functions which, besides having suitable integrability properties, are continuously differentiable in a neighborhood of $z = z_0$.

If φ is a test function admissible for both T and the derivative T', then it can be easily shown that

$$\langle T', \varphi\rangle = -\langle T, \varphi'\rangle \tag{3.7}$$

where φ' is the derivative of φ. This relation can also be assumed as a definition of the derivative of T. From Eqs. (3.7) and (2.8) one has

$$\int \delta'(z - z_0)\varphi(z)\,dz = -\int \delta(z - z_0)\varphi'(z)\,dz = -\varphi'(z_0) \tag{3.8}$$

which gives the practical meaning of δ'.

In the case of more than one independent variable, one can take the partial derivative with respect to any of the variables. In particular, one can consider the gradient of the delta function, i.e., a vector-valued generalized function, which can be written as $\frac{\partial \delta}{\partial \mathbf{z}}(\mathbf{z} - \mathbf{z}_0)$ and is a vector with components $\partial\delta/\partial z_1, \ldots, \partial\delta/\partial z_n$.

One can obviously define addition and subtraction for generalized functions. However, one cannot define, in general, the product of two generalized functions in such a way that scalar products as in Eq. (2.3) have a meaning when T is the product of two generalized functions; in particular, e.g., no simple meaning is attached to the square of the delta function.

One can, however, define the product of a generalized function $T(\mathbf{z})$ times an ordinary smooth function $f(\mathbf{z})$: such a product is associated with the sequence $f(\mathbf{z})T_n(\mathbf{z})$, and the test functions $\varphi(\mathbf{z})$ are such that $f(\mathbf{z})\varphi(\mathbf{z})$ is a test function for $T(\mathbf{z})$.

For instance, in the one-dimensional case one can define the product $z\,\delta(z)$, or more generally $(z - z_0)\delta(z - z_0)$; the test functions φ can be simply chosen by requiring boundedness at z_0 and suitable integrability properties. However, it turns out that

$$(z - z_0)\delta(z - z_0) = 0 \tag{3.9}$$

As a matter of fact, if we take any of the above sequences introduced to define $\delta(z - z_0)$ and multiply it by $(z - z_0)$, we find a sequence of functions which converges in the usual sense to an ordinary function equal to zero everywhere. It can also be shown that $C\delta(z - z_0)$, where C is an arbitrary constant, is the most general solution of the equation $(z - z_0)T(z - z_0) = 0$. We note that Eq. (3.9) also gives

$$z\delta(z - z_0) = z_0\delta(z - z_0) \tag{3.10}$$

Another case when multiplication of two or more generalized functions can be easily defined is when they are functions of different variables. For instance, if z_k $(k = 1, \ldots, n)$ are the components of $\mathbf{z}$, one can define $\delta(z_1 - z_1^0)\delta(z_2 - z_2^0)\cdots\delta(z_n - z_n^0)$, where z_k^0 $(k = 1, \ldots, n)$ are arbitrary numbers, as the ideal limit of sequences of products of corresponding terms in sequences defining the various one-dimensional deltas. In particular, one can take

$$\delta(z_1 - z_1^0)\delta(z_2 - z_2^0)\cdots\delta(z_n - z_n^0) = \lim_{m\to\infty}\prod_{k=1}^{n}(m/\sqrt{\pi})\exp[-(z_k - z_k^0)^2 m^2]$$
$$= \lim_{m\to\infty}(m/\sqrt{\pi})^n\exp[-m^2(\mathbf{z} - \mathbf{z}_0)^2] \tag{3.11}$$

where $\mathbf{z}_0$ is the n-dimensional vector with components z_k^0 $(k = 1, \ldots, n)$. Equations (3.11) and (2.5) show that

$$\delta(z_1 - z_1^0)\delta(z_2 - z_2^0)\cdots\delta(z_n - z_n^0) = \delta(\mathbf{z} - \mathbf{z}_0) \tag{3.12}$$

It can be checked that this relation is in agreement with all the properties of the delta functions, in particular Eq. (2.7). Finally, one sometimes meets the problem of division by an ordinary function. If the function $f(\mathbf{z})$ is different from zero everywhere, then we just have to multiply by the ordinary inverse $1/f(\mathbf{z})$. However, if the function $f(\mathbf{z})$ is zero at some point, $1/f(\mathbf{z})$ fails to exist there, and usually scalar products involving $1/f$ do not exist in the ordinary sense. Nevertheless, division by such functions can be defined in many important cases. We shall consider a particular but very significant case, that of finding the most general distribution T such that

$$(z - z_0)T(z) = 1 \tag{3.13}$$

for $z \neq z_0$ we find $T(z) = 1/(z - z_0)$. But this has no meaning at $z = z_0$, and, in particular, integrals of the kind shown in Eq. (2.3) do not exist in the ordinary sense. It is possible, however, to define a generalized function, e.g., as limit of the sequence

$$T_m(z) = m^2(z - z_0)/[m^2(z - z_0)^2 + 1] \tag{3.14}$$

which satisfies Eq. (3.13) and has a meaningful scalar product with sufficiently smooth functions (they are required to satisfy a Hölder condition in the neighborhood of $z = z_0$). Such a generalized function is denoted by

$$P\frac{1}{z - z_0} = \lim_{m\to\infty}\frac{m^2(z - z_0)}{1 + m^2(z - z_0)^2} \tag{3.15}$$

where P can be read "principal part of," and the scalar product $\left\langle P\frac{1}{z - z_0}, \varphi\right\rangle$ is to be interpreted as the Cauchy principal value integral of $\varphi(z)/(z - z_0)$:

$$\left\langle P\frac{1}{z - z_0}, \varphi\right\rangle = P\int\frac{\varphi(z)}{z - z_0}\,dz = \lim_{\epsilon\to 0}\int_{|z - z_0| > \epsilon}\frac{\varphi(z)}{z - z_0}\,dz \tag{3.16}$$

We can ask now whether or not $P\dfrac{1}{z - z_0}$ is the only solution of Eq. (3.13). The answer is no. As a matter of fact, the most general solution will be the sum of $P\dfrac{1}{z - z_0}$ and the general solution of the homogeneous equation

$$(z - z_0)T(z) = 0 \tag{3.17}$$

As noted before, the most general solution of this equation is a multiple of $\delta(z - z_0)$; therefore the general solution of Eq. (3.13) reads as follows:

$$T(z) = P\frac{1}{z - z_0} + C\delta(z - z_0) \tag{3.18}$$

where C is an arbitrary constant.

We can utilize the above results to evaluate the following limit, which occurs frequently in applied mathematics (where i is the imaginary unit):

$$\lim_{\epsilon \to 0+} \frac{1}{z - z_0 + i\varepsilon} = \lim_{\epsilon \to 0+} \frac{z - z_0 - i\epsilon}{(z - z_0)^2 + \epsilon^2} = \lim_{\epsilon \to 0+} \frac{z - z_0}{(z - z_0)^2 + \epsilon^2}$$

$$-i \lim_{\epsilon \to 0+} \frac{\epsilon}{(z - z_0)^2 + \epsilon^2} = P\frac{1}{z - z_0} - i\pi\delta(z - z_0) \tag{3.19}$$

Here Eqs. (3.5) and (3.15) (with $m = \epsilon^{-1}$) have been used. We note also that if, instead of letting ϵ go to zero through positive values, we let it go through negative values, then, since this is equivalent to changing i into $-i$, we obtain as a result the complex conjugate of Eq. (3.19), i.e.,

$$\lim_{\epsilon \to 0-} \frac{1}{z - z_0 + i\epsilon} = P\frac{1}{z - z_0} + i\pi\delta(z - z_0) \tag{3.20}$$

We close this section with a brief comment about the usefulness of the generalized functions. The main advantage in their use is that they obey the same formal rules as the ordinary (smooth) functions, and therefore the field of application of known procedures is enormously enlarged. By properly handling their formal properties, one can avoid many difficulties concerning the correctness of limiting procedures. The problem of a rigorous justification of the formal steps is always present, but is now shifted to a proper definition of the class of test functions to be considered. Usually, one can easily find large classes for which the formal steps are justified, and in the following, whenever necessary, we shall assume, as is usual in applied mathematics, that the admissible test functions have such properties.

4. Phase Space and the Liouville Equation

We now go back to our system of material particles. Let us assume for a moment that we know the initial data, and, correspondingly, that the solution of the dynamical equations, Eq. (1.1), can be found. Then we know that

$$\mathbf{x}_i = \mathbf{x}_i(t); \qquad \boldsymbol{\xi}_i = \dot{\mathbf{x}}_i(t) \tag{4.1}$$

where $\mathbf{x}_i(t)$ and its time derivative $\dot{\mathbf{x}}_i(t)$ are known functions of time. Let us now ask for the joint probability density of finding simultaneously (at some fixed instant of time t) the first molecule at a point of position vector $\mathbf{x}_1$ with velocity $\boldsymbol{\xi}_1$, the second molecule at $\mathbf{x}_2$ with velocity $\boldsymbol{\xi}_2, \ldots,$ the Nth molecule at $\mathbf{x}_N$ with velocity $\boldsymbol{\xi}_N$. This is to be understood in the sense explained in Section 2, i.e., "at $\mathbf{x}_i$ with velocity $\boldsymbol{\xi}_i$" actually means "between $\mathbf{x}_i$ and $\mathbf{x}_i + d\mathbf{x}_i$ with velocity between $\boldsymbol{\xi}_i$ and $\boldsymbol{\xi}_i + d\boldsymbol{\xi}_i$."

Now, since we assume the trajectories of the particles to be known, the mentioned probability density is zero unless $\mathbf{x}_i = \mathbf{x}_i(t)$ and $\boldsymbol{\xi}_i = \dot{\mathbf{x}}_i(t)$ for every value of i, i.e., we have the case where the probability reduces to certainty. Therefore, if we denote this "certainty" density by $C(\mathbf{x}_1, \ldots, \mathbf{x}_N; \boldsymbol{\xi}_1, \boldsymbol{\xi}_2, \ldots, \boldsymbol{\xi}_N; t)$ or, briefly, $C(\mathbf{x}_k, \boldsymbol{\xi}_k, t)$, we can write it as

$$\begin{aligned} C(\mathbf{x}_k, \boldsymbol{\xi}_k, t) &= \delta(\mathbf{x}_1 - \mathbf{x}_1(t))\delta(\mathbf{x}_2 - \mathbf{x}_2(t)) \cdots \delta(\mathbf{x}_N - \mathbf{x}_N(t))\delta(\boldsymbol{\xi}_1 - \dot{\mathbf{x}}_1(t)) \\ &\times \delta(\boldsymbol{\xi}_2 - \dot{\mathbf{x}}_2(t)) \cdots \delta(\boldsymbol{\xi}_N - \dot{\mathbf{x}}_N(t)) = \prod_{k=1}^{N} \delta(\mathbf{x}_k - \mathbf{x}_k(t))\delta(\boldsymbol{\xi}_k - \dot{\mathbf{x}}_k(t)) \end{aligned} \tag{4.2}$$

where δ denotes the delta function. It must be noted that here the same letter $\mathbf{x}_k$ denotes two basically different things: while $\mathbf{x}_k(t)$ is a given constant at a given time t, as z_0 in $\delta(z - z_0)$, $\mathbf{x}_k$ is an independent variable ranging over the volume occupied by the system, as $\mathbf{z}$ in $\delta(\mathbf{z} - \mathbf{z}_0)$. In order to avoid confusion, the t can be thought of as performing the same office as the subscript in $\mathbf{z}_0$. Analogously, we could write $\boldsymbol{\xi}_k(t)$ in place of $\dot{\mathbf{x}}_k(t)$, but we think that the notation adopted helps to distinguish between $\mathbf{x}_k$ and $\mathbf{x}_k(t)$.

The certainty density $C(\mathbf{x}_k, \boldsymbol{\xi}_k, t)$ is a generalized function of $6N + 1$ variables which contains exactly the same information as Eqs. (4.1). In fact, if one evaluates the average value of any $\mathbf{x}_i$ or $\boldsymbol{\xi}_i$ ($i = 1, \ldots, N$), he finds the values given by Eq. (4.1), and the deviation from such values is rigorously zero (the argument is the same as at the end of Section 2). Therefore the $\mathbf{x}_i$ and $\boldsymbol{\xi}_i$ are rigorously given by Eqs. (4.1), as was to be shown.

Let us now see the advantages and disadvantages of using a description based upon the single function of $6N + 1$ variables $C(\mathbf{x}_k, \boldsymbol{\xi}_k, t)$ instead of the $6N$ functions of one variable, $\mathbf{x}_i(t)$ and $\boldsymbol{\xi}_i(t)$. As we have seen, the two descriptions of the system are equivalent. In addition, there is, in principle, neither more nor less trouble in considering one independent variable and $6N$ dependent variables than in considering one dependent variable and

$6N + 1$ independent variables. From the intuitive point of view, it is true that one can easily imagine a swarm of particles with given positions and velocities at any instant t, according to the description based upon Eq. (4.1), while one undoubtedly has trouble in visualizing at any given time a point of a $6N$-dimensional space, where the coordinates are $\mathbf{x}_k$, $\boldsymbol{\xi}_k$ $(k = 1, \ldots, N)$. However, this $6N$-dimensional space, which is called the phase space of the system, has the advantage that a given state of the system (specified by the positions and velocities of *all* the particles) is just a point in the phase space; accordingly, if one forgets about the troubles of visualizing a space with such a large number of dimensions, the state of a system is a much simpler concept in the second picture than in the first. However, there is a really good reason for going from the first picture to the second one: the latter (with a probability density in place of a certainty density) retains a meaning even when our description of the system is incomplete, while the former does not. As a matter of fact, with an uncertainty in the definition of positions and velocities of the particles the picture based on the particle trajectories loses its meaning (especially from the point of view of predicting future evolution from the initial data), while the same uncertainty can be easily accounted for in the phase space description by simply spreading the density given by Eq. (4.2) in such a way that only averages of positions and velocities have a meaning, the deviations from average values no longer being zero.

It is therefore easy to foresee that the concept of the N-particle distribution function (this is the current name for the probability density of finding simultaneously the first particle at $\mathbf{x}_1$ with velocity $\boldsymbol{\xi}_1, \ldots,$ the Nth at $\mathbf{x}_N$ with velocity $\boldsymbol{\xi}_N$) will play a significant role in statistical mechanics.

It is to be noted, however, that until now we have introduced the function $C(\mathbf{x}_k, \boldsymbol{\xi}_k, t)$ under the assumption that we are able to solve the initial value problem, Eqs. (1.1) and (1.2), in order to write down Eq. (4.1); this is not satisfactory because it makes our second description depend heavily upon the first one, and therefore we lose any advantage in using the former. In other words, we have not made the second description completely independent of the first one from the mathematical standpoint. In order to do this, we need an equation which allows us to evaluate $C(\mathbf{x}_k, \boldsymbol{\xi}_k, t)$ at any time, given that we know it at $t = 0$, without going through the system of Eqs. (1.1). This can be done, and we shall eventually obtain a partial differential equation for $C(\mathbf{x}_k, \boldsymbol{\xi}_k, t)$ which is known as the Liouville equation.

In order to derive Liouville's equation, we form the time derivative of $C(\mathbf{x}_k, \boldsymbol{\xi}_k, t)$; since C is a function of $6N + 1$ variables, we have, of course, to take the partial derivative. We have

$$\frac{\partial C}{\partial t} = -\sum_{j=1}^{N}\left[\prod_{\substack{k=1\\k\neq j}}^{N}\delta(\mathbf{x}_k - \mathbf{x}_k(t))\delta(\boldsymbol{\xi}_k - \dot{\mathbf{x}}_k(t))\right]\dot{\mathbf{x}}_j\cdot\frac{\partial\delta}{\partial\mathbf{x}_j}(\mathbf{x}_j - \mathbf{x}_j(t))\,\delta(\boldsymbol{\xi}_j - \dot{\mathbf{x}}_j(t))$$

$$- \sum_{j=1}^{N} \left[\prod_{\substack{k=1 \\ k \neq j}}^{N} \delta(\mathbf{x}_k - \mathbf{x}_k(t))\delta(\boldsymbol{\xi}_k - \dot{\mathbf{x}}_k(t)) \right] \delta(\mathbf{x}_j - \mathbf{x}_j(t))\ddot{\mathbf{x}}_j \cdot \frac{\partial \delta}{\partial \boldsymbol{\xi}_j}(\boldsymbol{\xi}_j - \dot{\mathbf{x}}_j(t)) \tag{4.3}$$

Here we have applied the rule for differentiating a product to Eq. (4.2) and used the notation used in Section 3 for the gradient of each three-dimensional δ-function. If we now apply Eq. (3.10) (read from right to left), we have

$$\dot{\mathbf{x}}_j(t)\delta(\boldsymbol{\xi}_j - \dot{\mathbf{x}}_j(t)) = \boldsymbol{\xi}_j\delta(\boldsymbol{\xi}_j - \dot{\mathbf{x}}_j(t)) \tag{4.4}$$

Besides, $\ddot{\mathbf{x}}_j(t)$ is equal to $\mathbf{X}_j$, the force per unit mass acting on the jth particle, since $\mathbf{x}_j(t)$ satisfies Eq. (1.1b) by definition. Then

$$\begin{aligned} \frac{\partial C}{\partial t} = &- \sum_{j=1}^{N} \boldsymbol{\xi}_j \cdot \frac{\partial \delta}{\partial \mathbf{x}_j}(\mathbf{x}_j - \mathbf{x}_j(t))\delta(\boldsymbol{\xi}_j - \dot{\mathbf{x}}_j(t)) \left[\prod_{\substack{k=1 \\ k \neq j}}^{N} \delta(\mathbf{x}_k - \mathbf{x}_k(t))\delta(\boldsymbol{\xi}_k - \dot{\mathbf{x}}_k(t)) \right] \\ &- \sum_{j=1}^{N} \mathbf{X}_j \cdot \frac{\partial \delta}{\partial \boldsymbol{\xi}_j}(\boldsymbol{\xi}_j - \dot{\mathbf{x}}_j(t))\delta(\mathbf{x}_j - \mathbf{x}_j(t)) \left[\prod_{\substack{k=1 \\ k \neq j}}^{N} \delta(\mathbf{x}_k - \mathbf{x}_k(t))\delta(\boldsymbol{\xi}_k - \dot{\mathbf{x}}_k(t)) \right] \end{aligned} \tag{4.5}$$

However, the factor scalarly multiplying $\boldsymbol{\xi}_j$ is just $\partial C/\partial \mathbf{x}_j$, and the factor scalarly multiplying $\mathbf{x}_j$ is just $\partial C/\partial \boldsymbol{\xi}_j$; consequently,

$$\frac{\partial C}{\partial t} + \sum_{j=1}^{N} \boldsymbol{\xi}_j \cdot \frac{\partial C}{\partial \mathbf{x}_j} + \sum_{j=1}^{N} \mathbf{X}_j \cdot \frac{\partial C}{\partial \boldsymbol{\xi}_j} = 0 \tag{4.6}$$

This is a linear, homogeneous, first-order partial differential equation satisfied by C, the Liouville equation. Everything appearing in Eq. (4.6) is known (except the unknown C, of course): $\mathbf{x}_j$, $\boldsymbol{\xi}_j$ are independent variables which vary in some given region of the phase space (possibly the whole $6N$-dimensional phase space), t is the time variable, and $\mathbf{X}_j$ is a given function of the various $\mathbf{x}_k$. Therefore, if $C(\mathbf{x}_k, \boldsymbol{\xi}_k, 0)$ is given, we can find $C(\mathbf{x}_k, \boldsymbol{\xi}_k, t)$ without any reference to the first description in ordinary space. Now, how is the initial datum to be prescribed? If we maintain that we know the initial state of the mechanical system, i.e., we know exactly the initial positions, $\mathbf{x}_i^0$, and velocities, $\boldsymbol{\xi}_i^0$, of the N particles, then the initial datum for C is simply

$$C(\mathbf{x}_k, \boldsymbol{\xi}_k, 0) = \prod_{k=1}^{N} \delta(\mathbf{x}_k - \mathbf{x}_k^0)\delta(\boldsymbol{\xi}_k - \boldsymbol{\xi}_k^0) \tag{4.7}$$

Then the solution of Eq. (4.6) is given exactly by Eq. (4.2) where $\mathbf{x}_k(t)$ and $\dot{\mathbf{x}}_k(t)$ are obtained by solving Eq. (4.1) with the initial conditions (1.2). As a matter of fact, the function C is constant along the characteristic lines of Eq. (4.6), which are exactly given by Eqs. (1.1a): therefore C will be, at any time t, a delta centered at a point which, as t varies, moves along the characteristic line $\mathbf{x}_i = \mathbf{x}_i(t)$, $\boldsymbol{\xi}_i = \dot{\mathbf{x}}_i(t)$ corresponding to the initial data $\mathbf{x}_i^0$, $\boldsymbol{\xi}_i^0$, as was to be shown. In such a way we have shown that the two pictures considered, the one in ordinary space and the other in phase space, are not only completely equivalent, but are also autonomous, i.e., one can adopt the second picture and develop the necessary mathematics without borrowing any physical concept or result from the first one.

As noted before, the phase space picture becomes important when we forget about the idea of a complete description of a mechanical system and begin to talk about probabilities and averages. In fact, if we average over the possible initial values which give the same macroscopic behavior (at a certain level of description, arbitrarily chosen by us in agreement with the kinds of measurements that we want to consider as possible in our theoretical description), we obtain in place of $C(\mathbf{x}_k, \boldsymbol{\xi}_k, t)$ a smoother function $P(\mathbf{x}_k, \boldsymbol{\xi}_k, t)$ (yet P can be a generalized function and not an ordinary function, in case we have some partial but accurate information about the state of the system, e.g., its total energy). The first effect of the averaging process is that $P(\mathbf{x}_k, \boldsymbol{\xi}_k, t)$ will be a symmetric function of the positions and velocities of the N particles (if the particles are identical, i.e., have equal mass, and if any pair interacts with the same force law), since a macroscopic measurement cannot distinguish the particles from each other.

In spite of the conceptual difference between C and P, both satisfy the Liouville equation, since the latter is a linear homogeneous equation and the averaging procedure which transforms C into P is over the initial data, which do not appear in the coefficients of Eq. (4.6). Accordingly,

$$\frac{\partial P}{\partial t} + \sum_{j=1}^{N} \boldsymbol{\xi}_j \cdot \frac{\partial P}{\partial \mathbf{x}_j} + \sum_{j=1}^{N} \mathbf{X}_j \cdot \frac{\partial P}{\partial \boldsymbol{\xi}_j} = 0 \tag{4.8}$$

Although it is difficult to imagine in what sense the average over the initial data is to be performed and although the actual performance of such procedures is never required, one can consider, from a conceptual point of view, the following situation: Imagine having a large ensemble of replicas of the original system of N particles. These replicas have different initial data for the microscopic variables $\mathbf{x}_j^0$ and $\boldsymbol{\xi}_j^0$, but the macroscopic state (specified by such variables as density, pressure, temperature, mass velocity, etc.) is the same for all the replicas of the ensemble. Then one can construct, in principle and in the limit of infinitely many replicas, the probability density of a certain initial microscopic state as the "frequency" of its appearance in the set of the possible initial states.

5. An Example: Thermal Equilibrium of a Monatomic Ideal Gas

In this section we shall consider an elementary but significant application of the reasoning which is typical of statistical mechanics; this example also introduces the concept of the monatomic ideal gases, which we shall study in detail throughout this book. Let us consider a monatomic ideal gas in thermal equilibrium. Simple as it is, this sentence requires at least four definitions to be understood. We have to explain what we mean by "monatomic," "ideal," "gas," and "thermal equilibrium."

From our point of view, a "gas" is a dynamical system consisting of a large number N of particles having a rather small mass m. The statement that the gas is "ideal" means that the potential energy of intermolecular forces is negligible unless we consider particles which are closer than some distance σ (the molecular diameter), which is negligibly small with respect to any other length of interest. These two facts (negligible potential energy outside some regions, negligible extension of these regions) does not mean negligible potential energy, and hence negligible interactions, as is sometimes stated; rather, it means that (at our level of description) the only important effects of the strong repulsive force, which arises when two molecules are closer than a certain distance (the molecular diameter), are the deviation of the molecules from their rectilinear paths and the change of their speeds (which are not negligible effects at all). This suggests that we can introduce another kind of averaging, this time over the details of an interaction, the only interesting information being the probability that two interacting molecules having at the beginning of the interaction velocities $\boldsymbol{\xi}_1'$ and $\boldsymbol{\xi}_2'$, respectively, emerge from the interaction with velocities $\boldsymbol{\xi}_1$ and $\boldsymbol{\xi}_2$, respectively. We shall return to this new kind of average, but at present it will suffice to have it stated in this rudimentary form.

By "monatomic" we mean that our molecules have no internal degrees of freedom, i.e., the state of each molecule is completely described by three space coordinates and three velocity components.

Finally, by "thermal equilibrium" we mean that the probability density of a certain microscopic state, averaged over the details of the interactions as mentioned before, does not change with time ($\partial P/\partial t = 0$) and is homogeneous in space ($\partial P/\partial \mathbf{x}_j = 0$) (the latter assumption does not need to be made, as homogeneity could be deduced from the remaining assumptions; since, however, it is correct and avoids an overdetailed discussion, we shall regard it as part of the definition); in addition, the system is enclosed in a box at rest and does not exchange energy with other systems or parts of systems. This is to be understood to mean not only that the total energy is rigorously constant in time, but also that the balance is locally verified at each point of the boundaries which separate the system from the outside. It is clear, however, that an additional averaging over the local details of the interactions with the particles constituting the enclosing box is to be introduced, the only interesting information being the probability that a molecule arriving at the boundary with velocity $\boldsymbol{\xi}'$ emerges from the interaction with velocity $\boldsymbol{\xi}$. This information constitutes a boundary condition which is to be used in connection with the partial differential equation used to describe our gas [which can be Eq. (4.8) or some other equation which already takes into account the average over the details of the intermolecular interactions]. In so far as we are interested in thermal equilibrium only, this boundary condition can be given a very idealized form: we can assume that molecules are specularly reflected at the boundary; in such a case we have

$$P(\mathbf{x}_1, \ldots, \mathbf{x}_i, \ldots, \mathbf{x}_N; \boldsymbol{\xi}_1, \ldots, \boldsymbol{\xi}_i, \ldots, \boldsymbol{\xi}_N; t)$$
$$= P(\mathbf{x}_1, \ldots, \mathbf{x}_i, \ldots, \mathbf{x}_N; \boldsymbol{\xi}_1, \ldots, \boldsymbol{\xi}_i - 2\mathbf{n}(\mathbf{n}\cdot\boldsymbol{\xi}_i), \ldots, \boldsymbol{\xi}_N; t) \tag{5.1}$$
$$(\mathbf{x}_i \in \partial R; \boldsymbol{\xi}_i \cdot \mathbf{n} > 0)$$

where ∂R is the boundary of the region R of ordinary physical space occupied by the gas and $\mathbf{n}$ the inward normal at the point $\mathbf{x}_i$ of the boundary. We note that the boundary condition in Eq. (5.1) is consistent with the above definition of thermal equilibrium, since it conserves energy; as a matter of fact, $\boldsymbol{\xi}_i$ enters in the expression of energy only through its square, and

$$[\boldsymbol{\xi}_i - 2\mathbf{n}(\boldsymbol{\xi}_i\cdot\mathbf{n})]^2 = \xi_i^2 - 4(\boldsymbol{\xi}_i\cdot\mathbf{n})(\boldsymbol{\xi}_i\cdot\mathbf{n}) + 4(\boldsymbol{\xi}_i\cdot\mathbf{n})^2 = \xi_i^2 \tag{5.2}$$

It must be noted, however, that the boundary condition of specular reflection is too idealized: it is sufficient for treating equilibrium but is absolutely meaningless in many nonequilibrium situations which we shall have occasion to consider later. The assumption that the walls of the enclosing box are at rest is also embodied in Eq. (5.1).

We have now described completely (in a statistical sense) the gas and we must be able to find the probability density $P(\mathbf{x}_k, \boldsymbol{\xi}_k, t)$ of the system. As mentioned incidentally in the above discussion, P will not satisfy the Liouville equation, because of our averaging procedure over the details of interactions; this average differs from the average over the initial data because these details appear in the coefficients $\mathbf{X}_j$ of the Liouville equation. However, if, according to our definition of an ideal gas, we stay outside the region of phase space defined as the union of the sets where $|\mathbf{x}_i - \mathbf{x}_j| < \sigma$ $(i \neq j;$ $i, j = 1, \ldots, N)$, the Liouville equation will hold with $\mathbf{X}_j = 0$. However, since $\partial P/\partial t = \partial P/\partial x_j = 0$, it is trivially satisfied. Therefore, insofar as we consider the behavior in the region of the phase space bounded by the union of the boundaries ∂R_k $(k = 1, \ldots, N)$ (defined by $\mathbf{x}_k \in \partial R$, where ∂R is the boundary of the region occupied by the gas in ordinary space) and by the boundaries σ_{ij} $(i, j = 1, \ldots, N; i \neq j)$ defined by $|\mathbf{x}_i - \mathbf{x}_j| = \sigma$, the solution could be any function of the N velocity variables $\boldsymbol{\xi}_k$, $P(\boldsymbol{\xi}_k)$. However, the boundary conditions are also to be satisfied. No matter what kind of central interaction we choose to compute the probability that two interacting molecules having initial velocities ξ_1' and ξ_2', respectively, emerge from the interaction with velocities ξ_1 and ξ_2, respectively, the functions of these velocities which are conserved must depend only on the total momentum ($m\xi_1' + m\xi_2' = m\xi_1 + m\xi_2$) and the total kinetic energy ($\frac{1}{2}m\xi_1'^2 + \frac{1}{2}m\xi_2'^2 = \frac{1}{2}m\xi_1^2 + \frac{1}{2}m\xi_2^2$) of the pair, because these quantities are conserved in the interaction. Since this is true for any pair i, j ($i, j = 1, \ldots, N$), the only admissible P must be a function of the total momentum $m\sum_{j=1}^N \xi_j$ and the total kinetic energy $\frac{1}{2}m\sum_{i=1}^N \xi_j^2$. On the other hand, the boundary conditions, Eq. (5.1), do not conserve total momentum, and therefore P must depend only upon the total kinetic energy $\frac{1}{2}m\sum_{j=1}^N \xi_j^2$. However, because of the definition of an ideal gas, the contribution of the potential energy of intermolecular forces to the total energy E of the gas can be neglected, and we can write

$$\frac{m}{2}\sum_{j=1}^{N} \xi_j^2 = E = Nme \tag{5.3}$$

where e is the energy per unit mass of gas, i.e., a quantity which has a macroscopic significance, and therefore we can assume it is known exactly.

Therefore our P must depend only on $\sum_{j=1}^N \xi_j^2$ and express the certainty that this quantity has the value $2Ne$. Accordingly, it must be a multiple of a δ-function:

$$P_N = A_N\delta\left(\sum_{j=1}^{N} \xi_j^2 - 2Ne\right) \tag{5.4}$$

where we have explicitly indicated, by the subscript N, the dependence of P upon N, and A_N is a constant to be determined by the normalization condition, Eq. (2.1). In this case $\mathbf{z}$ is a $6N$-dimensional vector having as components the components of all the $\mathbf{x}$'s and ξ's, and Z is a region of the phase space which is the direct product ($R_{3N} \otimes R^{\otimes N}$) of the Nth direct power ($R^{\otimes N}$) of the three-dimensional region R occupied by the gas in physical space times the whole $3N$-dimensional space R_{3N} spanned by the $3N$ components of the N vectors ξ_k. Therefore

$$A_N \int_{R_{3N}} \delta\left(\sum_{j=1}^{N} \xi_j^2 - 2Ne\right) d\xi_1 \cdots d\xi_N \int_{R^{\otimes N}} d\mathbf{x}_1 \cdots d\mathbf{x}_N = 1 \tag{5.5}$$

Now, the integral over the space variables simply gives V^N, where V is the volume of the enclosing box, while the integral over R_{3N} is better evaluated if one introduces polar coordinates in this space; then

$$\sum_{j=1}^{N} \xi_j^2 = \rho^2; \qquad d\xi_1 \cdots d\xi_N = \rho^{3N-1}\, d\rho\, d^{3N}\boldsymbol{\Omega}$$

where $d^{3N}\mathbf{\Omega}$ is the surface element of the unit sphere in $3N$ dimensions. Consequently,

$$A_N V^N \int_0^\infty \delta(\rho^2 - 2Ne)\rho^{3N-1}\,d\rho \int d^{3N}\mathbf{\Omega} = 1$$

Denoting by ω_{3N} the surface of the unit sphere in $3N$ dimensions (see Section 3) and using Eq. (2.7) in one dimension with $z = \rho^2$, we have:

$$A_N V^N \tfrac{1}{2}(2Ne)^{(3N-2)/2}\omega_{3N} = 1 \tag{5.6}$$

Therefore

$$P_N = \frac{2}{\omega_{3N}(2Ne)^{(3N-2)/2}V^N}\delta\left(\sum_{j=1}^{N} \xi_j^2 - 2Ne\right) \tag{5.7}$$

This formula solves our problem; however, it does not look very illuminating in this form; we have to manipulate it in order to extract useful information. Suppose that we select a molecule at random and ask for the probability density $P_N^{(1)}$ that it has velocity between $\boldsymbol{\xi}$ and $\boldsymbol{\xi} + d\boldsymbol{\xi}$ and position between $\mathbf{x}$ and $\mathbf{x} + d\mathbf{x}$, without caring about the other molecules; this means that we have to "sum" P_N over all possible positions and velocities of all the molecules except one. We can assume that this molecule is specified by $\mathbf{x}_1$ and $\boldsymbol{\xi}_1$, since the labels j ($j = 1, \ldots, N$) can be interchanged at will. Then

$$\begin{aligned} P_N^{(1)} &= \int_{R_{3N-3}} P_N\, d\boldsymbol{\xi}_2 \cdots d\boldsymbol{\xi}_N \int_{R^{\otimes(N-1)}} d\mathbf{x}_2 \cdots d\mathbf{x}_N \\ &= V^{N-1} A_N \int_{R_{3N-3}} \delta\left(\sum_{j=1}^{N} \xi_j^2 - 2Ne\right) d\boldsymbol{\xi}_2 \cdots d\boldsymbol{\xi}_N \end{aligned} \tag{5.8}$$

We take now polar coordinates in the $(3N - 3)$-dimensional space spanned by the $3N - 3$ components of the $N - 1$ three-dimensional vectors $\boldsymbol{\xi}_k$ ($k = 2, \ldots, N$). We have

$$\sum_{j=2}^{N} \xi_j^2 = \rho^2 \qquad d\boldsymbol{\xi}_2 \cdots d\boldsymbol{\xi}_N = \rho^{3N-4}\,d\rho\, d^{3N-3}\Omega$$

and consequently we obtain for $P_N^{(1)}$ (the so-called one-particle distribution function):

$$\begin{aligned} P_N^{(1)} &= V^{N-1} A_N \omega_{3N-3} \int \delta[\rho^2 - (2Ne - \xi_1^2)]\rho^{3N-4}\,d\rho \\ &= \frac{\omega_{3N-3}}{\omega_{3N}} \frac{1}{V}(2Ne)^{-(3N-2)/2}(2Ne - \xi_1^2)^{(3N-5)/2} \end{aligned} \tag{5.9}$$

$$= \left(\frac{4\pi e}{3}\right)^{-3/2} V^{-1}\left(1 - \frac{\xi_1^2}{2Ne}\right)^{(3N-5)/2} \frac{\Gamma(\frac{3}{2}N)}{(\frac{3}{2}N)^{3/2}\Gamma((3N-3)/2)},$$

$$\text{for} \quad \xi_1^2 < 2Ne$$

$$P_N^{(1)} = 0 \qquad \text{for} \quad \xi_1^2 > 2Ne$$

Equation (3.2) has been used in the above. The result we have found means that although a random molecule is likely to be at any point of the region occupied by the gas ($P_N^{(1)}$ does not depend upon $\mathbf{x}_1$), the distribution of velocities is by no means uniform. If we open a little hole at any point of the boundary and let a very few molecules (few with respect to the total number) escape, when we measure their velocities and plot the distribution of the molecules according to their velocities, in the limit of infinitely repeated sampling, we find that the occupation density of the interval $(\xi_1, \xi_1 + d\xi_1)$ does depend upon ξ_1 and is given by Eq. (5.9). Actually, one has to be a little bit more careful, since if one extracts molecules from the gas, Eq. (5.9) cannot be applied, since it refers to a gas of exactly N molecules. If the sample is a small one, however, this should not give trouble; i.e., we expect that if N is very large, Eq. (5.9) is not essentially altered if N is changed by a small fraction. It is therefore customary to consider the limit of Eq. (5.9) as $N \to \infty$, which is quite justified by the huge numbers of molecules which are usually considered. But it is easily verified that

$$P_\infty^{(1)} = \lim_{N\to\infty} P_N^{(1)} = (\tfrac{4}{3}\pi e)^{-3/2} V^{-1} \exp[-3\xi_1^2/4e] \tag{5.10}$$

An elementary result which we shall find later is the connection between e and the temperature T of the ideal monatomic gas in thermal equilibrium:

$$e = \tfrac{3}{2}RT \tag{5.11}$$

where R is the Boltzmann constant related to the molecular mass m and the Boltzmann universal constant k ($k = 1.38 \times 10^{-23}$ J/°K) by

$$R = k/m \tag{5.12}$$

If we put Eq. (5.11) into Eq. (5.10), the latter reduces to the Maxwell formula for the velocity distribution which is frequently derived in elementary treatments of kinetic theory by a different kind of argument.

In addition to the one-particle distribution function $P_N^{(1)}$ (or its limit $P_\infty^{(1)}$), one has occasion to consider the two-particle distribution function $P_N^{(2)}$, defined as the probability density that two randomly chosen molecules have velocities between ξ_1 and $\xi_1 + d\xi_1$ and ξ_2 and $\xi_2 + d\xi_2$, respectively, and positions between $\mathbf{x}_1$ and $\mathbf{x}_1 + d\mathbf{x}_1$ and $\mathbf{x}_2$ and $\mathbf{x}_2 + d\mathbf{x}_2$, respectively, without caring about the $N - 2$ remaining molecules. This means that we have to "sum" P_N over all the possible positions and velocities of all the molecules except two. Then

$$\begin{aligned}
P_N^{(2)}(\mathbf{x}_1, \mathbf{x}_2; \boldsymbol{\xi}_1, \boldsymbol{\xi}_2) &= \int_{R_{3N-6}\otimes R^{\otimes(N-2)}} P_N \, d\boldsymbol{\xi}_3 \cdots d\boldsymbol{\xi}_N \, d\mathbf{x}_3 \cdots d\mathbf{x}_N \\
&= V^{N-2} A_N \omega_{3N-6} \frac{1}{2} [2ne - (\xi_1^2 + \xi_2^2)]^{(3N-8)/2} \\
&= \frac{\omega_{3N-6}}{\omega_{3N}} \frac{1}{V^2} (2Ne)^{-(3N/2-2)} [2Ne - (\xi_1^2 + \xi_2^2)]^{(3N-8)/2} \\
&= \left(\frac{4\pi}{3} e\right)^{-3} V^{-2} \left[1 - \frac{\xi_1^2 + \xi_2^2}{2Ne}\right]^{(3N-8)/2} \qquad (5.13) \\
&\quad \times \left(1 - \frac{2}{3N}\right)\left(1 - \frac{4}{3N}\right)\left(1 - \frac{6}{3N}\right) \\
&\quad \text{for} \quad \xi_1^2 + \xi_2^2 < 2Ne \\
P_N^{(2)}(\mathbf{x}_1, \mathbf{x}_2; \boldsymbol{\xi}_1, \boldsymbol{\xi}_2) &= 0 \qquad \text{for} \quad \xi_1^2 + \xi_2^2 > 2Ne
\end{aligned}$$

where calculations similar to the previous ones have been made and the recurrence formula for the Γ-function has been used. Now, if we let $N \to \infty$, we find

$$\begin{aligned}
P_\infty^{(2)}(\mathbf{x}_1, \mathbf{x}_2; \boldsymbol{\xi}_1, \boldsymbol{\xi}_2) &= (\tfrac{4}{3}\pi e)^{-3} V^{-2} \exp[-3(\xi_1^2 + \xi_2^2)/4e] \\
&= P_\infty^{(1)}(\mathbf{x}_1, \boldsymbol{\xi}_1) P_\infty^{(1)}(\mathbf{x}_2, \boldsymbol{\xi}_2) \qquad (5.14)
\end{aligned}$$

We have found, therefore, the remarkable result that in a gas of an extremely large number of molecules, two randomly selected molecules do not show any correlation, i.e., the probability density of finding the first molecule in a given place with a given velocity (i.e., in a state which we can call state onc) and simultaneously the second one at another given place with another given velocity (state two) is simply the product of the probability density of finding the first molecule in state one and the second one in any state, times the probability density of finding the second molecule in state two and the first one in any state. This means that the fact of finding one molecule in a given state has no influence upon the probability of finding the second one in another given state, and is expressed by saying that the states of the molecules (in the case of thermal equilibrium of a monatomic ideal gas) are statistically uncorrelated. An elementary example of random events, which are also statistically uncorrelated, is given by the results of throwing a pair of dice; the fact that we obtain a certain number from one die does not influence the number which is shown by the other die, and the probability of getting a given combination (say, two from the first die and five from the second one) is 1/36, i.e., the product of getting a given number from the first and any number from the second (which is 1/6) times the product of getting a given number from the second and any number from the first (which is also 1/6). (Note that in this example we make a distinction between the two dice—as if, e.g., one were red and the other green—and consequently we consider a "two and five" an event different from a "five and two.")

6. The Problem of Nonequilibrium States. The Boltzmann Equation

In Section 5 we saw that the problem of describing the state of thermal equilibrium of a monatomic ideal gas can be nicely solved; in particular, for the one-particle distribution function $P_{\infty}^{(1)}$ we found a very simple formula which has a large variety of applications in the statistical description of matter in the gaseous state.

It is clear, however, that the state of thermal equilibrium is a very special one, since in practice we usually have to deal with nonequilibrium gases, which can be, e.g., flowing in a channel or surrounding a body in flight in the atmosphere, something very different from being enclosed in a box and kept at uniform temperature and pressure. Can we handle these more general problems of nonequilibrium gases by the kind of argument used in Section 5? The answer can be "yes" or "no," depending on the meaning attached to the word "handle." It is certainly hopeless, e.g., to try to evaluate the distribution function $P_{\infty}^{(1)}$ for a gas in a general nonequilibrium situation by means of simple arguments. However, we can do something: we can derive an equation satisfied by this quantity $P_{\infty}^{(1)}$ under the same assumptions used in Section 5 except, of course, thermal equilibrium. Then we can hope to attack this equation (which is the already mentioned Boltzmann equation) on purely mathematical grounds, by means of standard procedures of approximate or exact nature. We must say from the beginning that the Boltzmann equation turns out to be particularly difficult to solve even for very simple nonequilibrium situations. But we shall have much more to say about this later, since the procedures for solving the Boltzmann equation are the main subject of this book.

What is the motivation for setting up and solving the Boltzmann equation? One can distinguish two main kinds of applications. The first one is concerned with deducing the macroscopic behavior of gases from their microscopic properties; these applications are therefore a particular instance of the basic problem of statistical mechanics, which is to bridge the gap between the atomic structure of matter and its continuum-like behavior at a macroscopic level. After the initial basic researches of Maxwell and Boltzmann in the last century this part of the subject developed mainly between the two world wars. Typical results of these researches are the explanation of the macroscopic behavior of gases and computations of viscosity and heat conduction coefficients from postulated laws of interaction between pairs of the constituent molecules. Besides their intrinsic importance, these researches constitute a model of what one should be able to do for other states of aggregation of matter (liquids, solids, many-phase systems).

In the last forty years, however, a new trend began, under the impetus of upper-atmosphere flight and of similar, more difficult problems encountered by plasma physicists in dealing with controlled thermonuclear reactions.

In this new situation the gas theory is expected, once more, to perform the double duty of giving intrinsically useful results (for aerodynamic purposes) and to show the way to another branch of research; but we must say that in this case the objectives are far from having been accomplished, and the whole subject is a matter of current research.

The aim of this modern aspect of kinetic theory, which will be of primary interest to us, is not that of deducing a macroscopic theory in the ordinary sense, although the final results are in terms of measurable and practical quantities, such as, e.g., the drag on an object moving in a rarefied atmosphere. As a matter of fact, the modern kinetic theory deals with situations where the gas is so rarefied that the average frequency of encounters between molecules is of the same magnitude as, or smaller than, the frequency of encounters with walls bounding the region of interest or the frequency of soundlike disturbances propagating through the gas. It is clear that in such conditions one cannot expect a "macroscopic behavior" easily described in terms of such quantities as density, pressure, temperature, or mass velocity, although all these concepts retain their meaning (in a statistical sense). In these conditions the use of the one-particle distribution function proves useful, and the Boltzmann equation becomes very important as an equation capable of encompassing the whole range of rarefactions and consequent behavior from the fluidlike regime of a moderately dense gas to the free-molecular regime, where the molecular encounters are practically negligible.

We want now to derive briefly the Boltzmann equation from the Liouville equation plus the assumption of averaging over the details of the molecular interactions. Actually, a small logical gap (covered only by an intuitive argument) will be left in the proof.

As we did in the case of thermal equilibrium, we assume that the force acting on a molecule is zero, unless the molecule itself is within the "action sphere" of another molecule, i.e.,

$$\mathbf{X}_j = 0 \quad \text{if} \quad |\mathbf{x}_i - \mathbf{x}_j| > \sigma \qquad (i, j = 1, \ldots, N; i \neq j) \tag{6.1}$$

Therefore, if we consider, as we did previously, the region, let us say U_{6N}, of the phase space bounded by the union of the ∂R_k and σ_{ij} (see Section 5), we have for Eq. (4.8)

$$\frac{\partial P_N}{\partial t} + \sum_{j=1}^{N} \boldsymbol{\xi}_j \cdot \frac{\partial P_N}{\partial \mathbf{x}_j} = 0 \tag{6.2}$$

Now we integrate both sides of this equation over $\mathbf{x}_j$ and $\boldsymbol{\xi}_j$ for $j = 2, \ldots, N$; $\boldsymbol{\xi}_j$ ranges over the whole set of values, while $\mathbf{x}_j$ ranges over the three-dimensional volume which is the intersection of U_{6N} with the three-dimensional space described by $\mathbf{x}_j$ itself. Then we have

$$\frac{\partial P_N^{(1)}}{\partial t} + \boldsymbol{\xi}_1 \cdot \frac{\partial P_N^{(1)}}{\partial \mathbf{x}_1} - \sum_{j=2}^{N} \int P_N^{(2)}(\mathbf{x}_1, \mathbf{x}_j; \boldsymbol{\xi}_1, \boldsymbol{\xi}_j)\boldsymbol{\xi}_1 \cdot \mathbf{n}_j \, dS_j \, d\boldsymbol{\xi}_j$$
$$+ \sum_{j=2}^{N} \int P_N^{(2)}(\mathbf{x}_1, \mathbf{x}_j; \boldsymbol{\xi}_1, \boldsymbol{\xi}_j)\boldsymbol{\xi}_j \cdot \mathbf{n}_j \, dS_j \, d\boldsymbol{\xi}_j = 0 \tag{6.3}$$

where $P_N^{(2)}$ is the two-particle distribution function, dS_j is the surface element of the sphere $|\mathbf{x}_1 - \mathbf{x}_j| = \sigma$ (where $\mathbf{x}_1$ is thought of as the fixed center and $\mathbf{x}_j$ as varying along the surface), and $\mathbf{n}_j$ is the normal directed toward the interior of the same sphere. The two surface integrals are extended to the whole surface of the mentioned sphere and have different origin: the second one simply arises from integrating the $N - 1$ terms with $j \neq 1$ in the sum appearing in Eq. (6.2) and applying the Gauss lemma, while the first surface integral arises from the fact that in order to exchange integration and differentiation in the remaining term ($j = 1$) of the mentioned sum, we have properly to take into account that the boundary of integration ($|\mathbf{x}_1 - \mathbf{x}_j| = \sigma$) depends upon $\mathbf{x}_1$. Additional integrals extended to the boundary of the region occupied by the gas are put equal to zero, since they express the difference between the probabilities that a molecule leaves and enters the region where the gas is contained, and we assume that no molecule leaves the mentioned region. Now, since $\mathbf{x}_j$ and $\boldsymbol{\xi}_j$ are integration variables in Eq. (6.3), all the terms of each sum are equal, and we have

$$\frac{\partial P_N^{(1)}}{\partial t} + \boldsymbol{\xi}_1 \cdot \frac{\partial P_N^{(1)}}{\partial \mathbf{x}_1} + (N-1) \int P_N^{(2)}(\mathbf{x}_1, \mathbf{x}_2; \boldsymbol{\xi}_1, \boldsymbol{\xi}_2; t)(\boldsymbol{\xi}_2 - \boldsymbol{\xi}_1) \cdot \mathbf{n}_2 \, dS_2 \, d\boldsymbol{\xi}_2 = 0 \tag{6.4}$$

Now we want to distinguish two hemispheres on the integration sphere: one is where $(\boldsymbol{\xi}_2 - \boldsymbol{\xi}_1) \cdot \mathbf{n}_2 > 0$, the other where $(\boldsymbol{\xi}_2 - \boldsymbol{\xi}_1) \cdot \mathbf{n}_2 < 0$. The first hemisphere (which we shall call the "plus" hemisphere) gives the contribution from the molecules which are entering into the "action sphere" of the first molecule; the second one (the "minus" hemisphere) gives the contribu-

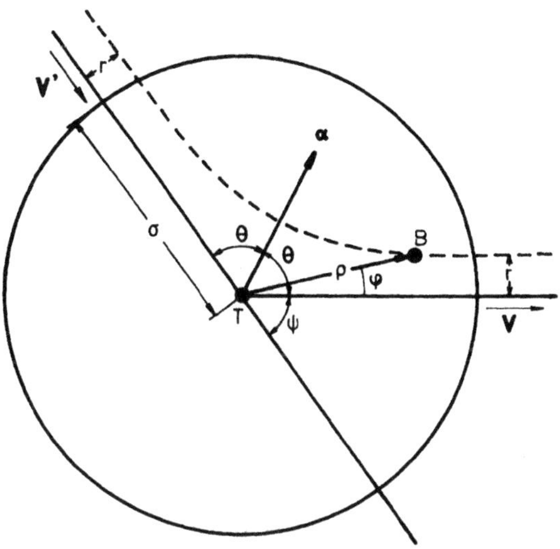

Fig. 1. Notation for a two-body collision. Here T is the "target" molecule, B the "bullet" molecule. The plane of the figure is the plane of motion, forming an angle ϵ with a fixed plane through **V**.

tion from the molecules which are leaving the same sphere. It is useful here to think of the first molecule ("target molecule") as if it were at rest and surrounded by a "protection sphere" having its radius equal to the molecular diameter, and the other molecules ("bullet molecules") as mass points entering and leaving the "protection sphere" (Fig. 1).

In order to treat the integrals over the two hemispheres properly, it is useful to consider the plane separating them and take as integration variables the polar coordinates (r, ϵ) of the point orthogonally projected from the sphere onto this plane; when a point describes the whole sphere the image point in the diameter plane describes the corresponding disk exactly twice, but one time it is the image of a point of the "plus" hemisphere and one time the image of a point of the "minus" hemisphere. If we also take into account an elementary geometrical argument that gives

$$(\boldsymbol{\xi}_2 - \boldsymbol{\xi}_1) \cdot \mathbf{n}_2 \, dS_2 = \pm |\boldsymbol{\xi}_2 - \boldsymbol{\xi}_1| r \, dr \, d\epsilon$$

where the double sign corresponds to the two possible signs of $(\boldsymbol{\xi}_2 - \boldsymbol{\xi}_1) \cdot \mathbf{n}_2$, we have

$$(N-1) \int P_N^{(2)}(\mathbf{x}_1, \mathbf{x}_2; \boldsymbol{\xi}_1, \boldsymbol{\xi}_2; t)(\boldsymbol{\xi}_2 - \boldsymbol{\xi}_1) \cdot \mathbf{n}_2 \, dS_2 \, d\boldsymbol{\xi}_2$$

$$= (N - 1) \int [P_N^{(2)}(\mathbf{x}_1, \mathbf{x}_2^+; \boldsymbol{\xi}_1, \boldsymbol{\xi}_2; t) - P_N^{(2)}(\mathbf{x}_1, \mathbf{x}_2^-; \boldsymbol{\xi}_1, \boldsymbol{\xi}_2; t)]$$

$$\times |\boldsymbol{\xi}_2 - \boldsymbol{\xi}_1| r \, dr \, d\epsilon \, d\boldsymbol{\xi}_1 \tag{6.5}$$

where $\mathbf{x}_2^+$ and $\mathbf{x}_2^-$ denote the points of the sphere having the point of polar coordinates r and ϵ as image, but located on the "plus" and "minus" hemispheres, respectively.

Now we must explain the reason for not being satisfied with Eq. (6.4) and for doing the preceding manipulations and the ones which will follow: the fact is that Eq. (6.4) is not an equation for our unknown $P_N^{(1)}$, since it contains the equally unknown $P_N^{(2)}$. Our aim is therefore to substitute for $P_N^{(2)}$ whenever it appears an equivalent expression, if available, in terms of $P_N^{(1)}$. Our present problem is therefore one of finding such an expression; it seems reasonable to expect that in order to get it, we should exploit the two assumptions of averaging over the details of intermolecular interactions and taking the limit of infinitely many particles, since these assumptions were used to obtain the neat result expressed by Eq. (5.10) in the particular case of thermal equilibrium. We remember here that in the case of thermal equilibrium we obtained the following interesting relation between $P_\infty^{(2)}$ and $P_\infty^{(1)}$:

$$P_\infty^{(2)}(\mathbf{x}_1, \mathbf{x}_2; \boldsymbol{\xi}_1, \boldsymbol{\xi}_2; t) = P_\infty^{(1)}(\mathbf{x}_1; \boldsymbol{\xi}_1; t) P_\infty^{(1)}(\mathbf{x}_2; \boldsymbol{\xi}_2; t) \tag{6.6}$$

If something of this kind could be shown to hold even in the case of nonequilibrium, our problem would be solved. However, we immediately realize that we *cannot* use Eq. (6.6) as it stands (quite apart from the problem of being able to prove it, if possible!); as a matter of fact, if we simply substitute Eq. (6.6) into Eq. (6.4), we find that our integral contains the difference

$$[P_\infty^{(1)}(\mathbf{x}_2^+; \boldsymbol{\xi}_2; t) - P_\infty^{(1)}(\mathbf{x}_2^-; \boldsymbol{\xi}_2; t)]$$

Now, this difference is practically zero, since $|\mathbf{x}_2^+ - \mathbf{x}_2^-| \leq 2\sigma$ and we want to average our distribution function over the details at such short distances; as a matter of fact, later we shall drop any distinction between different points of the protection sphere (we cannot do it now because $\mathbf{x}_2^+$ and $\mathbf{x}_2^-$ identify the hemisphere to which the molecules belong). Therefore the integral appearing in Eq. (6.4) should be cancelled, a particularly simple but also surely wrong result [it will be shown later that this integral is zero only in the case of thermal equilibrium, in which case Eq. (6.6), of course, applies]. This result warns us against applying Eq. (6.6) to Eq. (6.4) as it stands; if we think about it for a moment, we discover the physical reason for this. As pointed out at the end of Section 5, Eq. (6.6) means that the states of the two molecules considered are statistically uncorrelated; now, this makes sense for any two randomly chosen molecules of our gas, since they do not interact when they are far apart, and therefore behave independently. This is particularly true for two molecules which are going to interact, because they are just two random molecules whose paths happen to cross; but the same statistical independence is far from being true for two molecules which have just interacted, because the interaction creates a strong correlation between them. Accordingly, we are in the following situation: we can (in the limit of very large values of N) substitute Eq. (6.6) into one half of Eq. (6.5), i.e., only in place of $P_N^{(2)}(\mathbf{x}_1, \mathbf{x}_2^+; \boldsymbol{\xi}_1, \boldsymbol{\xi}_2; t)$, since $\mathbf{x}_2^+$ varies on the hemisphere of the molecules which are entering into the "action sphere" of the first molecule. We cannot do just the same thing for $P_N^{(2)}(\mathbf{x}_1, \mathbf{x}_2^-; \boldsymbol{\xi}_1, \boldsymbol{\xi}_2; t)$. However, we can do something a little more sophisticated. We are assuming that the molecular diameter is very small (in particular, that the total volume (of order $N\sigma^3$) of the molecules is negligible); therefore, during the interaction of the first and second molecules no other molecules will be around (i.e., the intersection of two sets $|\mathbf{x}_i - \mathbf{x}_j| < \sigma$ and $|\mathbf{x}_j - \mathbf{x}_k| < \sigma$, with $i \neq j$, $j \neq k$, $k \neq i$, is negligible). Then, inside the sphere we can write the Liouville equation as follows:

$$\frac{\partial P_N}{\partial t} + \boldsymbol{\xi}_1 \cdot \frac{\partial P_N}{\partial \mathbf{x}_1} + \boldsymbol{\xi}_2 \cdot \frac{\partial P_N}{\partial \mathbf{x}_2} + \sum_{j=3}^{N} \boldsymbol{\xi}_j \cdot \frac{\partial P_N}{\partial \mathbf{x}_j} + \mathbf{X}_{12} \cdot \frac{\partial P_N}{\partial \boldsymbol{\xi}_1} + \mathbf{X}_{21} \cdot \frac{\partial P_N}{\partial \boldsymbol{\xi}_2} = 0 \tag{6.7}$$

where $\mathbf{X}_{12}$ is the force (per unit mass) exerted by the second molecule on the first, and $\mathbf{X}_{21}$ the force (per unit mass) exerted by the first molecule on the second ($\mathbf{X}_{12} = -\mathbf{X}_{21}$, in accordance with Newton's third law). Now, if we integrate Eq. (6.7) with respect to the positions and velocities of the remaining $(N - 2)$ particles, we have

$$\frac{\partial P_N^{(2)}}{\partial t} + \boldsymbol{\xi}_1 \cdot \frac{\partial P_N^{(2)}}{\partial \mathbf{x}_1} + \boldsymbol{\xi}_2 \cdot \frac{\partial P_N^{(2)}}{\partial \mathbf{x}_2} + \mathbf{X}_{12} \cdot \frac{\partial P_N^{(2)}}{\partial \boldsymbol{\xi}_1} + \mathbf{X}_{21} \cdot \frac{\partial P_N^{(2)}}{\partial \boldsymbol{\xi}_2} = 0 \tag{6.8}$$

$$|\mathbf{x}_1 - \mathbf{x}_2| \leq \sigma$$

since no surface integrals arise now, because we decided to neglect the fact that in a small subset Eq. (6.7) does not hold and therefore we can integrate it over the whole phase space of the $(N - 2)$ molecules with $j \geq 3$.

Equation (6.8) says a rather obvious thing: the two-particle distribution function changes according to a Liouville equation for a system of two particles. Let τ be the duration of the interaction and $\mathbf{x}_1', \mathbf{x}_2', \boldsymbol{\xi}_1', \boldsymbol{\xi}_2'$ the positions and velocities of the pair of molecules at the beginning of the collision, which will take final values $\mathbf{x}_1, \mathbf{x}_2^-, \boldsymbol{\xi}_1, \boldsymbol{\xi}_2$ according to the laws of two-body dynamics:

$$\frac{d\mathbf{x}_1}{dt} = \boldsymbol{\xi}_1; \qquad \frac{d\mathbf{x}_2}{dt} = \boldsymbol{\xi}_2; \qquad \frac{d\boldsymbol{\xi}_1}{dt} = \mathbf{X}_{12}; \qquad \frac{d\boldsymbol{\xi}_2}{dt} = \mathbf{X}_{21} \tag{6.9}$$

then, since these equations are also the equations of the characteristic lines of the above partial differential equation, Eq. (6.8), we shall have

$$P_N^{(2)}(\mathbf{x}_1, \mathbf{x}_2^- ; \boldsymbol{\xi}_1, \boldsymbol{\xi}_2 ; t) = P_N^{(2)}(\mathbf{x}_1', \mathbf{x}_2' ; \boldsymbol{\xi}_1', \boldsymbol{\xi}_2' ; t - \tau)$$

since this is a consequence of the fact that $P_N^{(2)}$ is constant along the characteristics of Eq. (6.8). But now $\mathbf{x}_2'$ is the coordinate of a molecule entering the "action sphere" of the first molecule, and we can apply the idea of statistical independence, or "molecular chaos," Eq. (5.14) in the limit of very large values of N. Accordingly, we have

$$(N-1)\int P_N^{(2)}(\mathbf{x}_1, \mathbf{X}_2 ; \boldsymbol{\xi}_1, \boldsymbol{\xi}_2 ; t)(\boldsymbol{\xi}_2 - \boldsymbol{\xi}_1)\cdot \mathbf{n}_2\, dS_2\, d\boldsymbol{\xi}_2$$

$$\to N \int [P_\infty^{(1)}(\mathbf{x}_1 ; \boldsymbol{\xi}_1 ; t) P_\infty^{(1)}(\mathbf{x}_2 ; \boldsymbol{\xi}_2 ; t) - P_\infty^{(1)}(\mathbf{x}_1' ; \boldsymbol{\xi}_1' ; t - \tau)$$

$$\times P_\infty^{(1)}(\mathbf{x}_2' ; \boldsymbol{\xi}_2' ; t - \tau)] |\boldsymbol{\xi}_2 - \boldsymbol{\xi}_1| r\, dr\, d\epsilon\, d\boldsymbol{\xi}_2 \tag{6.10}$$

where we have dropped the superscript + of $\mathbf{x}_2$, since it is no longer necessary, and changed $N - 1$ to N, as is clearly permissible under our assumptions. The arrow in Eq. (6.10) means that we can effect the substitution expressed by this equation in the limit of very large N and very small σ (such that $N\sigma^2$ is a finite quantity; see below). Now, the points $\mathbf{x}_1, \mathbf{x}_1', \mathbf{x}_2, \mathbf{x}_2'$ are separated by distances of order σ; then we identify them ($\mathbf{x}_1' = \mathbf{x}_2 = \mathbf{x}_2' = \mathbf{x}_1$) in the argument of $P_\infty^{(1)}$, since taking into account their differences would give contributions of the same order of already neglected quantities. Since τ is of the same order of magnitude as σ divided by a typical molecular velocity, in order to be consistent, we have also to neglect differences of order τ in the argument of $P_\infty^{(1)}$, i.e., identify $t - \tau$ with t in Eq. (6.10). With such simplifications, substituting Eq. (6.10) into Eq. (6.4) gives (in the limit $N \to \infty$, $\sigma \to 0$, $N\sigma^2$ finite) the Boltzmann equation for P

$$\frac{\partial P}{\partial t} + \boldsymbol{\xi}\cdot\frac{\partial P}{\partial \mathbf{x}} = N \int [P(\boldsymbol{\xi}')P(\boldsymbol{\xi}_1') - P(\boldsymbol{\xi})P(\boldsymbol{\xi}_1)] V r\, dr\, d\epsilon\, d\boldsymbol{\xi}_1 \tag{6.11}$$

where the index 1 has been omitted and the index 2 changed into 1. We have also written P in place of $P_\infty^{(1)}$, V in place of $|\boldsymbol{\xi}_2 - \boldsymbol{\xi}_1|$ (the relative speed), and indicated only the dependence of P upon the velocity variables, since the other arguments are $\mathbf{x}$, t throughout. In addition, the integral term has been shifted to the right-hand side in order to conform to the usual way of writing the Boltzmann equation; this term is usually called the collision integral, since it takes into account the interactions between molecules.

As stated before, the above proof suffers from a logical gap; as a matter of fact, we didn't prove that Eq. (5.14) applies to molecules entering into an interaction. We only said that "it makes sense," since the two colliding molecules are just two randomly picked molecules in a set of infinitely many (in the limit $N \to \infty$).

This gap in the argument is not completely filled in the existing literature, although a lot of progress has been achieved in the last twenty years. When the first edition of this book appeared, there was only a conjecture, put forward by H. Grad in 1949, that, in the limit when N goes to infinity and σ goes to 0 in such a way that $N\sigma^2$ tends to a finite number (the so-called Boltzmann–Grad limit), a one-particle distribution obtained from the Liouville equation should tend, in a suitable sense, to a solution of the Boltzmann equation. We remark that $N\sigma^2$ (multiplied by a typical molecular velocity and divided by the volume of the region occupied by the gas) is the order of magnitude of the right-hand side of Eq. (6.11). In order to give an idea of the orders of magnitude involved in a typical case, we note that $N\sigma^2 = 10^4\ \text{cm}^2$ for $N = 10^{20}$ and $\sigma = 10^{-8}$ cm, while $N\sigma^3$ (a parameter measuring the importance of the terms that we have neglected) equals $10^{-10}\ \text{m}^3$. In 1972 a "proof" of the conjecture was given (ref. 11) under the proviso that: (1) the s-particle distributions (with a fixed s) obtained by solving the Liouville equation with uncorrelated initial data have a sufficiently smooth limit in the Boltzmann–Grad limit, (2) the equations that these limits turn out to satisfy admit a uniqueness theorem, and (3) the Boltzmann equation corresponding to the same initial state has an existence theorem. Then as long as the solution of the Boltzmann equation exists, it will agree with the Boltzmann–Grad limit of the one-particle distribution obtained from the Liouville equation. The proof was not mathematically rigorous, because many assumptions were made without showing that they were not contradictory. In 1975 O. Lanford (ref. 12) proved that if we restrict ourselves to a time interval of the order of one quarter of the mean free time (the time between two subsequent collisions of a molecule), then all the assumptions made above are in fact a consequence of the choice of sufficiently good initial data. The problem of a rigorous, globally valid justification of the Boltzmann equation is still open, except for the case of small data, for which the difficulties were overcome by R. Illner and M. Pulvirenti (refs. 13 and 14) in the last few years.

We now have to look more carefully at the equation which we have obtained, Eq. (6.11). We note that it is a nonlinear, integral, partial differential, functional equation, where the specification "functional" refers to the fact that the unknown function P appears in the integral term not only with the arguments ξ (the current velocity variable) and ξ_1 (the integration variable) but also with the arguments ξ' and ξ_1'. The latter are related to ξ and ξ_1 implicitly by the condition of being transformed into ξ and ξ_1 by effect of a collision, and explicitly by relations which will be investigated in the next section.

Frequently when dealing with the Boltzmann equation one introduces a different unknown function f which is related to P by

$$f = NmP \tag{6.12}$$

where N is the number of molecules and m the mass of a molecule. The meaning of f is an (expected) mass density in the phase space of a single particle, i.e., the (expected) "mass per unit volume" in the six-dimensional space described by $(\mathbf{x}, \boldsymbol{\xi})$. Sometimes the number density

$$F = NP = f/m \tag{6.13}$$

giving the expected number of molecules per unit volume in the phase space is also used. We note that because of the normalization condition

$$\int P \, d\mathbf{x} \, d\boldsymbol{\xi} = 1 \tag{6.14}$$

we have

$$\int f \, d\mathbf{x} \, d\boldsymbol{\xi} = Nm = M \tag{6.15}$$

where M is the total mass of the gas, and

$$\int F \, d\mathbf{x} \, d\boldsymbol{\xi} = N \tag{6.16}$$

It is clear that in terms of f one has

$$\frac{\partial f}{\partial t} + \boldsymbol{\xi} \cdot \frac{\partial f}{\partial \mathbf{x}} = \frac{1}{m} \int (f'f_1' - ff_1) Vr \, dr \, d\epsilon \, d\boldsymbol{\xi}_1 \tag{6.17}$$

where $f_1 = f(\boldsymbol{\xi}_1)$, $f' = f(\boldsymbol{\xi}')$, $f_1' = f(\boldsymbol{\xi}_1')$. This is the form of the Boltzmann equation which will be used in the following. We note also that the above considerations could be repeated if in addition to the small-range intermolecular force an external force per unit mass, $\mathbf{X}$, acts on the molecules, the only influence of this additional force being that one should add a term $\mathbf{X} \cdot \partial f / \partial \boldsymbol{\xi}$ to the left-hand side of Eq. (6.17). Since we shall usually consider cases when the external action on the gas is exerted through solid boundaries (surface forces), we shall not usually write the above-mentioned term describing body forces; however, it should be kept in mind that such simplification implies neglecting, *inter alia*, gravity.

We note also that there is no difficulty in extending the above treatment to the case when different gases are present, i.e., the molecules have different masses and different interaction laws. In such a case we obtain a system of

Boltzmann equations where the distribution functions f_i of the single species are coupled through collision terms describing interaction between molecules of different species.

Finally, we point out that the above derivation of the Boltzmann equation is not the standard one (see references at the end of this chapter); the skeleton of the above proof is taken from Grad's article (ref. 9).

7. Details of the Collision Operator

In order to specify completely the right-hand side of Eq. (6.17) we have to find the expressions for $\boldsymbol{\xi}'$ and $\boldsymbol{\xi}_1'$ in terms of $\boldsymbol{\xi}$, $\boldsymbol{\xi}_1$, r and ϵ. The terms $\boldsymbol{\xi}'$ and $\boldsymbol{\xi}_1'$ were defined as the velocities of two colliding molecules which after the collision have velocities $\boldsymbol{\xi}$ and $\boldsymbol{\xi}_1$, respectively. The evolution from $\boldsymbol{\xi}$ to $\boldsymbol{\xi}'$ and $\boldsymbol{\xi}_1$ to $\boldsymbol{\xi}_1'$ is described by Eqs. (6.9), which now we rewrite as follows:

$$\frac{d\mathbf{x}}{dt} = \mathbf{v}; \qquad \frac{d\mathbf{x}_1}{dt} = \mathbf{v}_1; \qquad \frac{d\mathbf{v}}{dt} = -\frac{1}{m}\frac{\partial U}{\partial \rho}\frac{\boldsymbol{\rho}}{\rho}; \qquad \frac{d\mathbf{v}_1}{dt} = \frac{1}{m}\frac{\partial U}{\partial \rho}\frac{\boldsymbol{\rho}}{\rho} \tag{7.1}$$

where $\mathbf{x}$ and $\mathbf{v}$ are, respectively, the position and velocity of the first molecule (whose velocity changes from $\boldsymbol{\xi}'$ to $\boldsymbol{\xi}$); $\mathbf{x}_1$ and $\mathbf{v}_1$ the position and velocity, respectively, of the second molecule (whose velocity changes from $\boldsymbol{\xi}_1'$ to $\boldsymbol{\xi}_1$); $\boldsymbol{\rho} = \mathbf{x} - \mathbf{x}_1$; $\rho = |\boldsymbol{\rho}|$; and $U(\rho)$ is the two-body potential such that $-\partial U/\partial \boldsymbol{\rho}$ is the force exerted by the second molecule on the first and $\partial U/\partial \boldsymbol{\rho} = -\partial U/\partial(-\boldsymbol{\rho})$ the force exerted by the first molecule on the second.

By simply adding the last two equations we obtain

$$\frac{d}{dt}(\mathbf{v} + \mathbf{v}_1) = 0 \tag{7.2}$$

which expresses the conservation of momentum. This means in particular that

$$\boldsymbol{\xi}' + \boldsymbol{\xi}_1' = \boldsymbol{\xi} + \boldsymbol{\xi}_1 \tag{7.3}$$

In addition, if we scalarly multiply the same equations by $\mathbf{v}$ and $\mathbf{v}_1$, respectively, and add we have

$$\begin{aligned}\frac{d}{dt}\left(\frac{1}{2}v^2 + \frac{1}{2}v_1^2\right) &= -\frac{1}{m}\frac{\partial U}{\partial \rho}\frac{\boldsymbol{\rho}}{\rho}\cdot(\mathbf{v} - \mathbf{v}_1) = -\frac{1}{m}\frac{\partial U}{\partial \rho}\frac{\boldsymbol{\rho}}{\rho}\cdot\left(\frac{d\mathbf{x}}{dt} - \frac{d\mathbf{x}_1}{dt}\right)\\ &= -\frac{1}{m}\frac{\partial U}{\partial \boldsymbol{\rho}}\cdot\frac{d\boldsymbol{\rho}}{dt} = -\frac{d}{dt}\left(\frac{U}{m}\right)\end{aligned} \tag{7.4}$$

i.e.,

$$\frac{d}{dt}\left[v^2 + v_1^2 + \frac{2}{m}U\right] = 0 \tag{7.5}$$

(In general, if a boldface letter denotes a vector, we shall use the same letter as lightface italic to denote the magnitude of the same vector). Equation (7.5) expresses the conservation of energy. In particular, since U has the same value at the beginning and at the end of the interaction (U is a constant outside the sphere $|\mathbf{x} - \mathbf{x}_1| = \sigma$), we have

$$\xi'^2 + \xi_1'^2 = \xi^2 + \xi_1^2 \tag{7.6}$$

Equations (7.3) and (7.6) relate $\boldsymbol{\xi}'$ and $\boldsymbol{\xi}_1'$ to $\boldsymbol{\xi}$ and $\boldsymbol{\xi}_1$. These equations are not sufficient, however, to give the complete expressions for $\boldsymbol{\xi}'$ and $\boldsymbol{\xi}_1'$ for which we are looking. In fact, we have four scalar equations for six scalar quantities. In order to exploit the information contained in Eqs. (7.3) and (7.6), we introduce the unit vector in the direction of $\boldsymbol{\xi}' - \boldsymbol{\xi}$:

$$\boldsymbol{\xi}' - \boldsymbol{\xi} = \boldsymbol{\alpha} C \qquad \alpha^2 = \boldsymbol{\alpha} \cdot \boldsymbol{\alpha} = 1 \tag{7.7}$$

where C is a scalar. We want now, as a first step, to express $\boldsymbol{\xi}'$ and $\boldsymbol{\xi}_1'$ in terms of $\boldsymbol{\xi}$, $\boldsymbol{\xi}_1$, and $\boldsymbol{\alpha}$. Our task is then reduced to finding a proper expression for C. We now have

$$\begin{aligned} \boldsymbol{\xi}' &= \boldsymbol{\xi} + \boldsymbol{\alpha} C \\ \boldsymbol{\xi}_1' &= \boldsymbol{\xi}_1 - \boldsymbol{\alpha} C \end{aligned} \tag{7.8}$$

where the first of these equations is a consequence of Eq. (7.7) and the second of both Eqs. (7.7) and (7.3). Conversely, Eqs. (7.8) imply Eq. (7.3); this means that we have already fully exploited the momentum conservation. Now we want to use conservation of energy to find C. Substituting Eqs. (7.8) into Eq. (7.6) gives

$$\xi^2 + 2\boldsymbol{\alpha} \cdot \boldsymbol{\xi} C + C^2 + \xi_1^2 - 2\boldsymbol{\alpha} \cdot \boldsymbol{\xi}_1 + C^2 = \xi^2 + \xi_1^2 \tag{7.9}$$

Cancelling equal terms from each side and putting $\mathbf{V} = \boldsymbol{\xi}_1 - \boldsymbol{\xi}$ (the relative velocity before the interaction), we have

$$-\boldsymbol{\alpha} \cdot \mathbf{V} C + C^2 = 0 \tag{7.10}$$

Hence, dismissing $C = 0$, which corresponds to a trivial solution of the conservation equations (absence of interaction), we have

$$C = \boldsymbol{\alpha} \cdot \mathbf{V} \tag{7.11}$$

and therefore

$$\boldsymbol{\xi}' = \boldsymbol{\xi} + \boldsymbol{\alpha}(\boldsymbol{\alpha} \cdot \mathbf{V}); \qquad \boldsymbol{\xi}_1' = \boldsymbol{\xi}_1 - \boldsymbol{\alpha}(\boldsymbol{\alpha} \cdot \mathbf{V}) \tag{7.12}$$

i.e., the desired expressions for $\boldsymbol{\xi}'$ and $\boldsymbol{\xi}_1'$ in terms of $\boldsymbol{\xi}$, $\boldsymbol{\xi}_1$, and $\boldsymbol{\alpha}$. We note a consequence of these equations: if we put $\mathbf{V}' = \boldsymbol{\xi}_1' - \boldsymbol{\xi}'$ (the relative velocity after the interaction), then

$$\mathbf{V}' = \mathbf{V} - 2\boldsymbol{\alpha}(\boldsymbol{\alpha}\cdot\mathbf{V}) \tag{7.13}$$

As a consequence,

$$V'^2 = V^2 - 4(\boldsymbol{\alpha}\cdot\mathbf{V})(\boldsymbol{\alpha}\cdot\mathbf{V}) + 4(\boldsymbol{\alpha}\cdot\mathbf{V})^2 = V^2 \tag{7.14}$$

i.e., the relative speed after the interaction is the same as before. Let us now see what the angle is between $\mathbf{V}'$ and $\mathbf{V}$ (which gives the deflection of a molecule in the center-of-mass system). We have from Eq. (7.13)

$$\mathbf{V}'\cdot\mathbf{V} = V^2 - 2(\boldsymbol{\alpha}\cdot\mathbf{V})^2 \tag{7.15}$$

Therefore if we denote by ψ the angle of deflection and θ the angle between $\boldsymbol{\alpha}$ and $\mathbf{V}$ (Fig. 1) we have, by means of Eqs. (7.14) and (7.15),

$$\cos\psi = 1 - 2\cos^2\theta = -\cos(2\theta) \tag{7.16}$$

i.e., the angle of deflection is $\psi = \pi - 2\theta$. Let us now take polar coordinates (θ, ϵ) for $\boldsymbol{\alpha}$ with $\mathbf{V}$ as pole; then

$$\begin{aligned} \boldsymbol{\alpha}\cdot\mathbf{V} &= V\cos\theta \\ \boldsymbol{\alpha} &= (\cos\theta, \sin\theta\cos\epsilon, \sin\theta\sin\epsilon) \end{aligned} \tag{7.17}$$

As is well known [and can be easily shown by means of Eqs. (7.1)] the relative motion of two bodies interacting with a central force takes place in a plane, which, of course, contains both $\mathbf{V}$ and $\boldsymbol{\alpha}$ and is specified by the angle ϵ formed by the plane of motion with an arbitrary fixed plane containing the polar axis V. This angle is, by definition, the coordinate ϵ of $\boldsymbol{\alpha}$ and also the angle ϵ in the diameter plane (orthogonal to $\mathbf{V}$) of the "protection sphere" introduced in Section 6. Therefore, if we find a relation between θ and the radial coordinate r in the diameter plane of the protection sphere $|\mathbf{x} - \mathbf{x}_1| = \sigma$, then our task of expressing $\boldsymbol{\xi}'$ and $\boldsymbol{\xi}_1'$ in terms of $\boldsymbol{\xi}, \boldsymbol{\xi}_1, \epsilon$, and r will be accomplished. To do this, however, we have to study the two-body interaction in the plane of motion more carefully.

It is a standard procedure to deduce from Eq. (7.1) the pair of equations

$$\begin{aligned} \tfrac{1}{4}m[\dot\rho^2 + \rho^2\dot\varphi^2] + U(\rho) &= \tfrac{1}{4}mV^2 + U(\sigma) \qquad (\rho \le \sigma) \\ \rho^2\dot\varphi &= rV \end{aligned} \tag{7.18}$$

where φ is the azimuthal angle in the plane of motion, ρ is the radial coordinate, and r the impact parameter, i.e., the distance of closest approach of the two particles had they continued their motion without interacting. It is clear (see Fig. 1) that this r is the same r used as polar coordinate in the diameter plane of the "protection sphere," since the latter r was defined as the distance between one mass point and the relative trajectory which the second mass point follows in absence of interaction. Equations (7.18) can also be written directly, since they express the conservation of energy and angular momentum in the reference system of one particle (with the reduced mass $m/2$ in place of the mass m of the second particle). We could omit $U(\sigma)$ in Eqs. (7.18) by stipulating $U(\sigma) = 0$, as is always possible; however, it is more instructive to retain the constant explicitly. We shall also restrict our considerations to repulsive potentials, which is the important case for close interactions between molecules in a gas.

Now, we can easily integrate the above equations (one can eliminate time derivatives, using φ as independent variable); the orbit is, of course, symmetric with respect to a straight line, the so-called apse line, on which both molecules lie at their closest approach. Accordingly, the apse line bisects the angle between the straight lines directed along $-\mathbf{V}'$ and $\mathbf{V}$, i.e., is directed along $\boldsymbol{\alpha}$. Therefore the angle θ can be easily evaluated since it is equal to the angle between $\mathbf{V}$ and the apse line; this angle can be easily obtained from the solution of Eqs. (7.18), and is given by

$$\theta = \frac{m^{1/2}Vr}{2}\int_{\rho_0}^{\sigma} \rho^{-2}\left[\frac{m}{4}V^2\left(1 - \frac{r^2}{\rho^2}\right) - U(\rho) + U(\sigma)\right]^{-1/2} d\rho + \sin^{-1}\left(\frac{r}{\sigma}\right) \tag{7.19}$$

where ρ_0 is the distance of closest approach, which satisfies

$$\frac{m}{4}V^2\left(1 - \frac{r^2}{\rho_0^2}\right) = U(\rho_0) - U(\sigma) \tag{7.20}$$

We note that $\rho_0 \leq \sigma$ (otherwise no deflection arises, since the molecules do not enter into an interaction); it is also clear that $r \leq \sigma$, but it is interesting to note that this follows from $\rho_0 \leq \sigma$ and the assumption of repulsive potential [which implies $U(\rho_0) - U(\sigma) \geq 0$].

What one should now do is insert Eq. (7.19) into Eq. (7.17) and then Eq. (7.17) into (7.12) in order to have the complete expression of $\boldsymbol{\xi}'$ and $\boldsymbol{\xi}_1'$ in terms of $\boldsymbol{\xi}$, $\boldsymbol{\xi}_1$, r, and ϵ. It is clear that after such a substitution the arguments of the unknown f in the collision integral would be extremely complicated. Therefore it is more expedient to adopt θ as integration variable in place of r. Then the final expressions of $\boldsymbol{\xi}'$ and $\boldsymbol{\xi}_1'$ are obtained by putting Eq. (7.17) into Eq. (7.12), Now, however, we have to make a change of variables from r to θ. The assumption of repulsive potential guarantees that $\theta = \theta(r)$ is a monotonic function, and therefore one can invert this relation to give $r = r(\theta)$. Then if we put

$$Vr\,\partial r/\partial\theta = B(\theta, V) \tag{7.21}$$

we can write the Boltzmann equation as follows:

$$\frac{\partial f}{\partial t} + \boldsymbol{\xi} \cdot \frac{\partial f}{\partial \mathbf{x}} = \frac{1}{m} \int_{\Xi} \int_{0}^{2\pi} \int_{0}^{\pi/2} (f'f_1' - ff_1)B(\theta, V)\, d\boldsymbol{\xi}\, d\epsilon\, d\theta \tag{7.22}$$

where the arguments of f' and f_1' are related to $\boldsymbol{\xi}$, $\boldsymbol{\xi}_1$, ϵ, and θ by Eqs. (7.12) and (7.17); the integration with respect to $\boldsymbol{\xi}_1$ extends to the whole three-dimensional velocity space Ξ, the integration with respect to ϵ goes, as obvious, from 0 to 2π, while the angle θ varies from 0 (head-on collisions, $r = 0$) to $\pi/2$ (grazing collisions, $r = \sigma$).

It is seen that all the complicated details of the two-body interactions are summarized by the quantity $B(\theta, V)$, which gives essentially the (unnormalized) probability density of a relative deflection equal to $\psi = \pi - 2\theta$ for a pair of molecules having relative speed V. It is true that $B(\theta, V)$ cannot be expressed in terms of elementary functions even for such simple potentials as inverse-power potentials ($U = kr^{1-n}, n \neq 2, 3$), but this will not prevent us from using Eq. (7.22) and deducing interesting results, as we shall see. The cases of inverse-square and inverse-cube force laws (with a cutoff, of course, for $r > \sigma$) are amenable to an analytic treatment, but describe too soft an interaction at small distances to be realistic. On the other hand, one can consider an extremely hard interaction by thinking of the molecules as rigid spheres of diameter σ ($U = 0, r \geq \sigma$; $U = \infty, r < \sigma$). In this case the calculation of $B(\theta, V)$ can be made easily: whenever the pointlike bullet molecule hits the "protection sphere" of the target molecule it is specularly reflected, and the angle θ formed by the initial trajectory with the normal to the sphere is related to the impact parameter r by the relation

$$r = \sigma \sin \theta \tag{7.23}$$

which is immediately obtained by elementary trigonometrical considerations. Accordingly, Eq. (7.21) gives

$$B(\theta, V) = V\sigma^2 \cos \theta \sin \theta \tag{7.24}$$

A short discussion about the meaning of the molecular diameter σ is not out of order here. We have assumed that the interaction force is zero for intermolecular distances larger than σ; it is known, however, that the interaction force between a pair of molecules is different from zero at any distance, being, in general, strongly repulsive at short distances and weakly attractive for large separations. We have neglected this weakly attractive part, as is justified for ideal gases; accordingly, we could define the molecular diameter σ as the distance at which repulsion changes into attraction. However, this definition is by no means standard. As a matter of fact, the classical treatments of the Boltzmann equation for force laws other than rigid spheres assume the radius σ of the protection sphere equal to ∞! This seems very odd at first sight, because the σ defined above is of the order 10^{-8} cm and we used the limit $\sigma \to 0$ when deriving the Boltzmann equation. However, there is a justification in putting $\sigma = \infty$; as a matter of fact, σ enters in Eq. (7.22) only through $B(\theta, V)$, and increasing σ means that we take more and more grazing collisions into account. Now, these added collisions are so grazing that they practically do not deflect the molecules from their initial paths; accordingly, a molecule which enters into such a grazing collision in a certain state of motion emerges in practically the same state, and therefore, it is argued, the contribution from such molecules to the integral in Eq. (7.22) is practically zero ($f'f_1' \approx ff_1$). In other words, if we accept this argument, we should say that, if we arbitrarily enlarge the value of σ, and, in particular, let $\sigma \to \infty$, we do not change anything, since we just add the same large number to the two terms of the subtraction

$$\left(\int_{\Xi}\int_0^{2\pi}\int_0^{\pi/2} f'f_1'B(\theta, V)\, d\xi_1\, d\epsilon\, d\theta\right) - \left(\int_{\Xi}\int_0^{2\pi}\int_0^{\pi/2} ff'B(\theta, V)\, d\xi_1\, d\epsilon\, d\theta\right) \tag{7.25}$$

How large these numbers are one can easily see, since, for any potential extending to infinity the above two integrals diverge! This is obvious if one remembers that, e.g., the second one is also given by

$$\int_{\Xi}\int_0^{2\pi}\int_0^{\sigma} ff_1 V\, d\xi_1\, r\, dr\, d\epsilon = f(\xi)\left[\int_{\Xi} f_1(\xi_1)|\xi_1 - \xi|\, d\xi_1\right]\pi\sigma^2 \tag{7.26}$$

which is plainly divergent when $\sigma \to \infty$.

It is true that, although both integrals diverge, if one does not separate them and writes the collision term as in Eq. (7.22), the final result is finite. However, this is not a justification for taking $\sigma = \infty$; it is just a vindication of the capital crime of introducing divergent integrals! A complete justification should be based upon a proof that the grazing collisions corresponding to very large values of the impact parameter make a negligible contribution to the collision integral. This, however, is not true, since there are special choices of f for which the contribution from large distances is overwhelming. Furthermore, the choice $\sigma = \infty$ imposes the restriction that one cannot make the splitting shown in Eq. (7.25), and this is inconvenient both in discussing properties of a general nature and in solving particular problems. On the other hand, one could object that taking a finite σ is an artificial procedure, since actual interaction potentials extend to infinity. Thus, why, instead of putting the correct potential extending to infinity into the equations, do we introduce the artificial molecular diameter σ? The answer is simple: because otherwise one could not even write the Boltzmann equation. As a matter of fact, the basic assumption of neglecting interactions among more than two molecules implies that calculations based on the two-body problem have no meaning for distances larger than $n^{-1/3}$, where n is the number density (number of molecules in the unit volume); i.e., consistency forces us to accept the restriction $\sigma < \sigma_0 \approx n^{-1/3}$. A first important step toward a justification of the standard treatment with $\sigma = \infty$ would be to show that there are no practical differences between taking $\sigma = \infty$ and $\sigma = \sigma_0 \approx n^{-1/3}$. But this seems unlikely to be verified when appreciable changes arise on distances of the order of the length of the average path of a molecule between two subsequent collisions (the so-called mean free path).

In the following we shall usually consider potentials with a cutoff at some unspecified distance σ. Since, however, potentials extending to infinity do offer some simplifications (especially in the case of power-law potentials), we shall sometimes consider the limiting case $\sigma = \infty$. In order to show where the mentioned simplifications arise, we note that in the case of power-law potentials we have $U(r) = kr^{-(n-1)}$ ($n > 3$, since for $n \leq 3$ the contribution from large distances is diverging even if the partial cancellation arising from subtraction is taken into account). Then if we take $\sigma = \infty$, $U(\sigma) = 0$, Eq. (7.19) becomes

$$\theta = \frac{m^{1/2}}{2} Vr \int_{\rho_0}^{\infty} \rho^{-2} \left[\frac{m}{4} V^2 \left(1 - \frac{r^2}{\rho^2}\right) - \frac{k}{\rho^{n-1}}\right]^{-1/2} d\rho \tag{7.27}$$

where ρ_0 satisfies

$$\frac{m}{4} V^2 \left(1 - \frac{r^2}{\rho_0^2}\right) - \frac{k}{\rho_0^{n-1}} = 0 \tag{7.28}$$

If we now put

$$b = r\left(\frac{m}{4k} V^2\right)^{1/(n-1)}; \qquad \frac{r}{\rho} = x; \qquad \frac{r}{\rho_0} = x_0 \tag{7.29}$$

we have

$$\theta = \int_0^{x_0} [1 - x^2 - (x/b)^{n-1}]^{-1/2}\, dx \tag{7.30}$$

where

$$1 - x_0^2 - (x_0/b)^{n-1} = 0 \tag{7.31}$$

Equations (7.30) and (7.31) give $\theta = \theta(b)$, and, as a consequence, $b = b(\theta)$; this relationship is purely mathematical and does not depend upon any other variable or parameter (in particular, not upon V). As a consequence,

$$B(\theta, V) = Vr\frac{\partial r}{\partial \theta} = V\left(\frac{mk}{4}V^2\right)^{-2/(n-1)} b\frac{db}{d\theta}$$

$$= \left(\frac{mk}{4}\right)^{-2/(n-1)} V^{(n-5)/(n-1)} b\frac{db}{d\theta} \tag{7.32}$$

The relevant simplification for inverse-power laws is therefore that $B(\theta, V)$ becomes the product of a function of θ alone times a fractional power of V.

A particularly noticeable simplification is present when $n = 5$, because then V disappears. This simplification was discovered by Maxwell and appealed so much to him that he endeavored to show, on the basis of the scanty experimental and theoretical information then available, that the real molecules interact with a force varying as the inverse fifth power of the distance. This is the reason why the fictitious molecules interacting in this way are usually called Maxwell molecules. Although today we know that actual molecules are not Maxwell molecules, yet the concept is extremely useful because the assumption of an inverse-fifth-power law frequently considerably simplifies the calculations and gives satisfactory answers, or, at least, first approximations to satisfactory answers.

References

Section 1——Among the many books dealing with equilibrium statistical mechanics (sometimes also called statistical thermodynamics) we shall quote only

1. E. W. Schrödinger, *Statistical Thermodynamics*, Cambridge University Press, Cambridge (1964),

which explains basic concepts both clearly and concisely (however, it does not deal with practical applications).

Sections 2 and 3——The standard references on the theory of generalized functions are

2. L. Schwartz, *Théorie des Distributions*, Hermann, Paris, Part I (1957), Part II (1959);
3. I. M. Gel'fand and co-authors, *Generalized Functions*, English translation from the Russian, Academic Press, New York (1964),

particularly the first of the five volumes of ref. 3. However, a short and very readable account, quite sufficient for the present purposes, is

4. J. M. Lighthill, *Fourier Analysis and Generalized Functions*, Cambridge University Press, Cambridge (1958).

Section 4——For the concept of characteristic lines of a partial differential equation, see, e.g.,

5. R. Courant and D. Hilbert, *Methods of Mathematical Physics*, Interscience Publishers, New York (1962).

Section 6——The standard derivation of the Boltzmann equation is due to Boltzmann himself, and can be found, e.g., in

6. L. Boltzmann, *Lectures on Gas Theory*, English translation from the German, University of California Press, Berkeley (1964);
7. S. Chapman and T. G. Cowling, *The Mathematical Theory of Non-Uniform Gases*, Cambridge University Press, Cambridge (1952);
8. T. Carleman, *Problèmes Mathématiques dans la Théorie Cinétique des Gaz*, Almqvist and Wiksells, Uppsala (1957);
9. H. Grad, in: *Handbuch der Physik*, Vol. XII, Springer, Berlin (1958);
10. M. N. Kogan, *Rarefied Gas Dynamics*, Plenum Press, New York (1969).

Reference 10 is also to be consulted for the Boltzmann equations describing gas mixtures and polyatomic gases. For recent developments in the justification of the Boltzmann equation see

11. C. Cercignani, *Transport Theory and Statistical Physics* **2**, 211 (1972);
12. O. E. Lanford, III, in: *Proceedings of the 1974 Battelle Rencontre on Dynamical Systems* (J. Moser, ed.), Lecture Notes in Physics, Volume 35, Springer, Berlin (1975);
13. R. Illner and M. Pulvirenti, *Commun. Math. Phys.* **105**, 189 (1986);
14. R. Illner and M. Pulvirenti, *Commun. Math. Phys.* **121**, 143 (1989);
15. C. Cercignani, *The Boltzmann Equation and Its Applications*, Springer, New York (1988).

Section 7——The content of this section is classical, except for the detailed discussion of the finiteness of σ, which appears to be new (see, however, ref. 7, p. 67). The standard treatment can be found in refs. 6–9. Maxwell molecules were introduced in

16. J. C. Maxwell, *Phil. Trans. R. Soc.* **157** (1866); reprinted in: *The Scientific Papers of J. C. Maxwell*, Dover Publications, New York (1965).

Chapter II

BASIC PROPERTIES

1. Elementary Manipulations of the Collision Operator. Collision Invariants

In this chapter we shall study the basic properties of the Boltzmann equation. It is clear from Eq. (7.22) of Chapter I that the left-hand side and the right-hand side are completely different in nature, from both a mathematical and a physical standpoint. The left-hand side contains a linear differential operator which acts on the space- and time-dependence of f; if we equate this side to zero, we obtain an equation for the time evolution in absence of collisions, and the differential operator is accordingly called the "free-streaming operator."

The right-hand side of Eq. (7.22) of Chapter I contains a quadratic integral operator

$$Q(f, f) = (1/m) \int (f'f'_1 - ff_1) B(\theta, V)\, d\xi_1\, d\epsilon\, d\theta \tag{1.1}$$

which acts on the velocity-dependence of f; it describes the effect of interactions, and is accordingly called the collision integral, or, equivalently, the collision operator, or, simply, the collision term. The limits of integration shown in Eq. (7.22) of Chapter I have been omitted here for simplicity in notation.

It is clear that although the free-streaming operator has its own interest and presents nontrivial problems when realistic boundary conditions are prescribed, the collision term is what characterizes the Boltzmann equation because of its unusual form. It therefore seems appropriate to study some properties which make the manipulation of Q possible in many problems of basic character, in spite of its complicated form.

Actually, we shall study here a slightly more general expression, the bilinear quantity

$$Q(f, g) = (1/2m) \int (f'g'_1 + f'_1 g' - fg_1 - f_1 g) B(\theta, V)\, d\xi_1\, d\epsilon\, d\theta \tag{1.2}$$

It is clear that when $g = f$ Eq. (1.2) reduces to Eq. (1.1); in addition,

$$Q(f, g) = Q(g, f) \tag{1.3}$$

To be precise, in this section we want to study some manipulations of the eightfold integral:

$$\int_{\Xi} Q(f, g)\varphi(\boldsymbol{\xi})\, d\boldsymbol{\xi} = (1/2m) \int (f'g_1' + f_1'g' - fg_1 - f_1 g) \times \varphi(\boldsymbol{\xi})B(\theta, V)\, d\boldsymbol{\xi}\, d\boldsymbol{\xi}_1\, d\epsilon\, d\theta \tag{1.4}$$

where $\varphi(\boldsymbol{\xi})$ is any function of $\boldsymbol{\xi}$ such that the indicated integrals exist.

We now perform the interchange of variables $\boldsymbol{\xi} \to \boldsymbol{\xi}_1$, $\boldsymbol{\xi}_1 \to \boldsymbol{\xi}$ (which implies also $\boldsymbol{\xi}' \to \boldsymbol{\xi}_1'$, $\boldsymbol{\xi}_1' \to \boldsymbol{\xi}'$ because of Eqs. (7.12) of Chapter I. Then, since both $B(\theta, V)$ and the quantity within brackets transform into themselves, and the Jacobian of the transformation is obviously 1, we have

$$\int Q(f, g)\varphi(\boldsymbol{\xi})\, d\boldsymbol{\xi} = (1/2m) \int (f'g_1' + f_1'g' - fg_1 - f_1 g) \times \varphi(\boldsymbol{\xi}_1)B(\theta, V)\, d\boldsymbol{\xi}\, d\boldsymbol{\xi}_1\, d\epsilon\, d\theta \tag{1.5}$$

This equation is identical to Eq. (1.4) except for having $\varphi(\boldsymbol{\xi}_1)$ in place of $\varphi(\boldsymbol{\xi})$. Now we consider another transformation of variables in Eq (1.4): $\boldsymbol{\xi} \to \boldsymbol{\xi}'$ and $\boldsymbol{\xi}_1 \to \boldsymbol{\xi}_1'$ (here, as above, $\boldsymbol{\alpha}$ can be considered as fixed). From Eq. (7.13) of Chapter I we have

$$\boldsymbol{\alpha} \cdot \mathbf{V}' = -\boldsymbol{\alpha} \cdot \mathbf{V} \tag{1.6}$$

and substituting this into Eqs. (7.12) of Chapter I gives

$$\boldsymbol{\xi} = \boldsymbol{\xi}' + \boldsymbol{\alpha}(\boldsymbol{\alpha} \cdot \mathbf{V}'); \qquad \boldsymbol{\xi}_1 = \boldsymbol{\xi}_1' - \boldsymbol{\alpha}(\boldsymbol{\alpha} \cdot \mathbf{V}') \tag{1.7}$$

These relations obviously invert Eq. (7.12) of Chapter I. If we regard Eqs. (7.12) of Chapter I as a linear transformation (with coefficients depending on $\boldsymbol{\alpha}$) of six variables into another six variables, comparison of Eqs. (7.12) of Chapter I and (1.7) here shows that the direct and the inverse transformation have the same coefficients; therefore, if A is the six-by-six matrix describing the direct transformation and A^{-1} its inverse, we have

$$\mathsf{A}^{-1} = \mathsf{A}, \qquad \text{i.e.,} \quad \mathsf{A}^2 = \mathsf{A} \cdot \mathsf{A} = \mathsf{I} \tag{1.8}$$

where I is the identity matrix and $\mathsf{A} \cdot \mathsf{A}$ denotes the usual rows-by-columns product. In accord with Binet's theorem on the determinant of a product, Eq. (1.8) implies

$$(\det \mathsf{A})^2 = 1 \tag{1.9}$$

i.e.,

$$|\det \mathsf{A}| = 1 \tag{1.10}$$

the vertical bars denoting absolute value. But det A is precisely the Jacobian of both the direct and inverse transformation; then Eq. (1.10) implies that $d\xi\, d\xi_1 = d\xi'\, d\xi_1'$ and Eq. (1.4) becomes

$$\int_{\Xi} Q(f, g)\varphi(\xi)\, d\xi = (1/2m) \int (f'g_1' + f_1'g' - fg_1 - f_1 g) \times \varphi(\xi)B(\theta, V)\, d\xi'\, d\xi_1'\, d\epsilon\, d\theta \tag{1.11}$$

where now, since ξ' and ξ_1' are the integration variables, we must express ξ and ξ_1 in terms of them by means of Eqs. (1.7). However, we can also change the names of the integration variables and call ξ and ξ_1 what we called ξ' and ξ_1' before, as a consequence, and because of the involutory character of the transformation (1.7), we can consistently call ξ' and ξ_1' what we called ξ and ξ_1 before, and write Eq. (1.11) as follows:

$$\int_{\Xi} Q(f, g)\varphi(\xi)\, d\xi = \frac{1}{2m} \int (fg_1 + f_1 g - f'g_1' - f_1'g') \times \varphi(\xi')B(\theta, V)\, d\xi\, d\xi_1\, d\epsilon\, d\theta \tag{1.12}$$

where $B(\theta, V)$ is not affected by the change, since Eq. (7.14) of Chapter I implies $V' = V$.

We can rewrite Eq. (1.12) as follows:

$$\int_{\Xi} Q(f, g)\varphi(\xi)\, d\xi = -\frac{1}{2m} \int (f'g_1' + f_1'g' - fg_1 - f_1 g) \times \varphi(\xi')B(\theta, V)\, d\xi\, d\xi_1\, d\epsilon\, d\theta \tag{1.13}$$

This equation is identical to Eq. (1.4) except for a minus sign and for having $\varphi(\xi')$ in place of $\varphi(\xi)$.

Finally, let us interchange ξ and ξ_1 in Eq. (1.13) as we did in Eq. (1.4) to obtain Eq. (1.5). The result is

$$\int_{\Xi} Q(f, g)\varphi(\xi)\, d\xi = -\frac{1}{2m} \int (f'g_1' + f_1'g' - fg_1 - f_1 g) \times \varphi(\xi_1')B(\theta, V)\, d\xi\, d\xi_1\, d\epsilon\, d\theta \tag{1.14}$$

which is identical to Eq. (1.4) except for a minus sign and for having $\varphi(\xi_1')$ in place of $\varphi(\xi)$.

We have thus obtained four different expressions for the same quantity: Eqs. (1.4), (1.5), (1.12), and (1.13). We can now obtain more expressions by taking linear combinations of the four basic ones; we are particularly interested in the combination which is obtained by adding the above four expressions and dividing by four. The result is

$$\int_{\Xi} Q(f, g)\varphi(\boldsymbol{\xi})\, d\boldsymbol{\xi} = \frac{1}{8m}\int (f'g_1' + f_1'g' - fg_1 - f_1g) \tag{1.15}$$

$$\times (\varphi + \varphi_1 - \varphi' - \varphi_1')B(\theta, V)\, d\boldsymbol{\xi}\, d\boldsymbol{\xi}_1\, d\epsilon\, d\theta$$

This equation expresses a basic property of the collision term, which will be frequently used in the following. In the particular case of $g = f$, Eq. (1.15) reads:

$$\int_{\Xi} Q(f, f)\varphi(\boldsymbol{\xi})\, d\boldsymbol{\xi} = \frac{1}{4m}\int (f'f_1' - ff_1)(\varphi + \varphi_1 - \varphi' - \varphi_1')$$

$$\times B(\theta, V)\, d\boldsymbol{\xi}\, d\boldsymbol{\xi}_1\, d\epsilon\, d\theta \tag{1.16}$$

We now observe that the integral appearing in Eq. (1.15) is zero, independent of the particular f and g, if

$$\varphi + \varphi_1 = \varphi' + \varphi_1' \tag{1.17}$$

is valid almost everywhere.

Since the integral appearing in the left-hand side of Eq. (1.15) is the average change of the function $\varphi(\boldsymbol{\xi})$ in unit time by the effect of the collisions, the functions satisfying Eq. (1.17) are usually called "collision invariants." We now have the property that Eq. (1.17) is satisfied if and only if

$$\varphi(\boldsymbol{\xi}) = a + \mathbf{b}\cdot\boldsymbol{\xi} + c\xi^2 \tag{1.18}$$

where a and c are constant scalars and $\mathbf{b}$ a constant vector. The functions $\psi_0 = 1, \psi_1, \psi_2, \psi_3 = \boldsymbol{\xi}, \psi_4 = \xi^2$ are usually called the elementary collision invariants; thus a general collision invariant is a linear combination of the five ψ's.

In order to prove the above statement, one can proceed either mathematically or physically. The purely mathematical proofs are somewhat lengthy and require some qualitative assumption on φ (e.g., continuity). We shall not consider them here, and refer to the bibliography at the end of this chapter for further information. From the physical standpoint, we note that Eq. (1.17) is certainly satisfied by Eq. (1.18), since a obviously cancels, $\boldsymbol{\xi}' + \boldsymbol{\xi}_1' = \boldsymbol{\xi} + \boldsymbol{\xi}_1$ because of momentum conservation and $\xi'^2 + \xi_1'^2 = \xi^2 + \xi_1^2$ because of energy conservation. Besides, no additional φ's linearly independent of the ψ's can exist, since they would give an additional relation between the velocities before and after a collision; then we would have five scalar equations instead of four and we could express ξ_1' and ξ' in terms of ξ and ξ_1 plus one scalar parameter instead of two (θ and ϵ). This is obviously wrong, since, e.g., the direction of the line joining the molecules at the beginning of the interaction can be determined only by giving two additional scalars. Therefore Eq. (1.18) gives the most general solution, as was to be shown. The weak point in this argument is, of course, that there could be a new linearly independent collision invariant, which is functionally dependent on ψ_0, ψ_1, ψ_2, ψ_3, and ψ_4.

2. Solution of the Equation $Q(f, f) = 0$.

In this section we investigate the existence of positive functions f which give a vanishing collision integral:

$$Q(f,f) = \int (f'f_1' - ff_1)B(\theta, V)\, d\xi_1\, d\theta\, d\epsilon = 0 \tag{2.1}$$

We want to show that such functions exist and are all given by

$$f(\xi) = \exp[a + \mathbf{b}\cdot\xi + c\xi^2] \tag{2.2}$$

where a, $\mathbf{b}$, and c have the same meaning as in Eq. (1.18).

In order to show that this statement is true, we prove a preliminary result which will also be important later, i.e., that no matter what the distribution function f is, the following inequality holds:

$$\int (\log f)Q(f,f)\, d\xi \leq 0 \tag{2.3}$$

and the equality sign applies if and only if f is given by Eq. (2.2). Now it is seen that the first statement is a simple corollary of the second one; in fact, if Eq. (2.1) is satisfied, then multiplying it by $\log f$ and integrating gives Eq. (2.3) with the equality sign, which implies Eq. (2.2) if the second statement applies. Let us prove, therefore, that Eq. (2.3) always holds and the equality sign implies (and is implied by) Eq. (2.2). If we use Eq. (1.16) with $\varphi = \log f$, we have

$$\int (\log f)Q(f,f)\, d\xi$$

$$= \frac{1}{4m}\int (f'f_1' - ff_1)\left[\log\left(\frac{ff_1}{f'f_1'}\right)\right]B(\theta, V)\, d\xi\, d\xi_1\, d\theta\, d\epsilon \tag{2.4}$$

Now $B \geq 0$ (the equality sign possibly applying only at $\theta = 0$ and $\theta = \pi/2$);

besides, for any $x > 0$, $y > 0$, we have

$$(x - y)\log(y/x) \leq 0 \tag{2.5}$$

and the equality sign applies if and only if $y = x$ [Eq. (2.5) is an elementary inequality, which can be simply obtained by putting $y/x = z$ and noticing that $(1 - z)$ and $\log z$ are negative or positive together and both are zero only at $z = 1$]. Then, if we put $ff_1 = y$, $f'f_1' = x$, and use Eq. (2.5), Eq. (2.4) implies Eq. (2.3) and the equality sign applies if and only if

$$ff_1 = f'f_1' \tag{2.6}$$

applies almost everywhere. But taking the logarithms of both sides of this equation, we obtain that $\varphi = \log f$ satisfies Eq. (1.17), i.e., φ is given by Eq. (1.18); this implies Eq. (2.2), as was to be shown.

We note that in Eq. (2.2) c must be negative, since f must be integrable over the whole velocity space. If we put $c = -\alpha$, $\mathbf{b} = 2\alpha\mathbf{v}$, where $\mathbf{v}$ is another constant vector, Eq. (2.2) can be written as follows:

$$f(\boldsymbol{\xi}) = A\exp[-\alpha(\boldsymbol{\xi} - \mathbf{v})^2] \tag{2.7}$$

where A is a constant related to α, c, and v^2 (α, $\mathbf{v}$, A constitute a new set of arbitrary constants). Equation (2.7) is the familiar Maxwellian distribution; it is different from Eq. (5.10) of Chapter I because Eq. (2.7) describes a gas which is not at rest, but it reduces to it if we change the reference frame to one moving with velocity $\mathbf{v}$ with respect to the gas, and interpret the quantities A and α suitably in terms of internal energy and mass density in physical space. This interpretation will be shown to be correct in the next section.

3. Connection of the Microscopic Description with Macroscopic Gas Dynamics

Until now we have considered our distribution function without investigating how we can obtain macroscopic quantities from it. In this section we shall consider the problem of evaluating such quantities once the distribution function is known. Because of the significance of the distribution function we have to take averages with respect to all the possible velocities if we want local information (as, e.g., the density at some point of physical space), while an additional integration with respect to space coordinates is required to obtain global quantities (as, e.g., the total mass of the gas).

Because of the definition of f it is clear that the ordinary density in physical space is given by

$$\rho(\mathbf{x}, t) = \int f\,d\boldsymbol{\xi} \tag{3.1}$$

the integration being extended, here and in the following, to the whole velocity space Ξ.

The mass velocity $\mathbf{v}$ is given by the average of the molecular velocity $\boldsymbol{\xi}$, i.e.,

$$\mathbf{v} = \frac{\int \boldsymbol{\xi} P \, d\boldsymbol{\xi}}{\int P \, d\boldsymbol{\xi}} = \frac{\int \boldsymbol{\xi} f \, d\boldsymbol{\xi}}{\int f \, d\boldsymbol{\xi}} \tag{3.2}$$

where P is the probability density related to f by Eq. (6.12) of Chapter I; the integral in the denominator is due to the fact that P is not normalized to unity when we consider $\mathbf{x}$ as fixed and integrate only with respect to $\boldsymbol{\xi}$. Because of Eq. (3.1), Eq. (3.2) can also be written as follows:

$$\rho \mathbf{v} = \int \boldsymbol{\xi} f \, d\boldsymbol{\xi}$$

or, using components,

$$\rho v_i = \int \xi_i f \, d\boldsymbol{\xi} \tag{3.3}$$

Here and in the following, unless otherwise stated, Latin subscripts take the values 1, 2, 3.

The mass velocity $\mathbf{v}$ is what we can perceive macroscopically of the molecular motion; it is zero for a gas at equilibrium in a box at rest. The single molecules have their own velocity $\boldsymbol{\xi}$ which can be decomposed into the sum of $\mathbf{v}$ and another velocity:

$$\mathbf{c} = \boldsymbol{\xi} - \mathbf{v} \tag{3.4}$$

which describes the random deviation of the molecular velocity from the ordered motion with velocity $\mathbf{v}$. The velocity $\mathbf{c}$ is usually called the peculiar velocity or the random velocity; it coincides with $\boldsymbol{\xi}$ when the gas is macroscopically at rest. We note that because of Eqs. (3.4), (3.3), and (3.1) we have

$$\int c_i f \, d\boldsymbol{\xi} = \int \xi_i f \, d\boldsymbol{\xi} - v_i \int f \, d\boldsymbol{\xi} = \rho v_i - v_i \rho = 0 \tag{3.5}$$

The quantity ρv_i which appears in Eq. (3.3) can be interpreted as the momentum density or, alternatively, as the mass flow (in the i direction). Other quantities which will be needed in the following are the momentum flow, the energy density, and the energy flow. Since momentum is a vector quantity, we have to consider the flow of the j component of momentum in the i direction; this is given by

$$\int \xi_i(\xi_j f) \, d\boldsymbol{\xi} = \int \xi_i \xi_j f \, d\boldsymbol{\xi} \tag{3.6}$$

where $\xi_j f$ is obviously the momentum density in phase space. Equation (3.6) shows that the momentum flow is described by a symmetric tensor of second order. It is to be expected that in a macroscopic description only part of the microscopically evaluated momentum flow will be identified as such, because the integral in Eq. (3.6) will be in general different from zero even if the gas is macroscopically at rest (absence of macroscopic momentum flow). In order to find out how the above momentum flow will appear in a macroscopic description, we have to use the splitting of $\boldsymbol{\xi}$ into mass velocity and peculiar velocity. Then, according to Eqs. (3.4), (3.5), and (3.1),

$$\begin{aligned}\int \xi_i \xi_j f\, d\boldsymbol{\xi} &= \int (v_i + c_i)(v_j + c_j) f\, d\boldsymbol{\xi} \\ &= v_i v_j \int f\, d\boldsymbol{\xi} + v_i \int c_j f\, d\boldsymbol{\xi} + v_j \int c_i f\, d\boldsymbol{\xi} + \int c_i c_j f\, d\boldsymbol{\xi} \\ &= \rho v_i v_j + \int c_i c_j f\, d\boldsymbol{\xi} \qquad (3.7)\end{aligned}$$

Therefore, the momentum flow decomposes into two parts, one of which is recognized as the macroscopic momentum flow (momentum density times velocity), while the second part is a hidden momentum flow due to the random motion of the molecules. How will this second part manifest itself in a macroscopic description? If we take a fixed region of the gas and observe the change of momentum inside it, we shall find that (in the absence of external body forces) the changes can only in part be attributed to the matter which enters and leaves the region, leaving a second part which has no macroscopic explanation unless we admit that the contiguous regions of the gas exert a force on the region of interest, which is thus subject to stresses. We shall therefore identify the second part of the momentum flow with the stress tensor

$$p_{ij} = \int c_i c_j f\, d\boldsymbol{\xi} \qquad (3.8)$$

(a complete identification is correctly justified by the fact that, as we shall see later, p_{ij} plays, in the macroscopic equations to be derived from the Boltzmann equation, the same role as the stress tensor in the conservation equations derived from macroscopic considerations).

An analogous decomposition is to be introduced for the energy density and energy flow. The energy density is given by

$$\tfrac{1}{2} \int \xi^2 f\, d\boldsymbol{\xi}$$

and we have only to take $j = i$ and sum from $i = 1$ to $i = 3$ in Eq. (3.7) to deduce

$$\tfrac{1}{2}\int \xi^2 f\, d\boldsymbol{\xi} = \tfrac{1}{2}\rho v^2 + \tfrac{1}{2}\int c^2 f\, d\boldsymbol{\xi} \tag{3.9}$$

Again the first term in the right-hand side will be macroscopically identified with the kinetic energy density, while the second term will be ascribed to the internal energy of the gas. Therefore, if we introduce the internal energy per unit mass e, we have

$$\rho e = \tfrac{1}{2}\int c^2 f\, d\boldsymbol{\xi} \tag{3.10}$$

since the density of internal energy is obviously the product of the internal energy per unit mass times the mass density.

We note that a relation exists between the internal energy density and the spur or trace (i.e., the sum of the three diagonal terms) of the stress tensor. In fact, Eqs. (3.10) and (3.8) give

$$p_{ii} = \int c^2 f\, d\boldsymbol{\xi} = 2\rho e \tag{3.11}$$

(Here and in the following we shall use the convention of summing over repeated subscripts from 1 to 3, unless otherwise stated.) The spur divided by three gives the isotropic part of the stress tensor; it is therefore convenient to identify $p = p_{ii}/3$ with the gas pressure, which is known to be isotropic, at least in the case of equilibrium (Pascal's principle; the identification is also correct for nonequilibrium situations in the case of a monatomic ideal gas, but is generally incorrect). Therefore

$$p = \tfrac{2}{3}\rho e \tag{3.12}$$

Equation (3.12) is called the state equation of the gas and allows us to express any of the three quantities p, ρ, e in terms of the remaining two. Since the macroscopic state equation for an ideal gas is Boyle's law

$$p = R\rho T \tag{3.13}$$

where R is a constant (depending on the gas) and T the gas temperature, we obtain the identification

$$e = \tfrac{3}{2}RT \tag{3.14}$$

which was anticipated in Eq. (5.11) of Chapter I.

We now have to investigate the energy flow; the total energy flow is obviously given by

$$\int \xi_i(\tfrac{1}{2}\xi^2) f\, d\boldsymbol{\xi} = \tfrac{1}{2}\int \xi_i \xi^2 f\, d\boldsymbol{\xi}$$

Using Eq. (3.4) gives

$$\frac{1}{2}\int \xi_i\xi^2 f\,d\xi = \frac{1}{2}\int (v_i + c_i)(v^2 + 2v_jc_j + c^2)f\,d\xi$$

$$= \tfrac{1}{2}v_iv^2\int f\,d\xi + v_iv_j\int c_jf\,d\xi + \tfrac{1}{2}v_i\int c^2f\,d\xi$$

$$+\tfrac{1}{2}v^2\int c_if\,d\xi + v_j\int c_ic_jf\,d\xi + \tfrac{1}{2}\int c_ic^2f\,d\xi$$

i.e., using Eqs. (3.1), (3.5), (3.10), and (3.8),

$$\frac{1}{2}\int \xi_i\xi^2 f\,d\xi = v_i(\tfrac{1}{2}\rho v^2 + \rho e) + v_jp_{ij} + \frac{1}{2}\int c_ic^2f\,d\xi \tag{3.15}$$

We now have three terms: the first one is obviously the energy flow due to macroscopic convection, and the second one is macroscopically interpreted as due to the work done by the stresses in unit time. The third term is another kind of energy flow; the additional term is usually called the heat flow vector, and is denoted by q_i:

$$q_i = \frac{1}{2}\int c_ic^2f\,d\xi \tag{3.16}$$

As for the case of the stress tensor, the identification is justified, as we shall see later, by the fact that q_i plays the same role as the heat flow vector in the macroscopic equations. However, the name "heat flow" is somewhat misleading, because there are situations when $q_i \neq 0$ and the temperature is practically constant everywhere; in this case one has to speak of a heat flow vector at constant temperature, which sounds a little paradoxical. The name "nonconvective energy flow" would be more accurate for q_i, but is not used.

The above discussion links the distribution function with the quantities used in the macroscopic description; in particular, e.g., p_{ij} can be used to evaluate the drag on a body moving inside a gas and q_i to evaluate the heat transfer from a hot body to a colder one when they are separated by a region filled by the gas.

In order to complete the connection, we now derive five differential equations, satisfied by the macroscopic quantities considered above, as a simple mathematical consequence of the Boltzmann equation; these equations are usually called the conservation equations, since they can be physically interpreted as expressing conservation of mass, momentum, and energy.

In order to obtain these equations, we consider the Boltzmann equation:

$$\frac{\partial f}{\partial t} + \xi_i \frac{\partial f}{\partial x_i} + X_i \frac{\partial f}{\partial \xi_i} = Q(f,f) \tag{3.17}$$

where for the sake of generality we have introduced the body force term which is usually left out. We multiply both sides of Eq. (3.17) by the five collision invariants ψ_α ($\alpha = 0, 1, 2, 3, 4$) and integrate with respect to $\boldsymbol{\xi}$; in accord with the results of Section 1,

$$\int \psi_\alpha Q(f,f)\, d\boldsymbol{\xi} = 0 \tag{3.18}$$

for any f. Therefore, for any f

$$\frac{\partial}{\partial t}\int \psi_\alpha f\, d\boldsymbol{\xi} + \frac{\partial}{\partial x_i}\int \xi_i \psi_\alpha f\, d\boldsymbol{\xi} + X_i \int \psi_\alpha \frac{\partial f}{\partial \xi_i}\, d\boldsymbol{\xi} = 0 \tag{3.19}$$

If we take successively $\alpha = 0, 1, 2, 3, 4$ and use Eqs. (3.1), (3.3), (3.7)–(3.10), (3.15), and (3.16), we obtain

$$\begin{gathered}
\frac{\partial \rho}{\partial t} + \frac{\partial}{\partial x_i}(\rho v_i) = 0 \\
\frac{\partial}{\partial t}(\rho v_j) + \frac{\partial}{\partial x_i}(\rho v_i v_j + p_{ij}) = \rho X_j \\
\frac{\partial}{\partial t}\left[\rho\left(\frac{1}{2}v^2 + e\right)\right] + \frac{\partial}{\partial x_i}\left[\rho v_i\left(\frac{1}{2}v^2 + e\right) + p_{ij}v_j + q_i\right] = \rho X_i v_i
\end{gathered} \tag{3.20}$$

where we have also used the following relations:

$$\begin{gathered}
\int \partial f/\partial \xi_j\, d\boldsymbol{\xi} = 0; \qquad \int \xi_i\, \partial f/\partial \xi_j\, d\boldsymbol{\xi} = -\int \delta_{ij} f\, d\boldsymbol{\xi} = -\rho\delta_{ij} \\
\tfrac{1}{2}\int \xi^2\, \partial f/\partial \xi_i\, d\boldsymbol{\xi} = -\int \xi_i f\, d\boldsymbol{\xi} = -\rho v_i
\end{gathered} \tag{3.21}$$

which follow by partial integration and the conditions $\lim_{\boldsymbol{\xi}\to\infty}(\psi_\alpha f) = 0$, which are required in order that all the integrals considered in the equations of this section exist. In the above δ_{ij} is the Kronecker delta, equal to 1 for $i = j$ and 0 for $i \neq j$. Equations (3.20) are the basic equations of macroscopic gas dynamics; as they stand, however, they constitute an empty scheme, since there are five equations for 13 quantities [if Eq. (3.12) is taken into account]. In order to have useful equations, one must have some expressions for p_{ij} and q_i in terms of ρ, v_i, and e. Otherwise, one has to go back to the

Boltzmann equation and solve it; and once this has been done, everything is known and Eqs. (3.20) are useless!

In any macroscopic approach to fluid dynamics one has to postulate, either on the basis of experiments or by plausibility arguments, some phenomenological relations (the so-called "constitutive equations") between p_{ij} and q_i on one hand and ρ, v_i, and e on the other. In the case of a gas, or, more generally, a fluid, there are two models which are well known: the Euler (or perfect) fluid,

$$p_{ij} = p\delta_{ij}; \qquad q_i = 0 \tag{3.22}$$

and the Navier–Stokes (or viscous) fluid,

$$\begin{aligned} p_{ij} &= p\delta_{ij} - \mu\left(\frac{\partial v_i}{\partial x_j} + \frac{\partial v_j}{\partial x_i}\right) - \lambda\frac{\partial v_k}{\partial x_k}\delta_{ij} \\ q_i &= -k\,\partial T/\partial x_i \end{aligned} \tag{3.23}$$

where μ and λ are the viscosity coefficients [frequently one neglects the so-called bulk viscosity; then $\lambda = -(2\mu)/3$] and k is the heat conduction coefficient (μ, λ, and k can be functions of density ρ and temperature T).

No such relations are to be introduced in the microscopic description; the single unknown f already contains all the information about density, velocity, temperature, stresses, and heat flow! Of course, this is possible because f is a function of seven variables instead of four; the macroscopic approach (five functions of four variables) is simpler than the microscopic one (one function of seven variables) and is to be preferred, whenever it can be applied. Therefore one of the tasks of a theory based on the Boltzmann equation is to deduce, for a gas in ordinary conditions, some approximate macroscopic model and find out what the limits of application of this model are. We shall consider this part of the theory in Chapter V.

There are, however, regimes of such rarefaction that no general macroscopic theory in the usual sense is possible; in this case the Boltzmann equation must be solved and not used only to justify the macroscopic equations (and to give the values of the quantities μ, λ, and k in terms of molecular constants).

We end this section by noting that if we apply Eqs. (3.1), (3.3), and (3.10) to the Maxwellian given by Eq. (2.7), we find that the constant $\mathbf{v}$ appearing in the latter equation is actually the mass velocity, while

$$\alpha = 3(4e)^{-1}, \qquad A = \rho(\tfrac{4}{3}\pi e)^{-3/2} \tag{3.24}$$

Furthermore,

$$p_{ij} = p\delta_{ij} = \tfrac{2}{3}\rho e\delta_{ij}, \qquad q_i = 0 \tag{3.25}$$

i.e., a gas with a Maxwellian distribution satisfies the constitutive equations of the Euler fluid, Eq. (3.22).

4. Boundary Conditions

Until now we have made no attempt to investigate the conditions which must be satisfied by the distribution function of a gas in contact with a solid body. It is clear, however, that such conditions are required, since the Boltzmann equation contains the space derivatives of f; furthermore, these conditions are very important because they describe the interactions of the gas molecules with the molecules of the solid body, i.e., the interaction between the body and the gas, to which one can trace the origin of the drag exerted by the gas upon the body and the heat transfer between the gas and the solid boundary.

In agreement with our statistical point of view, what we require is the probability density $B(\boldsymbol{\xi}' \to \boldsymbol{\xi}; \mathbf{x})$ that a molecule hitting the solid boundary ∂R at some point $\mathbf{x}$ with some velocity $\boldsymbol{\xi}'$ reemerges at practically the same point with some other velocity $\boldsymbol{\xi}$; accordingly, if $\mathbf{n}$ is the inward normal and $\mathbf{u}_0$ the velocity of the wall, we shall have $\boldsymbol{\xi}' \cdot \mathbf{n} < \mathbf{u}_0 \cdot \mathbf{n}$, $\boldsymbol{\xi} \cdot \mathbf{n} > \mathbf{u}_0 \cdot \mathbf{n}$. Accordingly, the general form of our boundary conditions will be

$$|(\boldsymbol{\xi} - \mathbf{u}_0) \cdot \mathbf{n}| f(\mathbf{x}, \boldsymbol{\xi}) = \int_{\boldsymbol{\xi}' \cdot \mathbf{n} < \mathbf{u}_0 \cdot \mathbf{n}} B(\boldsymbol{\xi}' \to \boldsymbol{\xi}; \mathbf{x}) |(\boldsymbol{\xi}' - \mathbf{u}_0) \cdot \mathbf{n}| f(\mathbf{x}; \boldsymbol{\xi}') \, d\boldsymbol{\xi}' \tag{4.1}$$
$$(\boldsymbol{\xi} \cdot \mathbf{n} > \mathbf{u}_0 \cdot \mathbf{n}; \mathbf{x} \in \partial R)$$

In fact, we have to use the mass flow $|(\boldsymbol{\xi} - \mathbf{u}_0) \cdot \mathbf{n}| f(\mathbf{x}, \boldsymbol{\xi})$ (number of molecules leaving or arriving per unit time and unit area) in doing a surface balance. The quantity $B(\boldsymbol{\xi}' \to \boldsymbol{\xi}; \mathbf{x})$, being a probability density, is normalized to unity,

$$\int_{\boldsymbol{\xi} \cdot \mathbf{n} > \mathbf{u}_0 \cdot \mathbf{n}} B(\boldsymbol{\xi}' \to \boldsymbol{\xi}; \mathbf{x}) \, d\boldsymbol{\xi} = 1 \qquad (\boldsymbol{\xi}' \cdot \mathbf{n} < \mathbf{u}_0 \cdot \mathbf{n}) \tag{4.2}$$

this means that we assume that the boundary does not capture molecules. Equation (4.2) implies

$$\int (\boldsymbol{\xi} - \mathbf{u}_0) \cdot \mathbf{n} f(\mathbf{x}; \boldsymbol{\xi}) \, d\boldsymbol{\xi} = 0 \qquad (\mathbf{x} \in \partial R) \tag{4.3}$$

where now the integration is extended to the whole velocity space. In fact, if we use Eq. (4.1), we have

$$\int (\boldsymbol{\xi} - \mathbf{u}_0)\cdot \mathbf{n} f(\mathbf{x};\boldsymbol{\xi})\,d\boldsymbol{\xi} = \int_{\boldsymbol{\xi}\cdot\mathbf{n} > \mathbf{u}_0\cdot\mathbf{n}} |(\boldsymbol{\xi} - \mathbf{u}_0)\cdot\mathbf{n}| f(\mathbf{x};\boldsymbol{\xi})\,d\boldsymbol{\xi}$$
$$- \int_{\boldsymbol{\xi}\cdot\mathbf{n} < \mathbf{u}_0\cdot\mathbf{n}} |(\boldsymbol{\xi} - \mathbf{u}_0)\cdot\mathbf{n}| f(\mathbf{x};\boldsymbol{\xi})\,d\boldsymbol{\xi}$$
$$= \int_{\boldsymbol{\xi}'\cdot\mathbf{n} < \mathbf{u}_0\cdot\mathbf{n}} \left[\int_{\boldsymbol{\xi}\cdot\mathbf{n} > \mathbf{u}_0\cdot\mathbf{n}} B(\boldsymbol{\xi}' \to \boldsymbol{\xi};\mathbf{x})\,d\boldsymbol{\xi}\right]$$
$$\times |(\boldsymbol{\xi}' - \mathbf{u}_0)\cdot\mathbf{n}| f(\mathbf{x};\boldsymbol{\xi}')\,d\boldsymbol{\xi}'$$
$$- \int_{\boldsymbol{\xi}'\cdot\mathbf{n} < \mathbf{u}_0\cdot\mathbf{n}} |(\boldsymbol{\xi}' - \mathbf{u}_0)\cdot\mathbf{n}| f(\mathbf{x};\boldsymbol{\xi}')\,d\boldsymbol{\xi}' \quad (4.4)$$

where we interchanged the order of the integrations in the first integral and changed the name of the integration variable in the second. Since the quantity in square brackets is equal to 1 because of Eq. (4.2), Eq. (4.3) follows. If we denote the mass velocity of the gas by $\mathbf{v}$ and use Eqs. (3.1) and (3.3), Eq. (4.3) can be written as follows:

$$\mathbf{v}\cdot\mathbf{n} - \mathbf{u}_0\cdot\mathbf{n} = 0 \qquad (\mathbf{x}\in\partial R) \tag{4.5}$$

i.e., the normal component of the mass velocity of the gas at the wall is equal to the normal component of the velocity of the wall. This also gives

$$\boldsymbol{\xi}\cdot\mathbf{n} - \mathbf{u}_0\cdot\mathbf{n} = \boldsymbol{\xi}\cdot\mathbf{n} - \mathbf{v}\cdot\mathbf{n} = \mathbf{c}\cdot\mathbf{n} \tag{4.6}$$

where $\mathbf{c}$ is the peculiar velocity of the molecules, given by Eq. (3.4). Accordingly, Eq. (4.1) can be rewritten as follows:

$$|\mathbf{c}\cdot\mathbf{n}| f(\mathbf{x};\boldsymbol{\xi}) = \int_{\mathbf{c}'\cdot\mathbf{n} < 0} B(\boldsymbol{\xi}' \to \boldsymbol{\xi};\mathbf{x})|\mathbf{c}\cdot\mathbf{n}| f(\mathbf{x},\boldsymbol{\xi}')\,d\boldsymbol{\xi}' \quad (\mathbf{c}\cdot\mathbf{n} > 0; \mathbf{x}\in\partial R) \tag{4.7}$$

We have now to specify $B(\boldsymbol{\xi}' \to \boldsymbol{\xi};\mathbf{x})$; this is not an easy problem, since the situation at the wall is very complicated, due to the complex phenomena of adsorption and evaporation which take place at the wall. An early and still valuable discussion on this subject was given by Maxwell in 1879. In an appendix to a paper published in the *Philosophical Transactions of the Royal Society* (see ref. 1) he discussed the problem of finding a boundary condition for the distribution function. As a first hypothesis he supposes the surface of a physical wall to be a perfectly elastic, smooth, fixed surface having the apparent shape of the solid, without any minute asperities. In this case the gas molecules are specularly reflected; therefore the gas cannot exert any stress on the surface, except in the direction of the normal. Then Maxwell points out that since gases can actually exert oblique stresses against real surfaces, such surfaces cannot be represented as perfectly reflecting.

As a second model for a real wall Maxwell considers a stratum in which fixed elastic spheres are placed so far apart from one another that any one sphere is not sensibly protected by any other sphere from the impact of molecules; he also assumes the stratum to be so deep that no molecule can pass through it without striking one or more of the spheres. Every molecule which comes from the gas toward such a wall must strike one or more of the spheres; when at last it leaves the stratum of spheres and returns into the gas its velocity must of course be directed from the surface toward the gas, but the probability of any particular magnitude and direction of the velocity will be the same as in a gas in thermal and mechanical equilibrium with the solid.

Then Maxwell considers more complicated models of physical walls, and finally concludes by saying that he prefers to treat the surface as something intermediate between a perfectly reflecting and a perfectly absorbing surface, and, in particular, to suppose that a portion of every surface element absorbs all the incident molecules, and afterwards allows them to reevaporate with velocities corresponding to those in still gas at the temperature of the solid wall, while the remaining portion perfectly reflects all the molecules incident upon it.

If we call α the fraction of evaporated molecules, Maxwell's assumption is equivalent to choosing

$$B(\boldsymbol{\xi}' \to \boldsymbol{\xi}; \mathbf{x}) = (1 - \alpha)\delta(\boldsymbol{\xi}' - \boldsymbol{\xi} + 2\mathbf{n}[\mathbf{n} \cdot \mathbf{c}]) + \alpha N_0 (2\pi R T_0)^{-3/2} |\mathbf{c} \cdot \mathbf{n}| \times \exp[-(\boldsymbol{\xi} - \mathbf{u}_0)^2 (2RT_0)^{-1}] \qquad (\mathbf{c} \cdot \mathbf{n} > 0; \mathbf{c}' \cdot \mathbf{n} < 0) \tag{4.8}$$

where $\mathbf{u}_0$ and T_0 are respectively, the velocity and temperature of the wall (which might vary from point to point), while N_0 is to be determined in such a way that Eq. (4.2) is satisfied; accordingly,

$$N_0 = [2\pi/(RT_0)]^{1/2} \tag{4.9}$$

We note here that if the boundary condition depends upon the temperature of the wall, as in the case of Maxwell's boundary conditions, then two additional properties are to be satisfied by the kernel $B(\boldsymbol{\xi}' \to \boldsymbol{\xi}; \mathbf{x})$. To see this, we note that if we assume the gas to have a Maxwellian distribution with temperature and mass velocity equal to the temperature and mass velocity of the wall, this gas is in thermal and mechanical equilibrium with the wall (at least locally); therefore, in the wall reference frame, the number of molecules which change their velocity from $\boldsymbol{\xi}' - \mathbf{u}_0$ to $\boldsymbol{\xi} - \mathbf{u}_0$ because of the interaction with the walls must be equal to those which change their velocity from $-(\boldsymbol{\xi} - \mathbf{u}_0) = \boldsymbol{\xi}_R - \mathbf{u}_0$ to $-(\boldsymbol{\xi}' - \mathbf{u}_0) = \boldsymbol{\xi}'_R - \mathbf{u}_0$ (in a state of equilibrium). Accordingly,

$$|\mathbf{c}' \cdot \mathbf{n}| B(\boldsymbol{\xi}' \to \boldsymbol{\xi}; \mathbf{x}) \exp\left[-\frac{(\boldsymbol{\xi}' - \mathbf{u}_0)^2}{2RT_0}\right]$$

$$= |\mathbf{c} \cdot \mathbf{n}| B(\boldsymbol{\xi}_R \to \boldsymbol{\xi}'_R; \mathbf{x}) \exp\left[-\frac{(\boldsymbol{\xi} - \mathbf{u}_0)^2}{2RT_0}\right] \tag{4.10}$$

where we have used the fact that $(\boldsymbol{\xi}'_R - \mathbf{u}_0)^2 = (\boldsymbol{\xi}' - \mathbf{u}_0)^2$. This equation expresses what is sometimes called the "detailed balancing"; it is easily seen that Maxwell's $B(\boldsymbol{\xi}' \to \boldsymbol{\xi}; \mathbf{x})$ satisfies Eq. (4.10).

In addition, if we consider the same equilibrium situation, the corresponding Maxwellian distribution must satisfy the boundary conditions, Eq. (4.1); i.e.,

$$\int |\mathbf{c}' \cdot \mathbf{n}| B(\boldsymbol{\xi}' \to \boldsymbol{\xi}; \mathbf{x}) \exp\left[-\frac{(\boldsymbol{\xi}' - \mathbf{u}_0)^2}{2RT_0}\right] d\boldsymbol{\xi}' = |\mathbf{c} \cdot \mathbf{n}| \exp\left[-\frac{(\boldsymbol{\xi} - \mathbf{u}_0)^2}{2RT_0}\right] \tag{4.11}$$

$$\mathbf{c} \cdot \mathbf{n} > 0; \quad \mathbf{x} \in \partial R$$

Again, one easily checks that Maxwell's boundary conditions satisfy Eq. (4.11).

If the boundary conditions do not contain the temperature of the wall [as, e.g., Eq. (4.8) with $\alpha = 0$—case of a completely reflected gas], then a Maxwellian with *any* temperature satisfies both the detailed balancing and Eq. (4.5). If this is true for any point of the boundary (including possible boundaries at infinity), it is clear that these boundary conditions are quite unrealistic in general, since they would allow the gas to stay in thermal equilibrium at any given temperature, irrespective of the surrounding bodies. This fact in general rules out these boundary conditions (adiabatic walls), which can be retained, however, in particular cases.

What can we say today about Maxwell's boundary conditions? The available information is not much, but we can say that they give satisfactory results with values of α rather close to 1; besides, in problems where the dynamics is more interesting than the thermodynamics (negligible energy transfer but important momentum transfer), $\alpha = 1$ is a rather accurate assumption.

However, it is to be noted that Maxwell's boundary conditions are by no means the only possible ones, and, in our opinion, they are not very satisfactory in general; as a matter of fact, Maxwell's arguments imply that specular reflection is reasonable for grazing molecules, which are likely to hit the *pole* of the spheres constituting the wall, while Eq. (4.3) extends this circumstance to all the molecules (even those moving perpendicularly to the wall). A more satisfactory boundary condition would be:

$$B(\boldsymbol{\xi}' \to \boldsymbol{\xi}; \mathbf{x}) = (1 - \alpha) \int\limits_{\boldsymbol{\beta}\cdot\mathbf{n}>0} \delta(\boldsymbol{\xi}' - \boldsymbol{\xi} + 2\boldsymbol{\beta}[\boldsymbol{\beta}\cdot(\boldsymbol{\xi} - \mathbf{u}_0)]H(\boldsymbol{\Omega}, \boldsymbol{\beta}, \mathbf{n})\, d\boldsymbol{\beta}$$

$$+ \alpha N_0 |\boldsymbol{\xi}\cdot\mathbf{n}| (2\pi R T_0)^{-3/2} \exp\left[-\frac{(\boldsymbol{\xi} - \mathbf{u}_0)^2}{2RT_0}\right] \tag{4.12}$$

$$\mathbf{c}\cdot\mathbf{n} > 0; \quad \mathbf{c}'\cdot\mathbf{n} < 0$$

where $\boldsymbol{\beta}$ is a unit vector, $\boldsymbol{\Omega} = \boldsymbol{\xi}/\xi$, and H a function which describes the partial shielding of a sphere of the outer layer by the spheres which lie next to it. Note that Eq. (4.12) reduces to Eq. (4.8) if we take $H(\boldsymbol{\Omega}, \boldsymbol{\beta}, \mathbf{n}) = \delta(\boldsymbol{\beta} - \mathbf{n})$.

However, Maxwell's boundary conditions are frequently used for their simplicity and reasonable accurateness; we return to this question later and show that Eq. (4.3) with $\alpha = 1$ is a reasonable approximation to any kind of more complicated boundary condition.

5. The H-Theorem

We want to show now that if we define

$$\mathscr{H} = \int f \log f \, d\boldsymbol{\xi} \tag{5.1}$$

$$\mathscr{H}_i = \int \xi_i f \log f \, d\boldsymbol{\xi} \tag{5.2}$$

where f is any function which satisfies the Boltzmann equation; then

$$\frac{\partial \mathscr{H}}{\partial t} + \frac{\partial \mathscr{H}_i}{\partial x_i} \leq 0 \tag{5.3}$$

Equation (5.3) is an obvious consequence of the inequality given by Eq. (2.3) and of the fact that f satisfies the Boltzmann equation; it suffices to multiply both sides of the Boltzmann equation by $(1 + \log f)$ and integrate over all possible velocities, taking into account that $d(f \log f) = (1 + \log f)\, df$ and 1 is a collision invariant.

We can now interpret Eq. (5.3) by rewriting it as follows:

$$\frac{\partial \mathscr{H}}{\partial t} + \frac{\partial \mathscr{H}_i}{\partial x_i} = \mathscr{S}; \qquad \mathscr{S} = \int (\log f) Q(f,f)\, d\boldsymbol{\xi} \leq 0 \tag{5.4}$$

and introducing the quantity

$$H = \int_R \mathscr{H}\, d\mathbf{x} \tag{5.5}$$

where R is the region filled by gas. If we had $\mathscr{S} = 0$, then H would be a conserved quantity like mass, energy, and momentum, since we can interpret the vector $\boldsymbol{\mathscr{H}} = (\mathscr{H}_1, \mathscr{H}_2, \mathscr{H}_3)$ as the flow of H (note that here $\mathscr{H}$ is not the absolute value of the vector $\boldsymbol{\mathscr{H}}$, but the density corresponding to the flow $\boldsymbol{\mathscr{H}}$). Since $\mathscr{S}$ is in general different from zero, but always nonpositive, we can say that the molecular collisions act as a negative source for the quantity H.

We also know that the source term $\mathscr{S}$ is zero if and only if the distribution function is Maxwellian. Finally, we note that, as we did for momentum flow, $\boldsymbol{\mathscr{H}}$ can be split into a macroscopic flow of H ($\mathscr{H}\mathbf{u}$) and a microscopic flow ($\boldsymbol{\mathscr{H}} - \mathscr{H}\mathbf{u}$).

From Eq. (5.3) we can deduce the celebrated H-theorem of Boltzmann: if there is no microscopic flow of H through the boundary, or if the boundary acts as a negative source of H, then the quantity H never increases with time and it is steady only when the distribution function is a Maxwellian.

In fact, if we integrate both sides of Eq. (5.3) with respect to $\mathbf{x}$, we have:

$$\frac{dH}{dt} - \int_{\partial R} [\boldsymbol{\mathscr{H}} \cdot \mathbf{n} - \mathscr{H}\mathbf{u}_0 \cdot \mathbf{n}]\, dS \leq 0 \tag{5.6}$$

where dS is a surface element of the boundary ∂R and $\mathbf{n}$ is the inward normal. The second term in the integral comes from the fact that if the wall is moving, when forming the time derivative of H we have to take into account that the region of integration in Eq. (5.5) changes with time. If the boundary does not allow inward flow of H, then

$$\int [\boldsymbol{\mathscr{H}} - \mathscr{H}\mathbf{u}_0] \cdot \mathbf{n}\, dS \leq 0 \tag{5.7}$$

and Eq. (5.6) gives

$$dH/dt \leq 0 \tag{5.8}$$

as was to be shown. It is clear that the equality sign applies only if it applies in both Eq. (5.7) and (5.6), and the equality sign in the latter applies only if it applies in Eq. (2.3), i.e., if f is a Maxwellian.

Several comments are in order in connection with this result. First of all, we have to inquire what kind of boundary conditions allow us to derive Eq. (5.7). We easily see that in the case of specular reflection Eq. (5.7) holds with the equality sign:

$$\int_{\partial R} [\mathscr{H} - \mathscr{H}\mathbf{u}_0] \cdot \mathbf{n}\, dS = \int_{\mathbf{c}\cdot\mathbf{n}>0} f(\log f)|\mathbf{c}\cdot\mathbf{n}|\, d\boldsymbol{\xi}\, dS - \int_{\mathbf{c}\cdot\mathbf{n}<0} f(\log f)|\mathbf{c}\cdot\mathbf{n}|\, d\boldsymbol{\xi}\, dS \tag{5.9}$$

Since $f(\boldsymbol{\xi}) = f(\boldsymbol{\xi} - 2\mathbf{n}(\mathbf{n}\cdot\mathbf{c}))$ implies that the two integrals are equal, Eq. (5.7) with the equality sign follows.

The other case which we want to consider here is the case of a purely diffusing wall (Maxwell's boundary condition with $\alpha = 1$). Then the integral in Eq. (5.9) becomes

$$\int (\mathscr{H} - \mathscr{H}\mathbf{u}_0) \cdot \mathbf{n}\, dS = \int_{\mathbf{c}\cdot\mathbf{n}>0} f_0(\log f_0)|\mathbf{c}\cdot\mathbf{n}|\, d\boldsymbol{\xi}\, dS - \int_{\mathbf{c}\cdot\mathbf{n}<0} f(\log f)|\mathbf{c}\cdot\mathbf{n}|\, d\boldsymbol{\xi}\, dS \tag{5.10}$$

where f_0 is the Maxwellian (possibly varying from point to point along the boundary) which gives the distribution of the reemitted molecules. Now, it is easy to see that for any pair of real numbers x and y ($x > 0$), the following inequality holds:

$$x \log x \geq (1 - y)x - \exp[-y] \tag{5.11}$$

[consider $\psi(x) = x \log x - (1 - y)x$ and look for its minimum]. Then if we take $x = f$, $y = -\log f_0$, we have

$$f \log f \geq f - f_0 + f \log f_0 \tag{5.12}$$

and integrating,

$$\int_{\mathbf{c}\cdot\mathbf{n}<0} f(\log f)|\mathbf{c}\cdot\mathbf{n}|\, d\boldsymbol{\xi} \geq \int_{\mathbf{c}\cdot\mathbf{n}<0} f|\mathbf{c}\cdot\mathbf{n}|\, d\boldsymbol{\xi} - \int_{\mathbf{c}\cdot\mathbf{n}<0} f_0|\mathbf{c}\cdot\mathbf{n}|\, d\boldsymbol{\xi} + \int_{\mathbf{c}\cdot\mathbf{n}<0} f(\log f_0)|\mathbf{c}\cdot\mathbf{n}|\, d\boldsymbol{\xi} \tag{5.13}$$

The second integral on the right-hand side can be extended to $\mathbf{c}\cdot\mathbf{n} > 0$ instead of $\mathbf{c}\cdot\mathbf{n} < 0$, since f_0 is an even function of $\mathbf{c}\cdot\mathbf{n}$; then the difference between the first two integrals on the right-hand side becomes the left-hand side of Eq. (4.3) and therefore is zero. Accordingly,

$$\int (\mathscr{H} - \mathscr{H}\mathbf{u}_0) \cdot \mathbf{n}\, dS \leq \int_{\mathbf{c}\cdot\mathbf{n}>0} f_0(\log f_0)|\mathbf{c}\cdot\mathbf{n}|\, d\boldsymbol{\xi}\, dS - \int_{\mathbf{c}\cdot\mathbf{n}<0} f(\log f_0)|\mathbf{c}\cdot\mathbf{n}|\, d\boldsymbol{\xi}\, dS \tag{5.14}$$

and if $f_0 = a \exp[-\alpha(\boldsymbol{\xi} - \mathbf{u}_0)^2]$ (where a and $\alpha = (2RT_0)^{-1}$ can vary along the boundary),

$$\int (\mathcal{H} - \mathcal{H}\mathbf{u}_0)\cdot\mathbf{n}\, dS \leq \int_{\partial R} \Bigg[(\log a) \int_{\mathbf{c}\cdot\mathbf{n}>0} f_0|\mathbf{c}\cdot\mathbf{n}|\, d\boldsymbol{\xi} - \alpha \int_{\mathbf{c}\cdot\mathbf{n}>0} (\boldsymbol{\xi} - \mathbf{u}_0)^2 |\mathbf{c}\cdot\mathbf{n}| f_0\, d\boldsymbol{\xi} - (\log a) \int_{\mathbf{c}\cdot\mathbf{n}<0} f|\mathbf{c}\cdot\mathbf{n}|\, d\boldsymbol{\xi} + \alpha \int_{\mathbf{c}\cdot\mathbf{n}<0} (\boldsymbol{\xi} - \mathbf{u}_0)^2 |\mathbf{c}\cdot\mathbf{n}| f\, d\boldsymbol{\xi} \Bigg] dS \tag{5.15}$$

The first and third integrals on the right-hand side cancel again because of Eq. (4.3), and, remembering that $f_0 = f$ for $\mathbf{c}\cdot\mathbf{n} > 0$, we can write

$$\begin{aligned} \int (\mathcal{H} - \mathcal{H}\mathbf{u}_0)\cdot\mathbf{n}\, dS &\leq -\int \alpha(\boldsymbol{\xi} - \mathbf{u}_0)^2 f\mathbf{c}\cdot\mathbf{n}\, d\boldsymbol{\xi}\, dS \\ &= -\int \alpha[c^2 + 2(\mathbf{v} - \mathbf{u}_0)\cdot\mathbf{c} + (\mathbf{v} - \mathbf{u}_0)^2] f\mathbf{c}\cdot\mathbf{n}\, d\boldsymbol{\xi}\, dS \end{aligned} \tag{5.16}$$

The last term in brackets again gives no contribution because of Eq. (4.3), and we have

$$\int (\mathcal{H} - \mathcal{H}\mathbf{u}_0)\cdot\mathbf{n}\, dS \leq -\frac{1}{R}\int \mathbf{q}\cdot\mathbf{n}T_0^{-1}\, dS - \frac{1}{R}\int \mathbf{p}_n\cdot(\mathbf{v} - \mathbf{u}_0)T_0^{-1}\, dS \tag{5.17}$$

where Eqs. (3.16) and (3.8) defining heat flow and stress tensor have been used ($\mathbf{p}_n$ is here the vector having components $p_{ij}n_j$). Note that only oblique stresses enter, because $\mathbf{v} - \mathbf{u}_0$ is orthogonal to $\mathbf{n}$, Eq. (4.5). If the gas does not slip on the boundary ($\mathbf{v} = \mathbf{u}_0$) or oblique stresses are zero ($p_{nj} = 0, j \neq n$) then the last term in Eq. (5.17) vanishes. The first integral is positive or zero if the gas does not release heat to the surrounding bodies. These remarks give sufficient conditions for Eq. (5.7) to be valid. Note that these conditions also apply to the case of specular reflection [$\mathbf{q} = 0$, $p_{in} = 0$ $(i \neq n)$] and are valid for any acceptable boundary conditions, as shown in several papers indicated in the references.

Boltzmann's H-theorem is of basic importance because it shows that the Boltzmann equation has a basic feature of irreversibility: the quantity H always decreases even if it is not released to outside bodies [equality sign in Eq. (5.7)]. This seems to be in conflict with the fact that the molecules constituting the gas follow the laws of classical mechanics, which are reversible. Accordingly, given at $t = t_0$ a motion with velocities $\mathbf{v}_1, \ldots, \mathbf{v}_N$, we can always consider the motion with velocities $-\mathbf{v}_1, \ldots, -\mathbf{v}_N$ at $t = t_0$ (and some molecular positions at $t = t_0$); the backward evolution of the latter state ($t < t_0$) will be equal to the forward evolution of the original motion. Therefore if $dH/dt < 0$ in the first case, we shall have $dH/d(-t) < 0$ in the second case; i.e., $dH/dt > 0$, which contradicts Boltzmann's H-theorem. The answer to this paradox is roughly as follows: If one sticks to the laws of mechanics, then "before" and "after" have no strict meaning, i.e., one can use the equations of motion to "predict" either the future or the past. However, this is not so in other branches of physics (e.g., thermodynamics) and in our everyday life, since we continuously experience the presence of a "time arrow" which introduces an intrinsic distinction between past and future. In particular, at a macroscopic level nobody can predict the past, not even in probabilistic terms: we cannot inquire about the probability of a past event, since the event either has surely happened or has not, and when we compute a probability we have to use all the information available at a macroscopic level, in this case also the fact that the event has or has not happened. One can only use the incomplete information in his possession to predict the future in probabilistic terms; it is obvious that the information, if it is not complete at $t = 0$, will in general decrease for $t > 0$. In this sense one can interpret H (which decreases with time) as the quantity of information which we have about the microscopic dynamics of the system; when we say "we," we mean any macroscopic apparatus; therefore the law of decreasing information is not something subjective, but an objective property of the world at a macroscopic level. In other words, the quantity H must be something which can also be measured macroscopically; from this point of view, however, it will loose any relation with the concept of information, since complete information at a macroscopic level is different from complete information at a microscopic level (the former requires extensions, compositions, mass velocities, temperatures, etc., while the latter requires positions and velocities of a large number of particles). We must not be surprised, therefore, to find that H, i.e., the quantity of information at a microscopic level, is related to entropy, which in macroscopic thermodynamics is defined in such a way as to have nothing to do with the quantity of information and is indeed a part of the information at a macroscopic level. That this interpretation is correct is suggested by the irreversibility properties of entropy and proved by the fact that for equilibrium states, as we shall show in the next section, we have

$$H = -(1/R)\eta \tag{5.18}$$

where η is the entropy of the gas; since η is a state function defined only for equilibrium states, Eq. (5.18) can be regarded as always true and considered as a generalization of the definition of η to nonequilibrium states. We note that Eq. (5.17) is also in agreement with the above identification, Eq. (5.18), and the macroscopic properties of entropy. The above interpretation shows that the Boltzmann equation gives results in agreement with macroscopic thermodynamics; the only modification consists in replacing a strict statement with a probabilistic one (violations of the second principle of thermodynamics are now not impossible but extremely unprobable).

However, we must not forget that we started from a discussion about a reconciliation of the irreversible behavior described by the Boltzmann equation and the complete reversibility of the mechanical model upon which the Boltzmann equation is based. As we said before, the irreversibility is to be connected with the existence of a time arrow which distinguishes past from future, in such a way that if $t_2 > t_1$, then our information at time t_2 is smaller than at time t_1 (unless new information flows in), if it was not microscopically complete at time t_1. (Or equivalently, the entropy of the system at t_2 is larger than at t_1, in a thermodynamical description, unless some entropy flows out.) We can also reverse the connection by saying that if $H(t_2) < H(t_1)$, then $t_2 > t_1$; i.e., use the decreasing of H (or increasing of η) to define the time arrow without reference to any subjective element. This can be done since all the physical systems of the universe are interacting with each other, and therefore it cannot happen that what is "earlier" for one system is "later" for another (this, however, could possibly not be true for systems composed of antimatter).

The fact that a system is really never isolated is to be kept in mind also in connection with another paradoxical objection to any mechanical interpretation of the second principle of thermodynamics; this objection is much subtler than the argument based upon reversing the molecular velocities. The objection is based upon a theorem of Poincaré which says that any finite mechanical system obeying the laws of classical mechanics will return arbitrarily close to its initial state, for almost any choice of the latter, provided we wait long enough. This theorem implies that our molecules can have, after a "recurrence time," positions and velocities so close to the initial ones that the macroscopic properties (density, temperature, etc.) calculated from them should also be practically the same as the initial ones; therefore the entropy which can be calculated in terms of them should also be practically the same, and if it increased initially, must have decreased at some later time. The usual answer to this objection is that the recurrence time is so large that practically speaking one would never observe a significant portion of the recurrence cycle; in fact, the recurrence time for a typical amount of gas is a huge number even if the estimated age of the universe is taken as time unit. It is clear that on such an enormous scale, apart from the question of the applicability of our model of classical pointlike molecules, we don't feel uneasy in conceding that the Boltzmann equation or, equivalently, the second principle of thermodynamics can be violated (a possible connection of the time arrow and the second principle of thermodynamics with the expansion of the Universe also presents itself as possible from this point of view).

6. Equilibrium States and Maxwellian Distributions

The decreasing of H shows that the Boltzmann equation describes an evolution toward a state of minimum H (compatible with the assigned constraints, such as volume of the region where the gas is contained, number of molecules, temperatures of the solid bodies surrounding the gas, etc.), provided no additional H flows in from the exterior. The final state (to be reached as $t \to \infty$) will presumably be an equilibrium state. If we define an equilibrium state as a state where the gas has a homogeneous and time-independent distribution ($\partial f/\partial t = \partial f/\partial \mathbf{x} = 0$), it follows from the Boltzmann equation that

$$Q(f, f) = 0 \tag{6.1}$$

and therefore the distribution is Maxwellian, in accord with the results of Section 2.

More generally, if Eq. (5.7) is satisfied, then the simple assumption of steadiness implies that the distribution function is Maxwellian; i.e., if the walls are at rest (or, more generally, move with a uniform motion perpendicular to themselves, arbitrarily in a parallel direction), f is time-independent, and Eq. (5.7) is satisfied, then f is a Maxwellian distribution. In fact, if Eq. (5.7) is satisfied, the H-theorem applies:

$$dH/dt \leq 0 \tag{6.2}$$

On the other hand, the steadiness assumptions (or their generalizations) imply that the equality sign holds in Eq. (6.2), and this can happen only if f is Maxwellian, because of the results of Section 5.

In particular, we note that if the boundaries are specularly reflecting, Eq. (5.7) applies, and, consequently, the distribution function must be a Maxwellian for any steady situation (this is not true for more realistic boundary conditions!).

Let us now look for the most general Maxwellian which satisfies the Boltzmann equation in the steady case; i.e., we want to find what dependence upon space variables is allowed for the parameters ρ, $\mathbf{v}$, T which appear in

$$f = \rho(2\pi RT)^{-3/2} \exp[-(\boldsymbol{\xi} - \mathbf{v})^2/(2RT)] \tag{6.3}$$

Since $\partial f/\partial t = 0$ and $Q(f, f) = 0$, we have to satisfy

$$\boldsymbol{\xi}\cdot\partial f/\partial \mathbf{x} = 0 \tag{6.4}$$

where f is given by Eq. (6.3), and ρ, $\mathbf{v}$, T are functions of $\mathbf{x}$ alone. Substituting Eq. (6.3) into Eq. (6.4) gives

$$\begin{aligned}&\frac{1}{\rho}c_i\frac{\partial\rho}{\partial x_i} + \frac{c_ic_j}{RT}\frac{\partial v_j}{\partial x_i} + \frac{c_jc_j}{2RT^2}c_i\frac{\partial T}{\partial x_i} - \frac{3}{2T}c_i\frac{\partial T}{\partial x_i}\\ &+\frac{1}{\rho}v_i\frac{\partial\rho}{\partial x_i} + \frac{c_jv_i}{RT}\frac{\partial v_j}{\partial x_i} + \frac{c_jc_j}{2RT^2}v_r\frac{\partial T}{\partial x_r} - \frac{3}{2T}v_i\frac{\partial T}{\partial x_i} = 0\end{aligned} \tag{6.5}$$

The left-hand side is a polynomial of third degree in the components of $\mathbf{c}$, which must be zero for all the values of these variables. If we equate the coefficients of the different terms to zero (paying attention to the fact that c_ic_j is not different from c_jc_i), we obtain

$$\begin{aligned}&\frac{1}{\rho}v_i\frac{\partial\rho}{\partial x_i} - \frac{1}{T}\frac{3}{2}v_i\frac{\partial T}{\partial x_i} = 0; \qquad \frac{1}{\rho}\frac{\partial\rho}{\partial x_j} - \frac{1}{T}\frac{3}{2}\frac{\partial T}{\partial x_j} + \frac{v_i}{RT}\frac{\partial v_j}{\partial x_i} = 0\\ &\frac{\partial v_i}{\partial x_j} + \frac{\partial v_j}{\partial x_i} + \frac{1}{T}v_r\frac{\partial T}{\partial x_r}\delta_{ij} = 0; \qquad \frac{\partial T}{\partial x_i} = 0\end{aligned} \tag{6.6}$$

The last equation gives $T = \text{const}$; the third one then becomes

$$\frac{\partial v_i}{\partial x_j} + \frac{\partial v_j}{\partial x_i} = 0 \tag{6.7}$$

These are the well-known equations for a motion without deformation; the general solution is well known:

$$\mathbf{v} = \mathbf{v}^0 + \boldsymbol{\omega}\times\mathbf{x} \tag{6.8}$$

or in terms of components

$$v_i = v_i^0 + \epsilon_{ijk}\omega_jx_k \tag{6.9}$$

where ϵ_{ijk} is the completely antisymmetric tensor equal to 1 if (ijk) is an even permutation of (123), to -1 if it is an odd permutation, to 0 if any two subscripts are equal. The vectors $\mathbf{v}^0$ and $\boldsymbol{\omega}$ are two constant vectors which define a rigid motion. As is well known, we can make $\mathbf{v}^0$ and $\boldsymbol{\omega}$ parallel to each other,

$$\epsilon_{jri}\omega_rv_i^0 = 0 \tag{6.10}$$

by suitably choosing the origin for $\mathbf{x}$.

Substituting Eq. (6.9) into the second of Eqs. (6.6) (with $T = \text{const}$), we have

$$\frac{1}{\rho}\frac{\partial \rho}{\partial x_j} + \frac{v_i^0}{RT}\epsilon_{jri}\omega_r + \frac{\epsilon_{isk}\omega_s x_k}{RT}\epsilon_{jri}\omega_r = 0 \tag{6.11}$$

Choosing the origin of $\mathbf{x}$ in such a way that Eq. (6.10) is satisfied, the second term disappears; if we use the identity

$$\epsilon_{isk}\epsilon_{jri} = \delta_{js}\delta_{kr} - \delta_{jk}\delta_{sr},$$

we have

$$\frac{1}{\rho}\frac{\partial \rho}{\partial x_j} + \frac{1}{RT}(\omega_j\omega_k x_k - \omega^2 x_j) = 0 \tag{6.12}$$

and integrating,

$$\rho = \rho_0 \exp[(\omega^2|\mathbf{x}|^2 - \omega_j\omega_k x_j x_k)/2RT] \tag{6.13}$$

where ρ_0 is a constant. If we take $\boldsymbol{\omega}$ along the z axis of a Cartesian reference frame and $\mathbf{x} = (x, y, z)$, we have

$$\rho = \rho_0 \exp[(\omega^2/2RT)(x^2 + y^2)] \tag{6.14}$$

Therefore the most general steady Maxwellian describes a gas with constant temperature, rigid motion, and density given by Eq. (6.14); the space variation of density in this equation reflects a centrifugation of the gas by means of the rotation with constant angular velocity.

The present results, together with the fact proved above that the only steady solution for specularly reflecting walls is the Maxwellian distribution, shows how small is the class of steady problems to which the assumption of specular reflection can be applied.

The function f given by Eq. (6.3) with constant T, ρ given by Eq. (6.14), and $\mathbf{v}$ by Eq. (6.8), must of course satisfy the boundary conditions. This imposes further restrictions: the only possible solutions for realistic boundary conditions are translations at constant velocity, the solid-body rotation being allowed only if the vessel is rotationally invariant about an axis directed as $\boldsymbol{\omega}$.

Thus we have investigated all the possible Maxwellian distributions which solve the Boltzmann equation under the assumption of steady state and absence of external forces; more general Maxwellians solving the Boltzmann equation can be found if external conservation forces are present and/or the situation is time-dependent. In any case, they form a very restricted class of solutions of the Boltzmann equation.

We end this section with the evaluation of H in the equilibrium state, in order to show that Eq. (5.18) applies. Equations (5.5), (5.1), and (6.3) give (assuming that ρ and T are constant)

$$H = \int f(\log f)\, d\boldsymbol{\xi}\, d\mathbf{x}$$

$$= \log[\rho(2\pi RT)^{-3/2}] \int f\, d\boldsymbol{\xi}\, d\mathbf{x} - (2RT)^{-1} \int c^2 f\, d\boldsymbol{\xi}\, d\mathbf{x} \tag{6.15}$$

If we now use Eq. (6.16) of Chapter I and Eqs. (3.10) and (3.14) of this chapter, we have

$$H = M \log[\rho(2\pi RT)^{-3/2}] - \tfrac{3}{2}M \tag{6.16}$$

where M is the total mass of gas. The entropy η of a gas is such that

$$d\eta = (1/T)[Mc_\rho\, dT + p d(M/\rho)] \tag{6.17}$$

where c_ρ is the specific heat at constant volume. Since $p = R\rho T$ and $c_\rho = \frac{3}{2}R$ for a monatomic gas, we have

$$d\eta = RM[\tfrac{3}{2}(dT/T) + \rho d(1/\rho)] \tag{6.18}$$

Hence

$$\eta = RM \log(T^{3/2}/\rho) + \text{const} \tag{6.19}$$

where the constant can depend on the total mass but not on the state of the gas (specified by T and ρ). Now Eq. (6.16) can be written

$$H = -M \log(T^{3/2}/\rho) + [-\tfrac{3}{2}M \log(2\pi R) - \tfrac{3}{2}M] \tag{6.20}$$

Since η is defined up to an additive constant, we can identify the constant in Eq. (6.19) with the quantity in square brackets in Eq. (6.20). Then Eq. (5.18) follows, as was to be shown.

References

The content of this chapter is generally standard. References 6 10 of Chapter I can be consulted as general references.

Section 1——For a mathematical proof that Eq. (1.18) gives all the solutions of Eq. (1.17) see ref. 8 of Chapter I. See also ref. 1 of Chapter III.

Section 4——Maxwell's discussion about boundary conditions can be found in:

1. J. C. Maxwell, *Phil. Trans. R. Soc.* I, Appendix (1879); reprinted in: *The Scientific Papers of J. C. Maxwell*, Dover Publications (1965).

Section 5——For the dispute on the H-theorem see the papers by Boltzmann, Zermelo, and Poincaré in the collection

2. S. G. Brush (ed.), *Kinetic Theory*, Vol. II, Pergamon Press, New York and London (1966).

For a modern proof of Poincaré's theorem see

3. M. Kac, *Probability and Related Topics in Physical Sciences*, Interscience Publishers, New York (1959).

The H-theorem in the presence of boundary conditions different from pure diffusion and specular reflection is discussed in ref. 15 of Chapter I and in the following papers:

4. C. Cercignani and M. Lampis, *Transport Theory and Statistical Physics* **1**, 101 (1971);
5. C. Cercignani, *Transport Theory and Statistical Physics* **2**, 27 (1972);
6. J. Darrozès and J. P. Guiraud, *C. R. Acad. Sci.* (*Paris*) **A262**, 1368 (1966).

Section 6——The Maxwellians solving the Boltzmann equation in time-dependent situations and in the presence of external forces are discussed in another book by the author (ref. 15 of Chapter I) and in Kogan's book (ref. 10 of Chapter I).

Chapter III

THE LINEARIZED COLLISION OPERATOR

1. General Comments on Perturbation Methods for the Boltzmann Equation

In Chapter II we found a solution of the Boltzmann equation, i.e., the Maxwellian. It is an exact solution of the Boltzmann equation and the most significant and widely used solution [other interesting solutions have been found but there is not sufficient space to discuss them here; some of them are discussed in detail in a book by Truesdell and Muncaster (ref. 1); see also Chapter VII, Section 14]. The meaning of the Maxwellian distribution is clear: it describes equilibrium states (or slight generalizations of them), characterized by the fact that neither heat flow nor stresses other than isotropic pressure are present. If we want to describe more realistic non-equilibrium situations, when oblique stresses are present and heat transfers take place, we have to rely upon approximation methods.

Some of the most useful methods are based upon perturbation techniques: we choose a parameter ϵ which can be small in some situations and expand f in a series of powers of ϵ [or, more generally, of functions $\sigma_n(\epsilon)$, such that $\lim_{\epsilon \to 0} \sigma_{n+1}(\epsilon)/\sigma_n(\epsilon) = 0$]. The resulting expansion, which in general cannot be expected to be convergent but only asymptotic to a solution of the Boltzmann equation, gives useful information for a certain range of small values of ϵ (sometimes larger than would be expected).

There are many different perturbation methods, corresponding to different choices of ϵ; however, in this section we want to study the general features of any perturbation method with respect to the collision operator $Q(f, f)$. We shall restrict ourselves to power series in ϵ:

$$f = \sum_{n=0}^{\infty} \epsilon^n f_n \tag{1.1}$$

If we insert this expansion into $Q(f, f)$ and take into account both the quadratic nature of the collision operator and the Cauchy rule for the product of two series, we find

$$Q(f, f) = \sum_{n=0}^{\infty} \epsilon^n \sum_{k=0}^{n} Q(f_k, f_{n-k}) \tag{1.2}$$

where $Q(f_k, f_{n-k})$ is the bilinear operator defined by Eq. (1.2) of Chapter II. The fact that a symmetrized expression appears depends upon the fact that we can combine the terms with $k = k_0$ and $k = n - k_0$ (for any k_0 between 0 and n), since both values of k appear in the coefficient of ϵ^n in Eq. (1.2). This equation shows that expanding f in a power series in the parameter ϵ implies an analogous expansion for the collision term Q, the coefficients being as follows:

$$Q_n = \sum_{k=0}^{n} Q(f_k, f_{n-k}) \qquad (n \geq 0) \tag{1.3}$$

A significant number of perturbation expansions which are used in connection with the Boltzmann equation have the following feature: either as a consequence of the zeroth-order equation or because of the assumptions underlying the perturbation method, the zeroth-order term in the expansion is a Maxwellian. We shall restrict our attention to this case at present; we note, however, that the parameters appearing in the Maxwellian (density, mass velocity, temperature) can depend arbitrarily upon the time and space variables (the Maxwellian is, in general, not required to satisfy the Boltzmann equation), but this will not concern us insofar as we deal with the collision operator which does not act on the space-time dependence of f.

Under the present assumptions we have

$$Q(f_0, f_0) = 0, \qquad \text{i.e.,} \qquad Q_0 = 0 \tag{1.4}$$

If we now consider Q_n $(n \geq 1)$ as given by Eq. (1.3), we observe that the first and last terms in the sum contain f_0, which is known to be a Maxwellian, and f_n, which is the nth-order coefficient in the expansion of f; the remaining terms $(1 \leq k \leq n - 1)$ contain only f_k of order less than n. If we apply our perturbation expansion to the Boltzmann equation, then we have to solve a sequence of equations; if we do this by recurrence, it is obvious that at the nth step f_k is known for $k \leq n - 1$. Therefore Q_n splits into the sum

$$Q_n = 2Q(f_0, f_n) + \sum_{k=1}^{n-1} Q(f_k, f_{n-k}) \qquad (n \geq 1) \tag{1.5}$$

where the second term is known at the nth step of the approximation (in particular, it is zero for $n = 1$) and can accordingly be written as a source term S_n; as a consequence, the relevant operator to be considered at each step is the linear operator $2Q(f_0, f_n)$ acting upon the unknown f_n. It is usual to put $f_n = f_0 h_n$ and consider h_n as unknown; then we can write as follows:

$$Q_n = f_0 L h_n + S_n \qquad (n \geq 1) \tag{1.6}$$

where, by definition, the linearized Boltzmann operator L is given by

$$Lh = 2f_0^{-1}Q(f_0, f_0h) \tag{1.7}$$

The above discussion suggests that the linearized collision operator is of basic importance in dealing with perturbation procedures for solving the Boltzmann equation; accordingly, we shall devote this chapter to the study of the properties of L.

2. Basic Properties of the Linearized Collision Operator

The definition, Eq. (1.7), of the linearized collision operator L, together with Eq. (1.2) of Chapter II, allows us to write the following expression for L:

$$Lh = (mf_0)^{-1}\int [f'_0 f'_{01}(h'_1 + h') - f_0 f_{01}(h_1 + h)]B(\theta, V)\, d\xi_1\, d\epsilon\, d\theta \tag{2.1}$$

f_0, being a Maxwellian, satisfies Eq. (2.6) of Chapter II; accordingly

$$Lh = (1/m)\int f_{01}(h'_1 + h' - h_1 - h)B(\theta, V)\, d\xi_1\, d\epsilon\, d\theta \tag{2.2}$$

since f_0 does not depend upon the integration variables and can be taken out of the integral to cancel f_0^{-1}. Equation (2.2) is the standard expression of the linearized collision operator.

We can use now the properties of $Q(f, g)$ and the definition, Eq. (1.7), of L to deduce some basic properties of the latter. If we consider Eq. (1.15) of Chapter II with $f = f_0$, $g = f_0h$, and $\varphi = g$, we have

$$\int f_0 g L h\, d\xi$$

$$= -\frac{1}{4m}\int f_0 f_{01}(h'_1 + h' - h_1 - h)(g'_1 + g' - g - g_1)$$

$$\times B(\theta, V)\, d\xi_1\, d\xi\, d\epsilon\, d\theta \tag{2.3}$$

The right-hand side shows that interchange of h and g does not change the value of the integral on the left-hand side. This suggests the introduction of a Hilbert space $\mathscr{H}$ where the scalar product is given by

$$(g, h) = \int f_0(\xi)g(\xi)h(\xi)\, d\xi \tag{2.4}$$

(When complex-valued functions are used the complex conjugate of g should appear; this, however, is not important here, and we restrict ourselves to real-valued functions.)

Equation (2.3) then shows that

$$(g, Lh) = (Lg, h) \tag{2.5}$$

i.e., L is formally self-adjoint (here both g and h belong to the domain of the operator L). If we put $g = h$ in Eq. (2.3), we obtain

$$(h, Lh) = -(1/4m) \int (h_1' + h' - h_1 - h)^2 B(\theta, V)\, d\boldsymbol{\xi}\, d\boldsymbol{\xi}_1\, d\epsilon\, d\theta \tag{2.6}$$

and since $B(\theta, V) > 0$, we have

$$(h, Lh) \leq 0 \tag{2.7}$$

and the equality sign holds if and only if the quantity which appears squared in Eq. (2.6) is zero, i.e., if h is a collision invariant. Equation (2.7) tells us that L is a nonpositive operator in $\mathscr{H}$. We note that when the equality sign holds in Eq. (2.7), i.e., h is a collision invariant, then Eq. (2.2) gives

$$Lh = 0 \tag{2.8}$$

and, conversely, if we scalarly multiply this equation by h, according to the scalar product in $\mathscr{H}$ defined above, we obtain

$$(h, Lh) = 0 \tag{2.9}$$

which implies that h is a collision invariant. Therefore the collision invariants are eigenfunctions of L corresponding to the eigenvalue $\lambda = 0$, and are the only eigenfunctions corresponding to such eigenvalue; all the other eigenvalues, if any, must be negative, because of Eq. (2.7). It can be verified that the latter equation is the linearized version of Eq. (2.3) of Chapter II; in fact, if we write $f = f_0(1 + h)$ in the latter equation and neglect terms of order higher than second (zeroth- and first-order terms cancel), we obtain Eq. (2.7). We note that it is sometimes useful to introduce the function

$$\hat{h} = [f_0]^{1/2} h \tag{2.10}$$

and the operator

$$\hat{L} = [f_0]^{1/2} L [f_0]^{-1/2} \tag{2.11}$$

since this avoids the weight function in the scalar product, Eq. (2.4); we shall not use this notation at present.

We observe now that if the range of the potential does not extend to infinity, or if it does but we artificially cut off the grazing collisions (see next section), we can split L into two parts,

$$Lh = Kh - \nu(\boldsymbol{\xi})h \tag{2.12}$$

It is also convenient to split K into two parts,

$$K = K_2 - K_1 \tag{2.13}$$

In these equations

$$\nu(\boldsymbol{\xi}) = (2\pi/m)\int f_{01}B(\theta, V)\, d\boldsymbol{\xi}_1\, d\theta \tag{2.14}$$

$$K_1 h = (2\pi/m)\int f_{01}h_1 B(\theta, V)\, d\boldsymbol{\xi}_1\, d\theta \tag{2.15}$$

$$K_2 h = (1/m)\int f_{01}(h_1' + h')B(\theta, V)\, d\boldsymbol{\xi}_1\, d\theta\, d\epsilon \tag{2.16}$$

It is clear that K_1 is an integral operator, whose kernel is picked out by inspection:

$$K_1(\boldsymbol{\xi}, \boldsymbol{\xi}_1) = (2\pi/m)f_0(\boldsymbol{\xi}_1)\int_0^{\pi/2} B(\theta, |\boldsymbol{\xi}_1 - \boldsymbol{\xi}|)\, d\theta \tag{2.17}$$

To find the kernel of K_2 requires some manipulation. We shall follow a procedure proposed by Grad in 1963. We recall (Chapter I, Section 7) that the range of definition of θ for repulsive potentials is $0 < \theta < \pi/2$ ($\boldsymbol{\alpha}$ is integrated over the hemisphere $\boldsymbol{\alpha} \cdot \mathbf{V} > 0$). Now a rotation of $\boldsymbol{\alpha}$ through $\pi/2$ ($\theta \to \frac{1}{2}\pi - \theta$, $\epsilon \to \epsilon \pm \pi$) interchanges the vectors $\boldsymbol{\xi}'$ and $\boldsymbol{\xi}_1'$ [note that if $\boldsymbol{\beta}$ lies in the plane of $\boldsymbol{\alpha}$ and $\mathbf{V}$ and makes a 90° angle with $\boldsymbol{\alpha}$, we can write $\mathbf{V} = \boldsymbol{\alpha}(\boldsymbol{\alpha} \cdot \mathbf{V}) + \boldsymbol{\beta}(\boldsymbol{\beta} \cdot \mathbf{V})$, which implies $\boldsymbol{\xi} + \boldsymbol{\alpha}(\boldsymbol{\alpha} \cdot \mathbf{V}) = \boldsymbol{\xi}_1 - \boldsymbol{\beta}(\boldsymbol{\beta} \cdot \mathbf{V})$, $\boldsymbol{\xi}_1 - \boldsymbol{\alpha}(\boldsymbol{\alpha} \cdot \mathbf{V}) = \boldsymbol{\xi} + \boldsymbol{\beta}(\boldsymbol{\beta} \cdot \mathbf{V})$]. Therefore if we introduce

$$B^*(\theta, V) = \tfrac{1}{2}[B(\theta, V) + B(\tfrac{1}{2}\pi - \theta, V)] \tag{2.18}$$

we can write

$$K_2 h = (2/m)\int f_0(\boldsymbol{\xi}_1)h(\boldsymbol{\xi}')B^*(\theta, V)\, d\theta\, d\epsilon\, d\boldsymbol{\xi}_1 \tag{2.19}$$

(the Jacobian is 1, since the above rotation of $\boldsymbol{\alpha}$ is a translation in terms of θ and ϵ). Next we notice that the transformation formulas, Eq. (7.12) of Chapter I, are invariant under $\boldsymbol{\alpha} \to -\boldsymbol{\alpha}$, i.e., $\theta \to \pi - \theta$, $\epsilon \to \epsilon \pm \pi$. Therefore if we extend the definition of $B^*(\theta, V)$ to the full range $0 < \theta < \pi$ by means of

$$B^*(\pi - \theta, V) = B^*(\theta, V) \tag{2.20}$$

we can integrate Eq. (2.19) over $0 < \theta < \pi$ after inserting a factor $\frac{1}{2}$. Now $\boldsymbol{\alpha}$ ranges over a full sphere, and in terms of

$$Q = B^*/\sin\theta \tag{2.21}$$

we may write

$$K_2h = (1/m)\int f_0(\boldsymbol{\xi}_1)h(\boldsymbol{\xi}')Q(V\cos\theta, V\sin\theta)\,d\boldsymbol{\alpha}\,d\mathbf{V} \tag{2.22}$$

where, following Grad, we have chosen the arguments of Q to be $V\cos\theta$ and $V\sin\theta$ (instead of V and θ) for later manipulation, and have introduced $\mathbf{V}$ in place of $\boldsymbol{\xi}_1$ as integration variable (they differ only by a translation $\boldsymbol{\xi}$).

Next we consider the components of $\mathbf{V}$ parallel and perpendicular to $\boldsymbol{\alpha}$:

$$\mathbf{V} = \mathbf{v} + \mathbf{w}; \qquad \mathbf{v} = \boldsymbol{\alpha}(\boldsymbol{\alpha}\cdot\mathbf{V}); \qquad \mathbf{w} = \mathbf{V} - \boldsymbol{\alpha}(\boldsymbol{\alpha}\cdot\mathbf{V}) \tag{2.23}$$

In these variables Q is a function of the magnitudes $v = V\cos\theta$ and $w = V\sin\theta$. We consider the following order of integration: first $\mathbf{w}$, then v, then $\boldsymbol{\alpha}$. With $\boldsymbol{\alpha}$ fixed the replacement of $\mathbf{V}$ by v and $\mathbf{w}$ amounts to a particular choice of the Cartesian reference frame for $\mathbf{V}$ and has unit Jacobian. After integrating over the plane $\mathbf{w}$ (which is perpendicular to $\boldsymbol{\alpha}$) we may combine the one-dimensional v integration in the direction $\boldsymbol{\alpha}$ with the integral of $\boldsymbol{\alpha}$ over a unit sphere to give a three-dimensional integration over the components of the vector $\mathbf{v} = \boldsymbol{\alpha}v$, introducing a factor 2 because the original range of v is $-\infty < v < +\infty$. The Jacobian of the transformation from $d\mathbf{v}$ to $d\boldsymbol{\alpha}\,dv$ (rectangular to polar coordinates) is simply v^2; summarizing,

$$d\boldsymbol{\alpha}\,dV = 2v^{-2}\,d\mathbf{v}\,d\mathbf{w} \tag{2.24}$$

In the new variables we have to integrate $\mathbf{w}$ over the plane perpendicular to $\mathbf{v}$, and $\mathbf{v}$ over the whole velocity space. Equations (2.23) also give

$$\boldsymbol{\xi}' = \boldsymbol{\xi} + \boldsymbol{\alpha}(\boldsymbol{\alpha}\cdot\mathbf{V}) = \boldsymbol{\xi} + \mathbf{v}; \qquad \boldsymbol{\xi}_1 = \boldsymbol{\xi} + \mathbf{V} = \boldsymbol{\xi} + \mathbf{v} + \mathbf{w} \tag{2.25}$$

and, consequently,

$$K_2h = \frac{2}{m}\int f_0(\boldsymbol{\xi} + \mathbf{v} + \mathbf{w})h(\boldsymbol{\xi} + \mathbf{v})Q(v, w)v^{-2}\,d\mathbf{w}\,d\mathbf{v} \tag{2.26}$$

Introducing the translated argument $\boldsymbol{\xi} + \mathbf{v} = \boldsymbol{\xi}_1$ instead of $\mathbf{v}$, we have

$$K_2h = (2/m)\int f_0(\boldsymbol{\xi}_1 + \mathbf{w})h(\boldsymbol{\xi}_1)Q(|\boldsymbol{\xi}_1 - \boldsymbol{\xi}|, w)|\boldsymbol{\xi} - \boldsymbol{\xi}_1|^{-2}\,d\mathbf{w}\,d\boldsymbol{\xi}_1 \tag{2.27}$$

Therefore the kernel of K_2 is

$$K_2(\boldsymbol{\xi}, \boldsymbol{\xi}_1) = (2/m)|\boldsymbol{\xi} - \boldsymbol{\xi}_1|^{-2}\int f_0(\boldsymbol{\xi}_1 + \mathbf{w})Q(|\boldsymbol{\xi}_1 - \boldsymbol{\xi}|, w)\,d\mathbf{w} \tag{2.28}$$

where the integration extends to the plane orthogonal to $\boldsymbol{\xi} - \boldsymbol{\xi}_1$.

For hard-sphere molecules the above equations reduce further; in fact, use of Eq. (7.24) of Chapter I gives

$$Q(V\cos\theta, V\sin\theta) = \sigma^2 V\cos\theta \tag{2.29}$$

and Eq. (2.28) reduces to

$$K_2(\boldsymbol{\xi}, \boldsymbol{\xi}_1) = (2\sigma^2/m)|\mathbf{c} - \mathbf{c}_1|^{-1} \int f_0(\boldsymbol{\xi}_1 + \mathbf{w})\, d\mathbf{w} \tag{2.30}$$

($\mathbf{c}$ is the peculiar velocity appearing in the Maxwellian, equal to $\boldsymbol{\xi}$ if the Maxwellian is taken to have zero mass velocity, and $\boldsymbol{\xi} - \boldsymbol{\xi}_1 = \mathbf{c} - \mathbf{c}_1$). Let us introduce f_0 explicitly:

$$f_0(\boldsymbol{\xi}) = \rho_0(2\pi RT_0)^{-3/2}\exp[-c^2(2RT_0)^{-1}] \tag{2.31}$$

then we have

$$\begin{aligned} f_0(\boldsymbol{\xi}_1 + \mathbf{w}) &= \rho_0(2\pi RT_0)^{-3/2}\exp[-(\mathbf{c}_1 + \mathbf{w})^2(2RT_0)^{-1}] \\ &= \rho_0(2\pi RT_0)^{-3/2}\exp\{-[c_1^2 + (\mathbf{c}_1 + \mathbf{c})\cdot\mathbf{w} + w^2](2RT_0)^{-1}\} \end{aligned} \tag{2.32}$$

where we have taken into account that $\mathbf{c}\cdot\mathbf{w} = \mathbf{c}_1\cdot\mathbf{w}$. Now, $\frac{1}{2}(\mathbf{c}_1 + \mathbf{c})$ can be decomposed into a part (say, $\boldsymbol{\zeta}$) in the $\mathbf{w}$ plane and a part along $\boldsymbol{\xi}_1 - \boldsymbol{\xi}$; the square of the latter part is $\frac{1}{4}(c_1^2 - c^2)^2|\mathbf{c}_1 - \mathbf{c}|^{-2}$, and

$$w^2 + (\mathbf{c}_1 + \mathbf{c})\cdot\mathbf{w} = (\mathbf{w} + \boldsymbol{\zeta})^2 - \tfrac{1}{4}(\mathbf{c}_1 + \mathbf{c})^2 + \tfrac{1}{4}(c_1^2 - c^2)|\mathbf{c}_1 - \mathbf{c}|^{-2} \tag{2.33}$$

Accordingly,

$$\begin{aligned} K_2(\boldsymbol{\xi}, \boldsymbol{\xi}_1) = {} & 2\sigma^2\rho_0(2\pi RT_0)^{-3/2}(m|\mathbf{c} - \mathbf{c}_1|)^{-1} \\ & \times\exp\{-[c_1^2 - \tfrac{1}{4}(\mathbf{c}_1 + \mathbf{c})^2 + \tfrac{1}{4}(c_1^2 - c^2)|\mathbf{c}_1 - \mathbf{c}|^{-2}](2RT_0)^{-1}\} \\ & \times\int \exp[-(\mathbf{w} + \boldsymbol{\zeta})^2(2RT_0)^{-1}]\, d\mathbf{w} \end{aligned} \tag{2.34}$$

(Here both $\mathbf{w}$ and

$$\boldsymbol{\zeta} = \tfrac{1}{2}(\mathbf{c}_1 + \mathbf{c}) - \tfrac{1}{2}(\mathbf{c}_1 - \mathbf{c})(c_1^2 - c^2)|\mathbf{c}_1 - \mathbf{c}|^{-2} \tag{2.35}$$

are two-dimensional vectors in the plane orthogonal to $\mathbf{c}_1 - \mathbf{c}$.)

The value of the integral appearing in Eq. (2.34) is trivially shown to be $2\pi RT_0$, since the integration extends to the whole plane $\mathbf{w}$; accordingly,

$$\begin{aligned} K_2(\boldsymbol{\xi}, \boldsymbol{\xi}_1) = {} & 2\sigma^2\rho_0(2\pi RT_0)^{-1/2}(m|\mathbf{c} - \mathbf{c}_1|)^{-1} \\ & \times\exp\{-[c_1^2 - \tfrac{1}{4}(\mathbf{c}_1 + \mathbf{c})^2 - \tfrac{1}{4}(c_1^2 - c^2)|\mathbf{c}_1 - \mathbf{c}|^{-2}](2RT_0)^{-1}\} \end{aligned} \tag{2.36}$$

This result for rigid spheres is due to Hilbert (ref. 3). It is also clear that the kernel of K_1 is very simple in this case, since, again using Eq. (7.24) of Chapter I in Eq. (2.17) of this chapter gives

$$K_1(\boldsymbol{\xi}, \boldsymbol{\xi}_1) = (2mRT_0)^{-1}(2\pi RT_0)^{-1/2}\rho_0\sigma^2|\mathbf{c} - \mathbf{c}_1| \exp[-c_1^2(2RT_0)^{-1}] \quad (2.37)$$

Finally, the collision frequency $\nu(\boldsymbol{\xi})$ defined by Eq. (2.14) can be written as

$$\nu(\boldsymbol{\xi}) = \rho_0\sigma^2(2mRT_0)^{-1}(2\pi RT_0)^{-1/2}\int \exp[-c_1^2(2RT_0)^{-1}]|\mathbf{c}_1 - \mathbf{c}|\, d\mathbf{c}_1 \quad (2.38)$$

The last integral can be easily evaluated in polar coordinates to give

$$\nu(c) = \rho_0\sigma^2 m^{-1}(2\pi RT_0)^{1/2}\Psi(c/(2RT_0)) \quad (2.39)$$

where

$$\Psi(x) = \exp(-x^2) + \left(2x + \frac{1}{x}\right)\int_0^x \exp(-y^2)\, dy \quad (2.40)$$

and we have written $\nu(c)$ in place of $\nu(\boldsymbol{\xi})$ in order to emphasize that the dependence is only upon the magnitude of $\mathbf{c}$.

3. Power-Law Potentials and Angular Cutoff

If we consider power-law potentials extending to infinity (see the end of Chapter I, Section 7), then the splitting shown in Eq. (2.12) is no longer possible, since the single integrals do not converge. Since, on the one hand, these potentials are the simplest from the computational point of view, but, on the other hand, the splitting in Eq. (2.12) makes a simple mathematical characterization of L possible, Grad has developed a theory based upon introducing a cutoff which excludes the grazing collisions. His definition of a cutoff power-law potential is to require that $B(\theta, V)$ is zero when θ exceeds a certain value $\theta_0 < \pi/2$. For inverse-power-law potentials we have [Eq. (7.32) of Chapter I]

$$B(\theta, V) = V^\gamma\beta(\theta) \qquad \gamma = (n-5)/(n-1) \quad (3.1)$$

where $\beta(\theta)$ is a function of θ, and has the following behavior:

$$\begin{aligned} \beta(\theta) &= O(\theta) & \theta &\to 0 \\ &= O([\tfrac{1}{2}\pi - \theta]^{-(n+1)/(n-1)}) & \theta &\to \pi/2 \end{aligned} \quad (3.2)$$

The behavior for $\theta \to \pi/2$ shows that $\beta(\theta)$ is not integrable, as we already know; this of course is no longer true if we introduce the above-mentioned cutoff, In fact, the cutoff allows us to introduce the quantity

$$\beta_0 = (2\pi/m) \int_0^{\pi/2} \beta(\theta)\, d\theta \tag{3.3}$$

and write

$$K_1(\xi, \xi_1) = \beta_0 f_0(\xi_1)|\xi_1 - \xi|^\gamma \tag{3.4}$$

$$\nu(c) = \beta_0 \int f_0(\xi_1)|\xi_1 - \xi|^\gamma\, d\xi_1 \tag{3.5}$$

Introducing

$$\tau = w/|\xi - \xi_1| = \tan\theta \tag{3.6}$$

as the argument of

$$q(\tau) = \tfrac{1}{2}[\beta(\theta) + \beta(\tfrac{1}{2}\pi - \theta)]/\sin\theta \tag{3.7}$$

we have

$$Q(v, w) = (v^2 + w^2)^{\gamma/2} q(\tau) \tag{3.8}$$

and

$$K_2(\xi, \xi_1) = (2/m)|\xi - \xi_1|^{-2} \int f_0(\xi_1 + \mathbf{w})(|\xi - \xi_1|^2 + w^2)^{\gamma/2} q(w/|\xi - \xi_1|)\, d\mathbf{w} \tag{3.9}$$

This expression can be manipulated further (see the paper by Grad, ref. 2) in order to deduce that $K_2(\xi, \xi_1)$, for power-law potentials with an angular cutoff, is bounded by a constant times the corresponding kernel for rigid spheres. It can be shown (see Grad, ref. 2) that the third iterate of the symmetrized form $f_0^{1/2} K_2 f_0^{-1/2}$ of the latter kernel is square integrable; therefore the operator K_2 is completely continuous in $\mathscr{H}$ (see ref. 4 for concepts of functional analysis) for rigid spheres and any power-law potential with angular cutoff. Since this is trivially true also for K_1 (in this case the kernel itself is square integrable), it follows that K is completely continuous. It is clear, however, that this property is useless if we want to go to the limit $\theta_0 \to \pi/2$, since the operator K has no meaning in this limit.

Another property which is very easy to prove is that the collision frequency $\nu(c)$ is a monotonic function of c. In fact, we have

$$\frac{\partial \nu}{\partial c} = \frac{\mathbf{c}}{c} \cdot \frac{\partial \nu}{\partial \mathbf{c}} = -\gamma \beta_0 c^{-1} \int f_0(\xi_1)\mathbf{c} \cdot \mathbf{V} V^{\gamma - 2}\, d\xi_1 \tag{3.10}$$

where $\mathbf{V} = \mathbf{c}_1 - \mathbf{c}$, and, consequently, $\partial V/\partial \mathbf{c} = -\mathbf{V}/V$. Since $f_0(\xi_1)$ is proportional to $\exp[-\alpha c_1^2]$, where $\alpha > 0$ and $c_1^2 = (\mathbf{c} + \mathbf{V})^2 = c^2 + V^2 + 2\mathbf{c} \cdot \mathbf{V}$, the contribution to the integral from the half-space $\mathbf{c} \cdot \mathbf{V} > 0$ is, in absolute value, smaller than the contribution from the half-space $\mathbf{c} \cdot \mathbf{V} < 0$; since the former integral is positive and the latter negative, the total integral is negative and $\partial v/\partial c$ is greater than or less than zero depending on whether γ is greater than or less than zero. Since $\gamma = (n - 5)/(n - 1)$, the collision frequency is monotonically increasing for $n > 5$ (including hard sphere as a limiting case, $n \to \infty$) and monotonically decreasing for $n < 5$; in the former case v is bounded from below,

$$v(c) \geq v_0 > 0 \tag{3.11}$$

whereas in the case $n < 5$ v is bounded from above,

$$0 < v(c) \leq v_0 \tag{3.12}$$

where $v_0 = v(0)$, which is a finite quantity.

The properties of K and $v(c)$ imply related properties of the operator L. The latter, being self-adjoint in $\mathscr{H}$, always admits a spectral decomposition, but the spectrum can be partly discrete and partly continuous; in correspondence with discrete points of the spectrum (eigenvalues), we obtain eigenfunctions belonging to $\mathscr{H}$, while in correspondence with points of the continuous spectrum (generalized eigenvalues), no square integrable eigenfunctions are obtained, although one can find generalized eigenfunctions not belonging to $\mathscr{H}$ (they are not ordinary functions, in general).

It is a trivial matter to verify that the operator "multiplication by $v(c)$" is self-adjoint and has a purely continuous spectrum which consists of just those values which are taken by the function $v(c)$ [in this case the generalized eigenfunction corresponding to the generalized eigenvalue λ is $\delta(\lambda - v(c))$]. For power potentials with $n > 5$ it extends from v_0 to infinity, while for $n < 5$ it covers the segment from v_0 to zero; for $n = 5$ the spectrum of $v = v_0$ reduces to the single point eigenvalue $\lambda = v_0$, infinitely many times degenerate. Now, the addition of a completely continuous operator to a self-adjoint operator does not alter the continuous spectrum (to be precise, it is the essential spectrum which is not altered, but the distinction is irrelevant for our purposes); thus L has as its continuous spectrum the values taken by $-v(c)$, for any potential with angular cutoff (with the exception of the inverse-fifth-power molecules, for which v_0 is a cluster point of eigenvalues) as well as for rigid spheres. In addition, L may have a discrete spectrum; this is the only spectrum which is present in the case $n = 5$. For $n \geq 5$ we have the result that a positive constant μ exists such that

$$-(h, Lh) > \mu(h, h) = \mu\|h\|^2 \tag{3.13}$$

for all functions h which are orthogonal to the collision invariants ψ_i:

$$(h, \psi_i) = 0 \qquad (i = 1, 2, 3, 4, 5) \tag{3.14}$$

[In (3.13) $\|h\|$ denotes the norm in the Hilbert space $\mathscr{H}$.] In fact, if there are any discrete eigenvalues between $\lambda = 0$ (to which the eigenfunctions ψ_i belong) and ν_0, we take μ equal to the smallest one, $\mu = \lambda_1$; otherwise we can take $\mu = \nu_0$.

An interesting question is whether there are indeed discrete values (other than $\lambda = 0$) besides the continuous spectrum for $n > 5$. The question was answered in the affirmative, for rigid spheres, by Kuščer and Williams in 1967. The idea, first introduced by Lehner and Wing in neutron transport theory, is to introduce an artificial parameter c as a multiplier of $[\nu(\xi) + \lambda]h$ in the eigenvalue equation, $Kg - \nu h = \lambda g$, and to study the eigenvalues $c = c_n(\lambda)$ as functions of the continuous parameter λ ranging from 0 to $-\nu(0)$. The eigenvalues of the original problem are then obtained as roots of $c_n(\lambda) = 1$. The result is that L has an infinite number of discrete eigenvalues in the interval $(-\nu(0), 0)$ which accumulate at the point $\lambda^* = -\nu(0)$. Kuščer and Williams restricted themselves to the case of isotropic eigenfunctions ($l = 0$ in the standard terminology of spherical functions), but Jenssen in 1972 generalized their method and showed that L has infinitely many discrete eigenvalues for $l = 0, 1, 2$, but at most a finite number of eigenvalues for each value of $l \geq 3$. Jenssen also gave a numerical evaluation of the largest eigenvalues and the corresponding eigenfunctions for $l = 0, 1, 2$. His numerical computations for $l = 3$ led him to conjecture that for $l \geq 3$, L has no eigenvalue at all.

For molecules interacting with central forces, Pao (1974) proved that the spectrum is purely discrete. This could be expected on the basis of a naive argument: in fact, if one cuts the grazing collisions off, finds the spectrum, and then lets the cutoff disappear, one is led to a purely discrete spectrum. Pao's proof was found to be wrong by Klaus in 1977, because of an error in the discussion of the essential self-adjointness of L. In order to avoid this difficulty, as well as the tedious estimates of the so-called symbols of the pseudodifferential operator used by Pao, Klaus constructs the collision operator for an infinite range potential as the limit of operators with a cutoff, thus transforming the intuitive argument given above into a rigorous proof. In his paper, the approximation is in the sense of strong resolvent convergence.

Another transformation which is useful in the case of inverse-power-law potentials will now be described. The Maxwellian is explicitly given by Eqs. (2.31) and (2.32); hence

$$f_0(\boldsymbol{\xi}_1 + \mathbf{w}) = \rho_0(2\pi R T_0)^{-3/2} \exp[-(c_1^2 + 2\boldsymbol{\zeta}\cdot\mathbf{w} + w^2)(2RT_0)^{-1}] \quad (3.15)$$

where $\boldsymbol{\zeta}$ is given by Eq. (2.35), or, alternatively, by

$$\boldsymbol{\zeta} = \pm\tfrac{1}{2}(\mathbf{c}_1 + \mathbf{c}) \times \frac{(\mathbf{c}_1 - \mathbf{c})}{|\mathbf{c}_1 - \mathbf{c}|} = \pm\frac{\mathbf{c} \times \mathbf{c}_1}{|\mathbf{c}_1 - \mathbf{c}|} \quad (3.16)$$

the choice of the sign depending only on $\mathbf{c}$ and $\mathbf{c}_1$.

Also, let w, ϵ be the polar coordinates of $\mathbf{w}$, with ϵ the angle between $\mathbf{w}$ and $\boldsymbol{\zeta}$. Then

$$K_2(\boldsymbol{\xi}, \boldsymbol{\xi}_1) = \frac{2\rho_0}{m}(2\pi RT_0)^{-3/2}|\mathbf{c} - \mathbf{c}_1|^{-2} \times \int \exp[-(c_1^2 + 2zw\cos\epsilon + w^2)(2RT_0)^{-1}] \times (|\boldsymbol{\xi} - \boldsymbol{\xi}_1|^2 + w^2)^{\gamma/2} q(w/|\boldsymbol{\xi} - \boldsymbol{\xi}_1|) w\, dw\, d\epsilon \qquad (3.17)$$

It is convenient to again introduce the variable θ related to w by Eq. (3.6) ($|\boldsymbol{\xi} - \boldsymbol{\xi}_1|$ is a constant during the integration over w). Then

$$K_2(\boldsymbol{\xi}, \boldsymbol{\xi}_1) = \frac{2\rho_0}{m}(2\pi RT_0)^{-3/2}|\mathbf{c} - \mathbf{c}_1|^{\gamma} \times \int (\sec\theta)^{\gamma+2} \tan\theta\, q(\tan\theta) \times \exp[-(c_1^2 + 2|\mathbf{c} \times \mathbf{c}_1|\cos\epsilon\tan\theta + |\mathbf{c} - \mathbf{c}_1|^2 \tan^2\theta)(2RT_0)^{-1}]\, d\theta\, d\epsilon \qquad (3.18)$$

where Eq. (3.16) has been taken into account.

The integration over ϵ can be handled in terms of the modified Bessel function I_0, giving

$$K_2(\boldsymbol{\xi}, \boldsymbol{\xi}_1) = \frac{4\pi\rho_0}{m}(2\pi RT_0)^{-3/2}|\mathbf{c} - \mathbf{c}_1|^{\gamma} \int (\sec\theta)^{\gamma+2} \tan\theta\, q(\tan\theta) \times \exp[-(c_1^2 + |\mathbf{c} - \mathbf{c}_1|^2 \tan^2\theta)(2RT_0)^{-1}] \times I_0(2|\mathbf{c} \times \mathbf{c}_1| \tan\theta\, (2RT_0)^{-1})\, d\theta \qquad (3.19)$$

Also, if we write

$$K_2(\boldsymbol{\xi}, \boldsymbol{\xi}_1; \theta) = \frac{4\pi\rho_0}{m}(2\pi RT_0)^{-3/2}|\mathbf{c} - \mathbf{c}_1|^{\gamma} \times \exp[-(c_1^2 + |\mathbf{c} - \mathbf{c}_1|^2 \tan^2\theta)(2RT_0)^{-1}] \times I_0(2|\mathbf{c} \times \mathbf{c}_1| \tan\theta\, (2RT_0)^{-1})(\sec\theta)^{\gamma+2} \tan\theta\, q(\tan\theta) \qquad (3.20)$$

and analogously,

$$K_1(\boldsymbol{\xi}, \boldsymbol{\xi}_1; \theta) = \frac{2\pi\rho_0}{m}(2\pi RT_0)^{-3/2}|\mathbf{c}_1 - \mathbf{c}|^{\gamma} \exp[-c_1^2(2RT_0)^{-1}]\beta(\theta) \qquad (3.21)$$

$$\nu(c;\theta) = \frac{2\pi\rho_0}{m}(2\pi R T_0)^{-3/2}\int \exp[-c_1^2(2RT_0)^{-1}]|\mathbf{c}_1 - \mathbf{c}|^{\gamma}\, d\mathbf{c}_1\, \beta(\theta) \quad (3.22)$$

$$K(\xi, \xi_1; \theta) = K_2(\xi, \xi_1; \theta) - K_1(\xi, \xi_1; \theta) \quad (3.23)$$

we can write

$$Lh = \int_0^{\pi/2}\left[\int K(\xi, \xi_1; \theta)h(\xi_1)\, d\xi_1 - \nu(c;\theta)h(\xi)\right] d\theta \quad (3.24)$$

This expression retains a meaning even when we let $\theta_0 \to \pi/2$, i.e., we let the cutoff disappear, and shows that for potentials extending to infinity we can write L as an integral, with respect to the parameter θ, of the difference between an integral operator K_θ and a multiplication operator ν_θ, both depending upon θ. However, since the dependence on θ of each operator is nonintegrable in the vicinity of $\theta = \pi/2$, we cannot integrate each term to yield Eq. (2.12).

4. Potentials with a Strictly Finite Range

The angular cutoff introduced in Section 3 has the advantage of yielding a rather simple mathematical theory of the collision operator without changing the dependence of the differential collision cross section [which is proportional to $B(\theta, V)$] upon the relative velocity. However, the angular cutoff is to be regarded as a mathematical trick, which would acquire a meaning only if one could pass to the limit $\theta_0 \to \pi/2$. On the other hand, as discussed in Chapter I, Section 7, a physically meaningful two-body interaction should be based on a potential with a strictly finite range; in this case an angle cutoff is not needed to obtain the splitting shown in Eq. (2.12). The disadvantage of using the cutoff potential instead of infinite range plus the angular cutoff procedure is that the operator K is now much tougher to handle; in particular, it is not easy either to prove or disprove that K is completely continuous in $\mathscr{H}$ (see ref. 5). However, one can prove that the integral operator with kernel $K(\xi, \xi_1)[\nu(\xi)\nu(\xi_1)]^{-\alpha}$ is completely continuous for suitable values of α (it is easy to show that this is true for any $\alpha \geq 2$). It proves difficult, however, to show, if possible, that the value of α can be reduced to zero; in this author's opinion, although complete continuity can fail to hold for $\alpha = 0$, it is highly probable that the value $\alpha = \frac{1}{2}$ can yield complete continuity. This result, as will be seen in a later chapter, would allow the construction of a theory which is both consistent and elegant.

Some useful properties of the linearized collision operator can be obtained even without proving the complete continuity of K. A highly interesting property is given by the following:

Theorem. Inequality (3.13) applies to any h such that Eq. (3.14) is verified and $(\nu h, h)$ exists, provided $U(\rho) \geq O(\rho^{-(n-1)})$ $(\rho \to 0; n \geq 5)$, $U(\rho)$ denoting the intermolecular potential.

To prove the theorem, let us consider the general expression for L, Eq. (2.2). Let us split L as follows:

$$L = L^{\delta} + L_{\delta} \tag{4.1}$$

here L^{δ} and L_{δ} are given by the same expression as L, Eq. (2.2), where, however, $B(\theta, V)$ is replaced, respectively, by the following cutoff expressions:

$$\begin{aligned} B^{\delta}(\theta, V) &= B(\theta, V) && \text{for} \quad \theta < \delta \\ &= 0 && \text{for} \quad \theta > \delta \\ B_{\delta}(\theta, V) &= 0 && \text{for} \quad \theta < \delta \\ &= B(\theta, V) && \text{for} \quad \theta > \delta \end{aligned} \tag{4.2}$$

where δ is any fixed angle between 0 and $\pi/2$. Now L^{δ} and L_{δ}, having the general expression given by Eq. (2.2), enjoy the property (2.7), i.e.,

$$(h, L^{\delta}h) \leq 0 \qquad (h, L_{\delta}h) \leq 0 \tag{4.3}$$

Let us now assume that the theorem is not true, i.e., for any $\mu > 0$, no matter how small, we can find a function orthogonal to the collision invariants and such that

$$-(h, Lh) \leq \mu(h, h) \tag{4.4}$$

However, because of the splitting (4.1) and the property (4.3) we have

$$-(h, L_{\delta}h) = -(h, Lh) + (h, L^{\delta}h) \leq -(h, Lh) \leq \mu(h, h) \tag{4.5}$$

i.e., given any $\mu > 0$, no matter how small, we can find a function h such that Eq. (4.5) is satisfied, i.e., the origin is not an isolated point eigenvalue for L_{δ}. But L_{δ} is an operator with angular cutoff, which is harder than an inverse-power-potential with $n \geq 5$; then we can apply Grad's reasoning to show that Eq. (4.5), and hence Eq. (4.4) as well, is not true, as was to be shown.

There is an assumption in the above theorem, that $(\nu h, h)$ exists, which has not been used explicitly, but is required in order that the scalar products appearing in the above proof, in particular (h, Lh), exist. In fact, because of Eq. (2.12), (h, Lh) certainly exists, if both (h, Kh) and $(h, \nu h)$ exist; now if we assume the latter to exist, then the existence of the former is a consequence of the following:

Theorem. The operator K satisfies the inequality

$$|(h, Kh)| \leq \lambda(\nu h, h) \qquad (h \in \mathcal{N}) \tag{4.6}$$

for some positive λ. Here $\mathcal{N}$ denotes the (Hilbert) space of the functions such that $(\nu h, h)$ exists.

A preliminary result is the trivial statement:

$$(h, Kh) \le (\nu h, h) \tag{4.7}$$

which follows from Eqs. (2.12) and (2.7). We note also that since the kernels of K_1 and K_2 are nonnegative, Eq. (2.13) implies

$$|(h, Kh)| \le (|h|, K_2|h|) + (|h|, K_1|h|) \tag{4.8}$$

But Eqs. (2.13) and (4.7) give

$$(|h|, K_2|h|) = (|h|, K_1|h|) + (|h|, K|h|) \le (|h|, K_1|h|) + (h, \nu h) \tag{4.9}$$

where the bars are not required in the last term. Substituting into Eq. (4.8) gives

$$|(h, Kh)| \le (\nu h, h) + 2(|h|, K_1|h|) \tag{4.10}$$

Now, because of Eq. (2.15) and the Schwarz inequality

$$\begin{aligned}(|h|, K_1|h|) &= \frac{2\pi}{m}\int d\theta \int d\xi_1\, d\xi\, \{[f_0(\xi)f_0(\xi_1)B(\theta, |\xi_1 - \xi|)]^{1/2}|h(\xi)|\} \\ &\quad \times \{[f_0(\xi)f_0(\xi_1)B(\theta, |\xi_1 - \xi|)]^{1/2}|h(\xi_1)|\} \\ &\le \frac{2\pi}{m}\int d\theta \left\{\int d\xi\, d\xi_1\, f_0(\xi)f_0(\xi_1)B(\theta, |\xi_1 - \xi|)|h(\xi)|^2\right\}^{1/2} \\ &\quad \times \left\{\int d\xi\, d\xi_1\, f_0(\xi)f_0(\xi_1)B(\theta, |\xi_1 - \xi|)|h(\xi_1)|^2\right\}^{1/2}\end{aligned} \tag{4.11}$$

Both integrals within brackets have the same value, as seen by interchanging ξ and ξ_1. Then, if Eq. (2.14) is also used we obtain

$$\begin{aligned}(|h|, K_1|h|) &\le \int d\xi \left\{\frac{2\pi}{m}\int d\theta\, d\xi_1\, f_0(\xi_1)B(\theta, |\xi_1 - \xi|)\right\} f_0(\xi)|h(\xi)|^2 \\ &= \int d\xi\, \nu(\xi)f_0(\xi)|h(\xi)|^2 = (\nu h, h)\end{aligned} \tag{4.12}$$

Putting this result into Eq. (4.10) gives Eq. (4.6) with $\lambda = 3$. Another property of the collision operator which is worthwhile noting is that the "collision frequency" $\nu(\xi)$ for any cutoff potential is equal to the collision frequency of rigid-sphere molecules having the diameter equal to the range σ of the intermolecular force. As a matter of fact

$$\nu(\xi) = \frac{2\pi}{m}\int B(\theta, |\xi_1 - \xi|) f_0(\xi_1)\, d\xi_1\, d\theta$$

$$= \frac{2\pi}{m}\int f_0(\xi_1)|\xi_1 - \xi| r \frac{\partial r}{\partial \theta}\, d\theta\, d\xi_1 = \frac{2\pi}{m}\int_0^\sigma r\, dr \int f_0(\xi_1)|\xi_1 - \xi|\, d\xi_1$$

$$= \frac{\pi\sigma^2}{m}\int f_0(\xi_1)|\xi_1 - \xi|\, d\xi_1 \tag{4.13}$$

and comparison with Eq. (2.38) gives the desired result. Therefore Eq. (2.39) holds; in particular, $\nu(c)$ increases linearly with c as $c \to \infty$.

In 1975 Drange introduced a new cutoff, called the radial integral cutoff, which simply consisted of cutting all the interactions with impact parameters larger than an assigned value. He also developed a spectral theory for the linearized collision operator with either the radial cutoff discussed above or the new one. In his paper of 1977, mentioned in the previous section, Klaus was also able to describe the spectrum in the case of a radial integral cutoff.

5. The Case of Maxwell's Molecules

In spite of the several interesting results discussed in Section 3, the knowledge of the spectrum and the corresponding eigenfunctions in the case of a potential without cutoff is rather scanty; an exception is offered by the case of Maxwell's molecules, for which we are able to evaluate explicitly both the eigenfunctions and the eigenvalues.

In order to do this, we observe that for Maxwell's molecules ($n = 5$; i.e., $\gamma = 0$), the dependence of the kernel $K_2(\xi, \xi_1)$ [Eqs. (3.19) or (3.20)] on $\mathbf{c}$ and $\mathbf{c}_1$ is contained in the factor

$$\varphi(\xi, \xi_1) = \exp\{-[c_1^2 + (c^2 + c_1^2)\tan^2\theta]\} \exp[2cc_1 \tan^2\theta \cos\varphi'] \times I_0(2cc_1 \sin\varphi' \tan\theta) \tag{5.1}$$

where we have temporarily taken velocity units such that $2RT_0 = 1$ and introduced the angle φ' between $\mathbf{c}$ and $\mathbf{c}_1$. We now use the identity

$$\exp[\alpha r \cos\psi \cos\varphi'] I_0(\alpha r \sin\psi \sin\varphi') = [\pi/(2r\alpha)]^{1/2} \sum_{l=0}^{\infty} (2l + 1) I_{l+(1/2)}(\alpha r) P_l(\cos\varphi') P_l(\cos\psi) \tag{5.2}$$

where I_ν is the modified Bessel function of the order ν and P_l the Legendre polynomial of degree l. The quickest way of proving Eq. (5.2) is to observe that if we consider r, φ', and, say, ϵ as polar coordinates of a point in three-dimensional space, then the left-hand side Ψ of Eq. (5.2), as a function of the coordinates of this point, satisfies the equation

$$\Delta\Psi = \alpha^2\Psi \tag{5.3}$$

Separating this equation in polar coordinates yields the following general solution regular at the origin:

$$\Psi = \sum_{l=0}^{\infty} Q_l r^{-1/2} I_{l+(1/2)}(\alpha r) P_l(\cos\varphi') \tag{5.4}$$

where the coefficients Q_l are arbitrary and must be chosen suitably in order to represent the left-hand side of Eq. (5.2). The coefficients Q_l can depend upon α and ψ, but not on r and φ', of course; however, the left-hand side of Eq. (5.2) is symmetric with respect to the interchange $\alpha \leftrightarrow r$, $\psi \leftrightarrow \varphi'$, and, consequently,

$$Q_l = b_l \alpha^{-1/2} P_l(\cos\psi) \tag{5.5}$$

where b_l no longer depends upon α and ψ. Then

$$\exp[\alpha r \cos\psi \cos\varphi'] I_0(\alpha r \sin\psi \sin\varphi') = \sum_{0}^{\infty} b_l (\alpha r)^{-1/2} I_{l+(1/2)}(\alpha r) P_l(\cos\varphi') P_l(\cos\psi) \tag{5.6}$$

Now putting $\psi = 0$ and comparing with the well-known identity

$$\exp(\alpha z) = (\pi/2)^{1/2} \sum_{0}^{\infty} (2l+1)(\alpha r)^{-1/2} I_{l+(1/2)}(\alpha r) P_l(\cos\varphi') \tag{5.7}$$

$$z = r\cos\varphi'$$

which is a particular case of Sonine's formula (see References), we obtain

$$b = (2l+1)(\pi/2)^{1/2} \tag{5.8}$$

and consequently Eq. (5.2).

In order to exploit Eq. (5.2), we put

$$\cot\psi = \tan\theta, \qquad \alpha r = 2cc_1 \tan\theta \sec\theta \tag{5.9}$$

and obtain

$$\varphi(\xi, \xi_1) = [\pi/(4cc_1 \tan\theta \sec\theta)]^{1/2} \exp\{-[c_1^2 + (c^2 + c_1^2)\tan^2\theta]\} \times \sum_{l=0}^{\infty} (2l+1) I_{l+(1/2)}(2cc_1 \tan\theta \sec\theta) P_l(\cos\varphi') P_l(\sin\theta) \tag{5.10}$$

Now we use another identity, the so-called Miller–Lebedev formula (see References), which can be written as

$$\sum_{n=0}^{\infty} \frac{n!}{\Gamma(n+\beta+1)} L_n^{\beta}(x) L_n^{\beta}(y) Z^n$$
$$= \frac{(xyZ)^{-\beta/2}}{1-Z} \exp[-Z(x+y)(1-Z)^{-1}] I_\beta(2(xyZ)^{1/2}(1-Z)^{-1})$$
$$(|Z| < 1) \tag{5.11}$$

where the L_n^α are the associated Laguerre polynomials of order α and degree n.

In order to exploit this equation, we put

$$x = c^2, \qquad y = c_1^2, \qquad Z = \sin^2\theta, \qquad \beta = l + \tfrac{1}{2} \tag{5.12}$$

and put the result into Eq. (5.10) to obtain

$$\varphi(\xi, \xi_1) = \frac{\sqrt{\pi}}{4} \cos^3\theta \exp[-c_1^2] \sum_{l,n=0}^{\infty} (2l+1) g_{nl}(c) g_{nl}(c_1) P_l(\cos\varphi')$$
$$\times P_l(\sin\theta) \sin^{2n+l}\theta \tag{5.13}$$

where

$$g_{nl}(c) = \left[\frac{n!2}{\Gamma(n+l+\frac{3}{2})}\right]^{1/2} c^l L_n^{l+(1/2)}(c^2) \tag{5.14}$$

Now, the summation formula for the spherical harmonics Y_l^m gives

$$P_l(\cos\varphi') = \frac{4\pi}{2l+1} \sum_{m=-l}^{l} Y_l^{*m}(\theta, \varphi) Y_l^m(\theta_1, \varphi_1) \tag{5.15}$$

where θ, φ and θ_1, φ_1 are the angles defining the vectors $\mathbf{c}$ and $\mathbf{c}_1$ in an arbitrary reference frame. Inserting Eq. (5.15) into Eq. (5.13) gives

$$\varphi(\xi, \xi_1) = \pi^{3/2} \cos^3\theta \exp[-c_1^2] \sum_{l,n,m} \psi_{nlm}(\mathbf{c}) \psi_{nlm}^*(\mathbf{c}_1) P_l(\sin\theta) \sin^{2n+l}\theta \tag{5.16}$$

where

$$\psi_{nlm}(\mathbf{c}) = g_{nl}(c) Y_l^m(\theta, \varphi) \tag{5.17}$$

Using the well-known properties of the associated Laguerre polynomials and the spherical harmonics, one easily verifies that

$$\int \psi_{nlm}^*(\mathbf{c}) \psi_{n'l'm'}(\mathbf{c}) \exp[-c^2]\, d\mathbf{c} = \delta_{nn'} \delta_{ll'} \delta_{mm'} \tag{5.18}$$

Inserting Eq. (5.16) into Eq. (3.20) [Eq. (5.1) has to be used, of course], we obtain

$$K_2(\boldsymbol{\xi}, \boldsymbol{\xi}_1; \theta) = (2RT_0)^{-3/2} \exp[-c_1^2(2RT_0)^{-1}] \times \sum_{l,n,m} \mu_{nl}(\theta)\psi_{nlm}[\mathbf{c}(2RT_0)^{-1}]\psi^*_{nlm}[\mathbf{c}_1(2RT_0)^{-1}] \quad (5.19)$$

where

$$\mu_{nl}(\theta) = (4\pi\rho_0/m)P_l(\sin\theta)\sin^{2n+l+1}\theta\, q(\tan\theta) \quad (5.20)$$

and we have restored general units for the molecular velocity vector.

On the other hand, if we use the fact that $\psi_{000} = \pi^{-3/4}$, $K_1(\boldsymbol{\xi}, \boldsymbol{\xi}_1; \theta)$ can be written as follows ($\gamma = 0$):

$$\frac{2\pi}{m}\rho_0\beta(\theta)(2RT_0)^{-3/2} \exp[-c_1^2(2RT_0)^{-1}]\psi_{000}[\mathbf{c}(2RT_0)^{-1}]\psi^*_{000}[\mathbf{c}_1(2RT_0)^{-1}] \quad (5.21)$$

$\nu(c;\theta)$ reduces to $(2\pi/m)\rho_0\beta(\theta)$ and $h(\boldsymbol{\xi})$ can be expanded according to the complete set $\{\psi_{nlm}\}$ as follows:

$$h(\boldsymbol{\xi}) = (2RT_0)^{-3/2} \sum_{n,l,m} \psi_{nlm}[\mathbf{c}(2RT_0)^{-1}] \int \exp[-c_1^2(2RT_0)^{-1}]h(\boldsymbol{\xi}_1) \times \psi^*_{nlm}[\mathbf{c}_1(2RT_0)^{-1}]\, d\boldsymbol{\xi}_1 \quad (5.22)$$

Accordingly, Eq. (5.24) gives

$$Lh = (2RT_0)^{-3/2} \int_0^{\pi/2} d\theta \Big\{ \sum_{n,l,m} \lambda_{nl}(\theta)\psi_{nlm}[\mathbf{c}(2RT_0)^{-1}] \times \int \exp[-c_1^2(2RT_0)^{-1}]h(\boldsymbol{\xi}_1)\psi^*_{nlm}[\mathbf{c}_1(2RT_0)^{-1}]\, d\boldsymbol{\xi}_1 \Big\} \quad (5.23)$$

where

$$\lambda_{nl}(\theta) = \mu_{nl}(\theta) - (2\pi/m)\rho_0\beta(\theta)(\delta_{n0}\delta_{l0} + 1) \quad (5.24)$$

We can now integrate term by term with respect to θ (convergence of the integrals holds even in the absence of cutoff) and obtain

$$Lh = \sum_{n,l,m} \lambda_{nl}\psi_{nlm}[\mathbf{c}(2RT_0)^{-1}] \times \int \psi^*_{nlm}[\mathbf{c}_1(2RT_0)^{-1}] \exp[-c_1^2(2RT_0)^{-1}]h(\boldsymbol{\xi}_1)(2RT_0)^{-3/2}\, d\boldsymbol{\xi}_1 \quad (5.25)$$

where

$$\lambda_{nl} = \int_0^{\pi/2} \lambda_{nl}(\theta)\, d\theta \tag{5.26}$$

If we put $h = \psi_{n'l'm'}$ into Eq. (5.25) and use Eq. (5.18), we obtain

$$L\psi_{nlm} = \lambda_{nl}\psi_{nlm} \tag{5.27}$$

i.e., the ψ's are eigenfunctions of L corresponding to the eigenvalue λ_{nl}. This basic result accounts for the preference given to Maxwell's molecules with respect to other, more realistic molecular models.

We note that the eigenvalues do not depend on m; accordingly, λ_{nl} is, at least, $(2l + 1)$ times degenerate.

If $\beta(\theta)$ were integrable, then the eigenvalues λ_{nl} would be bounded by some constant, since both $P_l(\sin\theta)$ and $\sin^{2n+l+1}\theta$ are uniformly bounded. The nonintegrable nature of $\beta(\theta)$ implies that λ_{nl} grows without bounds when $n \to \infty$; it is interesting to look for the asymptotic growth of λ_{nl}. We recall [Eq. (3.2) with $n = 5$] that

$$\beta(\theta) \approx A(\cos\theta)^{-3/2} \qquad (\theta \to \pi/2) \tag{5.28}$$

where A is a constant. It is clear that the singular behavior of $\beta(\theta)$ as $\theta \to \pi/2$ will dictate the behavior of λ_{nl} for large n because λ_{nl} would be bounded in the absence of singularities. Since $P_l[\sin(\pi/2)] = 1$ is finite, we have

$$\lambda_{nl}(\theta) \approx \frac{2\pi\rho_0}{m} A[\cos\theta]^{-3/2}[\sin^{2n+l+1}\theta - 1] \qquad (\theta \to \pi/2) \tag{5.29}$$

since the remaining terms are either zero or less singular. Therefore we have to find the asymptotic behavior as $n \to \infty$ of the integral

$$\int_0^{\pi/2} (\cos\theta)^{-3/2}(\sin^{2n+l+1}\theta - 1)\, d\theta \tag{5.30}$$

If we change the variables according to $\sin\theta = e^{-t}$, we can replace $\cos\theta$ by $(2t)^{1/2}$ and $d\theta$ with $dt\,(2t)^{-1/2}$, since the differences are of the order of previously neglected quantities; then

$$\lambda_{nl} \approx \frac{\pi\rho_0 A}{m} 2^{-1/4} \int_0^{\infty} t^{-5/4}\{[\exp - (2n + l + 1)t] - 1\}\, dt \qquad (n \to \infty) \tag{5.31}$$

The last integral is obviously proportional to $(2n + l + 1)^{1/4}$ (the numerical coefficient is easily evaluated in terms of the Γ-function, but does not matter at present). Accordingly, we obtain

$$\lambda_{nl} \approx -B(2n + l + 1)^{1/4} \qquad (n \to \infty) \tag{5.32}$$

where B is a new (positive) constant. Therefore the eigenvalues of the collision operator for Maxwell's molecules grow asymptotically as the fourth root of an integer; it can also be shown that there is an additional constant contribution, while the remaining part goes to zero as $n^{-1/4}$ when $n \to \infty$. Since the eigenfunctions of L are also eigenfunctions of the quantum-mechanical Hamiltonian M for a three-dimensional harmonic oscillator, the symmetrized operator $\hat{L}$ [Eq. (2.11)] can be written as the sum of the fourth root of M plus a constant plus a completely continuous operator.

References

Section 1——The exact solutions of the Boltzmann equation are discussed in some detail in the book

1. C. Truesdell and R. G. Muncaster, *Fundamentals of Maxwell's Kinetic Theory of a Simple Monatomic Gas*, Academic Press, New York (1980).

Section 2——The manipulations of L follow the treatment given by

2. H. Grad, in: *Rarefied Gas Dynamics* (J. A. Laurmann, ed.), Vol. I, p. 100, Academic Press, New York (1963);

   The result for rigid sphere was first given by

3. D. Hilbert, *Math. Ann.* **72**, 562 (1912).

   An alternative procedure for general molecular models can be found in ref. 7 of Chapter I.

Section 3—— The results in this section are mainly based on ref. 2. The basic concepts of functional analysis required for the spectral theory can be found in

4. F. Riesz and B. Sz-Nagy, *Functional Analysis* (translated by L. Boron), Frederick Ungar, New York (1955).

   The main results on the discrete spectrum for the sphere collision operator were given by

5. I. Kuščer and M. M. R. Williams, *Phys. Fluids* **10**, 1922 (1967);
6. O. O. Jenssen, *Phys. Norvegica* **6**, 179 (1972);
7. Y. P. Pao, in *Rarefied Gas Dynamics* (M. Becker and M. Fiebig, eds.), Vol. I, p. A.6-1, DFVLR Press, Porz-Wahn (1974);
8. Y. P. Pao, *Commun. Pure Appl. Math.* **27**, 407 (1974);
9. Y. P. Pao, *Commun. Pure Appl. Math.* **27**, 559 (1974);
10. M. Klaus, *Helv. Phys. Acta* **50**, 893 (1977).

Section 4——The treatment in this section is mainly based on

11. C. Cercignani, *Phys. Fluids* **10**, 2097 (1967)

    but the results on the spectrum can be found in ref. 10 and in

12. H. Drange, *SIAM J. Appl. Math.* **29**, 4 (1975).

Section 5——The fact that the Laguerre (or Sonine) polynomials times the spherical harmonics give the eigenfunctions for Maxwell's molecules was first pointed out in an explicit fashion by

13. C. S. Wang Chang and G. E. Uhlenbeck, *On the Propagation of Sound in Monatomic Gases*, University of Michigan Press, Project M999, Ann Arbor, Michigan (1952).

The Sonine and Miller–Lebedev formulas used in this section can be found in

14. Bateman Manuscript Project, *Higher Transcendental Functions*, Vol. II, McGraw-Hill, New York (1953), pp. 64 and 188.

The asymptotic behavior of the eigenvalues of the collision operator for Maxwell molecules was first pointed out by

15. H. M. Mott-Smith, "A New Approach in the Kinetic Theory of Gases," MIT Lincoln Laboratory Group Report V-2 (1954).

Chapter IV

MODEL EQUATIONS

1. Guessing Models: The Nonlinear Bhatnagar, Gross, and Krook Model and Generalizations

One of the major shortcomings in dealing with the Boltzmann equation is the complicated nature of the collision integral, in both the full nonlinear version, Eq. (1.1) of Chapter II, and the linearized form, Eq. (2.2) of Chapter III. It is therefore not surprising that alternative, simpler expressions have been proposed for the collision term; they are known as collision models, and any Boltzmann-like equation where the Boltzmann collision integral is replaced by a collision model is called a model equation or a kinetic model.

The idea behind this replacement is that the large amount of detail of the two-body interactions (which is contained in the collision term and reflected, e.g., in the details of the spectrum of the linearized operator) is not likely to influence significantly the values of many experimentally measured quantities. That is, unless very refined experiments are devised, the fine structure of the collision operator $Q(f, f)$ is blurred into a coarser description based upon a simpler operator $J(f)$, which retains only the qualitative and average properties of the true collision operator.

The most widely known collision model is usually called the Bhatnagar, Gross, and Krook (BGK) model, although Welander proposed it independently at about the same time. The idea behind the BGK model (retained by more sophisticated models) is that the main features of the collision operator are:

1. The true collision term $Q(f,f)$ satisfies Eq. (3.18) of Chapter II; hence the collision model $J(f)$ must satisfy

$$\int \psi_\alpha J(f)\, d\xi = 0 \qquad (\alpha = 0, 1, 2, 3, 4) \tag{1.1}$$

2. The collision term expresses the tendency to a Maxwellian distribution (H-theorem).

The simplest way of taking this second feature into account is to imagine that each collision changes the distribution function $f(\boldsymbol{\xi})$ by an amount proportional to the departure of f from a Maxwellian $\Phi(\boldsymbol{\xi})$; i.e., if ν is a constant with respect to $\boldsymbol{\xi}$, we introduce the following collision model:

$$J(f) = \nu[\Phi(\boldsymbol{\xi}) - f(\boldsymbol{\xi})] \tag{1.2}$$

The Maxwellian $\Phi(\boldsymbol{\xi})$ has five disposable scalar parameters $(\rho, \mathbf{v}, T)$; however, these are fixed by Eq. (1.1), which implies

$$\int \psi_\alpha \Phi(\boldsymbol{\xi})\, d\boldsymbol{\xi} = \int \psi_\alpha f(\boldsymbol{\xi})\, d\boldsymbol{\xi} \tag{1.3}$$

i.e., at any space point and time instant $\Phi(\boldsymbol{\xi})$ must have exactly the same density, velocity, and temperature of the gas given by the distribution function $f(\boldsymbol{\xi})$. Since the latter will in general vary with time and space coordinates, the same will be true for the parameters of $\Phi(\boldsymbol{\xi})$, which is accordingly called the local Maxwellian. The "collision frequency" ν is not restricted at this level and has to be fixed by means of additional considerations; we note, however, that ν can be a function of the local state of the gas and hence vary with both time and space coordinates.

We observe that the nonlinearity of the proposed $J(f)$ is much worse than the nonlinearity of the true collision term $Q(f, f)$; in fact, the latter is simply quadratic in f, while the former contains f in both the numerator and the denominator of an exponential (the $\mathbf{v}$ and T appearing in Φ are moments of f).

The main advantage in using the BGK operator is that for any given problem one can deduce integral equations for the macroscopic variables $\rho, \mathbf{v}, T$ (see Chapter VIII); these equations are strongly nonlinear, but simplify some iteration procedures and make the treatment of interesting problems feasible on a high-speed computer. Another advantage of the BGK model is offered by its linearized form:

$$Lh = \nu\left[\sum_{\alpha=0}^{4} \psi_\alpha(\psi_\alpha, h) - h\right] \tag{1.4}$$

where h is the perturbation of the distribution function considered in Chapter III and (,) denotes the scalar product in $\mathscr{H}$. The collision invariants ψ_α are here normalized in such a way that

$$(\psi_\alpha, \psi_\beta) = \delta_{\alpha\beta} \qquad (\alpha, \beta = 0, 1, 2, 3, 4) \tag{1.5}$$

It is obvious that Eq. (1.4) has a structure definitely simpler than the true linearized collision operator.

A problem which is easily solved with the nonlinear BGK model is the relaxation to equilibrium in the spatially homogeneous case. An arbitrary distribution function $g(\xi)$ depending only on the velocity vector ξ is given, and we want to find its time evolution according to kinetic theory; the problem cannot be solved analytically with the full Boltzmann equation, but is trivial with the BGK model. As a matter of fact, we have

$$\partial f/\partial t = \nu(\Phi - f), \qquad f(\xi, 0) = g(\xi) \tag{1.6}$$

where the space derivatives can be omitted because the distribution function will remain homogeneous in physical space at any time. But Eqs. (1.1), or, equivalently, Eqs. (3.20) of Chapter II, give

$$\partial \rho/\partial t = 0; \qquad \partial \mathbf{v}/\partial t = 0; \qquad \partial T/\partial t = 0 \tag{1.7}$$

i.e. ρ, $\mathbf{v}$, T are time-independent (mass, momentum, and energy are locally conserved in the homogeneous case). Then Eq. (1.6) is a very simple first-order differential equation with constant coefficients and constant source term; the solution is

$$f(\xi, t) = g(\xi)e^{-\nu t} + (1 - e^{-\nu t})\Phi(\xi) \tag{1.8}$$

where the ρ, $\mathbf{v}$, T in Φ can be calculated as moments of $g(\xi)$. The interpretation of Eq. (1.8) is quite simple: it describes an exponential approach to the equilibrium distribution $\Phi(\xi)$ with relaxation time $1/\nu$.

Another simple property of the BGK model is that it admits an easily proved H-theorem; i.e., for any distribution function f we have

$$\int (\log f) J(f)\, d\xi \le 0 \tag{1.9}$$

and the equality sign applies only if f is a Maxwellian. In fact, we have

$$\int (\log f) J(f)\, d\xi = \int \nu(\log f)(\Phi - f)\, d\xi$$

$$= \int \nu[\log(f/\Phi)](\Phi - f)\, d\xi + \int (\log \Phi) J(f)\, d\xi \tag{1.10}$$

Since $\log \Phi$ is a linear combination of the collision invariants, Eq. (1.1) shows that the last integral is zero. On the other hand,

$$(\Phi - f)\log(f/\Phi) \le 0 \tag{1.11}$$

[see Eq. (2.5) of Chapter II] and the equality sign applies only if $f = \Phi$; hence Eq. (1.9) follows.

The BGK model contains the most basic features of the Boltzmann collision integral, but presents some shortcomings. Some of them can be avoided by suitable modifications, at the expense, however, of the simplicity of the model. A first modification can be introduced in order to allow the collision frequency ν to depend on the molecular velocity instead of being locally constant; this modification is suggested by the circumstance that for rigid-sphere molecules, all the potentials with finite range, and power-law potentials with angular cutoff (except Maxwell's molecules), the collision frequency varies with the molecular velocity and this variation is expected to be important at high molecular velocities. Formally, the modification is quite simple: we have only to allow ν to depend on $\boldsymbol{\xi}$ (more precisely, on c) in Eq. (1.2), while requiring that Eq. (1.1) still holds. All the basic formal properties (including the H-theorem) are retained, but the density, velocity, and temperature which now appear in the Maxwellian Φ are not the local density, velocity, and temperature, but some fictitious local parameters, related to five moments of f weighted with $\nu(c)$; this follows from the fact that Eq. (1.1) now gives

$$\int \nu(c)\Phi\psi_\alpha \, d\boldsymbol{\xi} = \int \nu(c) f \psi_\alpha \, d\boldsymbol{\xi} \tag{1.12}$$

instead of Eq. (1.3). A consequence of this fact is that the solution of the initial value problem is no longer as simple as before.

A different kind of correction to the BGK model is obtained when we want to adjust the model to give the same Navier–Stokes equations as the full Boltzmann equation; in fact, as we shall see in the next chapter, the BGK model gives the value $\mathrm{Pr} = 1$ for the Prandtl number, a value which is not in agreement with both the true Boltzmann equation and the experimental data for a monatomic gas (which agree in giving $\mathrm{Pr} \approx \frac{2}{3}$). In order to have a correct Prandtl number, a further adjustable parameter is required beside the already available ν; accordingly, one is led (see References) to generalize the BGK model by substituting a locally anisotropic three-dimensional Gaussian in place of the local Maxwellian (which is an isotropic Gaussian):

$$\begin{aligned} \Phi &= \rho\pi^{-3/2}(\det A)^{1/2} \exp\left[-\sum_{i,j=1}^{3} \alpha_{ij}(\xi_j - v_j)(\xi_i - v_i)\right] \\ A &= \|\alpha_{ij}\| = \|(\mathrm{Pr})^{-1}(2RT)\delta_{ij} - 2(1-\mathrm{Pr})(\rho\,\mathrm{Pr})^{-1}p_{ij}\|^{-1} \end{aligned} \tag{1.13}$$

If we set the Prandtl number Pr equal to 1, we recover the BGK model.

A disadvantage of this model (to be called the ES, or ellipsoidal statistical, model) is that it has not been possible to prove or disprove that the H-theorem holds.

2. Deducing Models : The Gross and Jackson Procedure and Generalizations

One of the unsatisfactory features of the BGK model as discussed in Section 1 is that it is not derived from the Boltzmann equation by any kind of systematic procedure, but is just guessed at on the basis of some qualitative information. The same statement applies to the two variants which have been briefly discussed at the end of the preceding section.

In this section we shall investigate systematic procedures for deducing models of increasing accuracy. A satisfactory treatment is available only for the linearized operator, but something can also be said about nonlinear kinetic models.

The simplest model for the linearized operator is given by the linearized version of the BGK model, Eq. (1.4). This model has the following properties:

$$L\psi_\alpha = 0 \qquad (\alpha = 0, 1, 2, 3, 4) \tag{2.1}$$

$$(h, Lh) \leq 0 \qquad \left(= 0 \quad \text{only if} \quad h = \sum_{\alpha=0}^{4} c_\alpha \psi_\alpha\right) \tag{2.2}$$

$$(g, Lh) = (Lg, h) \tag{2.3}$$

Putting $g = \psi_\alpha$ in Eq. (2.3) and using Eq. (2.1), we also get

$$(\psi_\alpha, Lh) = 0 \qquad (\alpha = 0, 1, 2, 3, 4) \tag{2.4}$$

The three basic properties expressed by Eqs. (2.1)–(2.3) should be retained by any significant model. Equations (2.1) and (2.3) for the BGK model are almost evident, while Eq. (2.2) follows from the fact that Eq. (1.4) can also be written as follows:

$$Lh = \nu(Ph - h) = -\nu(I - P)h \tag{2.5}$$

where P is the projection operator onto the five-dimensional space $\mathscr{F}$ spanned by the ψ's, and I the identity operator (accordingly, $I - P$ is the projector onto $\mathscr{W}$, the orthogonal complement of $\mathscr{F}$); Eq. (2.5) implies

$$(h, Lh) = -\nu(h, (I - P)h) = -\nu\|(I - P)h\|^2 \leq 0$$

and equality obviously holds only if $(I - P)h = 0$.

A systematic procedure for improving the linearized BGK model and for characterizing the latter as the first step in a hierarchy of models approximating the collision operator for Maxwell molecules with arbitrary accuracy was proposed by Gross and Jackson (ref. 6). These authors started from the expansion of the collision operator for Maxwell molecules into a series of eigenfunctions, Eq. (5.26) of Chapter III, which we now rewrite as

$$Lh = \sum_{R=0}^{\infty} \lambda_R \psi_R(\psi_R, h) \tag{2.6}$$

where R is a single label which summarizes the triplet (n, l, m) in such a way that the collision invariants ψ_α correspond to $R = 0, 1, 2, 3, 4$. A systematic procedure for approximating L consists in partially destroying the fine structure in the spectrum of L by collapsing all the eigenvalues corresponding to $R > N$ into a single eigenvalue, which we shall denote by $-\nu_N$ (remember that $\lambda_R \leq 0$). This amounts to replacing L by an approximate operator L_N defined as follows:

$$L_N h = \sum_{R=0}^{N} \lambda_R \psi_R(\psi_R, h) - \nu_N \sum_{R=N+1}^{\infty} \psi_R(\psi_R, h) \tag{2.7}$$

Now, since the ψ_R constitute a complete set, we have

$$h = \sum_{R=0}^{\infty} \psi_R(\psi_R, h) \tag{2.8}$$

which is simply the Fourier expansion of h in terms of the ψ_R. Then

$$\begin{aligned} \sum_{R=N+1}^{\infty} \psi_R(\psi_R, h) &= \sum_{R=0}^{\infty} \psi_R(\psi_R, h) - \sum_{R=0}^{N} \psi_R(\psi_R, h) \\ &= h - \sum_{R=0}^{N} \psi_R(\psi_R, h) \end{aligned} \tag{2.9}$$

and substituting into Eq. (2.7),

$$L_N h = \sum_{R=0}^{N} (\nu_N + \lambda_R)\psi_R(\psi_R, h) - \nu_N h \tag{2.10}$$

In particular, if $N = 4$, $\lambda_R = 0$ for $0 \leq R \leq N$, and consequently, Eq. (2.10) reduces to Eq. (1.4) (with $\nu_4 = \nu$); by taking N larger and larger we include more and more details of the spectrum of L into the model. If we take $N = 9$ by including the five eigenfunctions corresponding to $n = 0$, $l = 2$ in Eq. (5.17) in Chapter III, we obtain the linearized version of the ES model, Eq. (1.13).

The above procedure applies only to the case of Maxwell's molecules. However, a slight generalization of the expansion (2.6) is capable of producing collision models in correspondence with any kind of linearized collision operator. In fact, nothing prevents us from expanding h in a series of the ψ_R (eigenfunctions for the Maxwell collision operator) even if we are considering a different molecular model; in this case we get

$$Lh = \sum_{Q=0}^{\infty} (\psi_Q, h)L\psi_Q \tag{2.11}$$

and, by expanding the result again in terms of the ψ_R

$$Lh = \sum_{R,Q=0}^{\infty} \beta_{QR}(\psi_Q, h)\psi_R \tag{2.12}$$

where

$$\beta_{QR} = (\psi_R, L\psi_Q) = \beta_{RQ} \tag{2.13}$$

Equation (2.12) generalizes Eq. (2.6) and reduces to the latter when $\beta_{RQ} = \lambda_R\delta_{RQ}$. If we now introduce the assumption $\beta_{QR} = -\nu_N\delta_{QR}$ for $Q, R > N$, we obtain the model

$$L_N h = \sum_{R,Q=0}^{N} (\nu_N\delta_{RQ} + \beta_{QR})\psi_R(\psi_Q, h) - \nu_N h \tag{2.14}$$

which generalizes Eq. (2.10) for linearized collision operators other than Maxwell's. Taking $N = 4$ gives again the BGK model. It is clear now how the procedure can be extended to the nonlinear collision operator: it suffices to expand f in a series of the ψ's corresponding to the local Maxwellian Φ:

$$Q(f,f) = \sum_{P,Q=0}^{\infty} (\psi_P, f)(\psi_Q, f)Q(\Phi\psi_P, \Phi\psi_Q) \tag{2.15}$$

where now the scalar products are *not* weighted with a Maxwellian, and expand the result in terms of the same ψ's:

$$Q(f,f) = \Phi \sum_{P,Q,R=0}^{\infty} \beta_{PQR}(\psi_P, f)(\psi_Q, f)\psi_R \tag{2.16}$$

where

$$\beta_{PQR} = (\psi_R, Q(\Phi\psi_P, \Phi\psi_Q)) \tag{2.17}$$

If we now assume

$$\beta_{PQR} = -\tfrac{1}{2}[\delta_{P0}\delta_{QR} + \delta_{Q0}\delta_{PR}]\nu_N(\psi_0, f)^{-1} \qquad (P, Q, R > N) \tag{2.18}$$

then $Q(f,f)$ is replaced by a nonlinear model $J_N(f)$, given by

$$J_N(f) = \Phi \sum_{P,Q,R=0}^{N} \beta_{PQR}(\psi_P, f)(\psi_Q, f)\psi_R + \nu_N\left[\sum_{Q=0}^{N} \Phi(\psi_Q, f)\psi_Q - f\right] \tag{2.19}$$

We note that, since Φ is the local Maxwellian, $(\psi_Q, f) = (\psi_Q, \Phi) = 0$ for $Q = 1, 2, 3, 4$; the latter equality follows from the orthogonality properties of the ψ's. Accordingly, if we take $N = 4$ and take into account that $\beta_{PQR} = 0$ for $R \leq 4$, we reobtain the nonlinear BGK model.

The linearized models for molecules other than Maxwell molecules and the nonlinear models discussed above were proposed by Sirovich (ref. 7).

3. Models with Velocity-Dependent Collision Frequency

A characteristic feature of the linearized collision models which have been surveyed in Section 2 is that they are based on bounded operators with a purely discrete spectrum (with an infinitely many-times degenerate eigenvalue). This follows from the fact that the above-mentioned collision operators can be written $L_N = K_N - \nu_N I$, where ν_N is a constant, I the identity operator, and K_N maps any function onto the finite-dimensional subspace spanned by the ψ_R ($R \leq N$). In particular, by taking the projections of the eigenvalue equation for L_N onto this space and its orthogonal complement, one can show that the eigenfunctions are linear combinations of the ψ_R, and hence are polynomials [in particular, the eigenfunctions are obviously the ψ_R themselves in the case of the model defined by Eq. (2.10)]. We recall now that the collision operator for hard spheres or hard potentials with angular cutoff is unbounded and displays a continuous spectrum (similar results should also hold for the case of a radial cutoff, but have never been proved rigorously); if these features of the operator have any influence on the solution of particular problems, this influence is lost when we adopt one of the models proposed in Section 3. It is therefore convenient to introduce and investigate models which retain the above-mentioned features of the linearized collision operator; this can be done in many different ways.

Conceptually, the simplest procedure is based upon exploiting the fact that we either know (rigid spheres and angular cutoff) or conjecture (radial cutoff) that the operator $K^* = \nu^{-1/2} K \nu^{-1/2}$ is self-adjoint and completely continuous in $\mathscr{H}$ (here we use terminology and results from Chapter III); accordingly, the kernel of K^* can be expanded in a series of its square summable eigenfunctions $\varphi_R \nu^{1/2}$ (such that $K\varphi_R = \lambda_R \nu \varphi_R$). In other words, we can write

$$Kh = \nu(c) \sum_{R=0}^{\infty} \lambda_R \varphi_R (\nu\varphi_R, h) \tag{3.1}$$

Truncating this series, we get the model

$$L_N h = \nu(c) \sum_{R=0}^{N} \lambda_R \varphi_R (\nu\varphi_R, h) - \nu(c) h \tag{3.2}$$

Since the first five φ_R are the collision invariants and $\lambda_R = 1$ $(0 \leq R \leq 4)$, if we take $N = 4$, we have

$$L_4 h = \nu(c) \sum_{\alpha=0}^{4} \psi_\alpha(\psi_\alpha, \nu h) - \nu(c)h \tag{3.3}$$

where the collision invariants are orthonormalized according to

$$(\psi_\alpha, \nu\psi_\beta) = \delta_{\alpha\beta} \tag{3.4}$$

Equation (3.3) is nothing other than the linearized version of the nonlinear model with velocity-dependent collision frequency which was briefly discussed in Section 1.

If we want to obtain models corresponding to $N \geq 5$ we are faced with the trouble that we don't have analytical expressions for the φ_R $(R \geq 5)$; it is true that we can compute them numerically, but this is obviously a notable complication. Besides, the procedure by which we have derived Eqs. (3.2) and (3.3) shows that $\nu(c)$ is fixed by the original molecular model; hence we do not have any parameters to be adjusted in order to reproduce the correct continuum limit (see the end of Section 1).

We can introduce an adjustable parameter by a slight modification of the above procedure; instead of simply truncating the series in Eq. (3.1) (i.e., putting $\lambda_R = 0$ for $R > N$), we can put $\lambda_R = k$ for $R > N$. The result is exactly the same model as in Eq. (3.2), except for the fact that $\nu(c)$ is replaced by $(1 - k)\nu(c)$ and λ_R by $(\lambda_R - k)/(1 - k)$; in particular, for $N = 4$ the only change is an adjustable factor in $\nu(c)$, which gives the model the same flexibility as the BGK model.

A greater flexibility can be obtained by using higher-order models; here we meet the above-mentioned trouble that we don't know the φ_R $(R \geq 5)$. This difficulty can be avoided by a procedure analogous to the non-Maxwell modeling considered in Section 2; i.e., we take a complete set of orthogonal functions and expand everything in terms of these functions, as we did in Section 2. Of course, if we want to retain the features of the models discussed in this section, we have to make a proper choice of the set; the simplest one is to take polynomials orthogonalized with respect to the weight $f_0\nu(c)$, f_0 being the basic Maxwellian. The simplest model again takes the form shown in Eq. (3.3), but now we can extend it to arbitrarily high orders without trouble (the only things to be computed numerically are coefficients, not functions); the procedure is obvious and will not be described in detail.

Another procedure for constructing models can be based upon the requirement that the model be able, in the continuum limit, to reproduce the behavior not only of some coefficients, but also of the distribution function; such models can be constructed (as will be clear from the results of

Chapter V) by insisting that the equations given by

$$Lh = \xi_i g - P(\xi_i g) \tag{3.5}$$

(where P is the projector onto the space spanned by the five collision invariants) have the same solution as the corresponding equations with L_N in place of L for a certain class of source functions g. To be precise, one starts with the five collision invariants for g and then constructs step by step the set of functions h, using the results of each step as sources for the next step and orthogonalizing the set when necessary; again the trouble is that, except for Maxwell molecules, we don't have analytical tools for solving Eq. (3.5), and, consequently, for constructing the above-mentioned models in an explicit fashion. Sometimes, however, the procedure can be worthwhile because it is capable of giving very accurate results for molecules which are not Maxwell molecules; this is particularly true when the final results can be expressed in terms of quadratures to be performed numerically.

4. Models for the Boundary Conditions

The importance of boundary conditions has been stressed repeatedly in the previous chapters. From the discussion in Section 4 of Chapter II we recall that theoretical investigations of the gas–wall interactions are exceedingly difficult because of the complexity of phenomena taking place at the wall. Even if we restrict ourselves to the linear boundary conditions expressed by Eq. (4.1) of Chapter II with a kernel independent of f, we are still faced with the problem of finding $B(\boldsymbol{\xi}' \to \boldsymbol{\xi}; \mathbf{x})$. General considerations can only succeed in putting restrictions upon this kernel, such as the detailed balancing, Eq. (4.10) of Chapter II, and Eq. (4.11) of Chapter II; otherwise we have to construct a physical model of the surface and try to evaluate the corresponding kernel $B(\boldsymbol{\xi}' \to \boldsymbol{\xi}; \mathbf{x})$. This is what Maxwell did in order to find his boundary conditions, Eq. (4.8) of Chapter II; it is to be noted, however, that he was not able to master the problem completely and had to resort to qualitative arguments and introduce a phenomenological parameter α $(0 \le \alpha \le 1)$ which is not directly related to the structure of the surface.

We note that according to Maxwell's boundary conditions the tangential momentum and the random kinetic energy of the reevaporated molecules are influenced partially by the velocity and temperature of the boundary, and partially by the momentum and random kinetic energy of the incoming stream; if $\alpha = 0$ (specular reflection), the reemitted stream does not feel the boundary (as far as the tangential momentum and kinetic energy are concerned), while if $\alpha = 1$ (completely diffuse reevaporation), the reemitted stream has completely lost its memory of the incoming stream (except for conservation of the number of molecules). For this reason the coefficient α

(originally defined as the fraction of the diffusively evaporated molecules) is sometimes called the "accommodation coefficient" because it expresses the tendency of the gas to accommodate to the state of the wall. It is to be stressed, however, that momentum and energy accommodate differently in physical interactions, momentum being lost or gained much faster than energy; this points out a basic inaccuracy of Maxwell's boundary conditions (of the same order of magnitude as the failure of the BGK model to give a correct Prandtl number). Although Maxwell's physical model of the wall suggests certain improvements, such as Eq. (4.12) of Chapter II, a different procedure will be discussed presently; this procedure is completely phenomenological in the sense that it is not based on a physical model of the wall, but only on the general properties of the wall interactions plus the introduction of numerical parameters to be determined experimentally or by means of other theoretical arguments.

Let us consider the following eigenvalue equation:

$$A\psi = \int_{\mathbf{c}'\cdot\mathbf{n}>0} K(\boldsymbol{\xi}, \boldsymbol{\xi}')\psi(\boldsymbol{\xi}')\,d\boldsymbol{\xi}' = \lambda\psi(\boldsymbol{\xi}) \qquad (\mathbf{c}\cdot\mathbf{n} > 0) \tag{4.1}$$

where, if $\boldsymbol{\xi}'_R = -\boldsymbol{\xi}' + 2\mathbf{u}_0$ (hence $\mathbf{c}'_R \cdot \mathbf{n} < 0$), we put

$$K(\boldsymbol{\xi}, \boldsymbol{\xi}') = [|\mathbf{c}'\cdot\mathbf{n}|F_0(\boldsymbol{\xi}')]^{1/2}[|\mathbf{c}\cdot\mathbf{n}|F_0(\boldsymbol{\xi})]^{-1/2}B(\boldsymbol{\xi}'_R \to \boldsymbol{\xi}; \mathbf{x}) \tag{4.2}$$

In accord with Eq. (4.10) of Chapter II, $K(\boldsymbol{\xi}', \boldsymbol{\xi})$ is symmetric; in addition, Eq. (4.11) of Chapter II shows that

$$\psi_0(\boldsymbol{\xi}) = [2\pi(RT_0)^2]^{-1/2}|\mathbf{c}\cdot\mathbf{n}|^{1/2}\exp[-(\boldsymbol{\xi} - \mathbf{u}_0)^2(4RT_0)^{-1}] \tag{4.3}$$

is an eigensolution of Eq. (4.1) corresponding to the eigenvalue $\lambda = 1$. If A has a purely discrete spectrum, we can write

$$K(\boldsymbol{\xi}, \boldsymbol{\xi}') = \sum_{n=0}^{\infty} \lambda_n\psi_n(\boldsymbol{\xi})\psi_n(\boldsymbol{\xi}') \tag{4.4}$$

where ψ_n denotes the normalized eigenfunction corresponding to the eigenvalue $\lambda = \lambda_n$. The generalization to cases when the spectrum is not purely discrete is possible, but is not interesting at present. Equation (4.4) implies a similar relation for $B(\boldsymbol{\xi}' \to \boldsymbol{\xi}; \mathbf{x})$, which can be obtained from Eq. (4.2). We also assume that all the eigenvalues λ_n satisfy

$$0 \le \lambda_n \le 1 \qquad (n = 0, 1, 2, \ldots) \tag{4.5}$$

and, in general, $\lambda_n = 1$ only for $n = 0$; the condition $\lambda_n < 1$ for $n \neq 0$ comes from the requirement that the Maxwellian corresponds to a stable equilibrium (hence to the maximum eigenvalue) and $\lambda \ge 0$ from the positive nature of the probability density $B(\boldsymbol{\xi}' \to \boldsymbol{\xi}; \mathbf{x})$.

We can now use the above relations to construct models for the boundary conditions in much the same way as we did for the collision term in the previous sections. The simplest assumption is complete degeneracy, i.e., $\lambda_n = 1$ for any n (note that this violates the general condition that $\lambda_n = 1$ can hold only for $n = 0$); then Eq. (4.4) shows that $K(\boldsymbol{\xi}', \boldsymbol{\xi}) = \delta(\boldsymbol{\xi}' - \boldsymbol{\xi})$, and we obtain the boundary condition of purely specular reflection.

The next possible assumption (which takes into account the above-mentioned restriction on the eigenvalue $\lambda = 1$) is that we have two distinct eigenvalues $\lambda_0 = 1$, $\lambda_n = 1 - \alpha$ $(0 < \alpha \leq 1; n = 1, 2, \ldots)$; then

$$\begin{aligned} K(\boldsymbol{\xi}, \boldsymbol{\xi}') &= \psi_0(\boldsymbol{\xi})\psi_0(\boldsymbol{\xi}') + (1 - \alpha) \sum_{n=1}^{\infty} \psi_n(\boldsymbol{\xi})\psi_n(\boldsymbol{\xi}') \\ &= \alpha\psi_0(\boldsymbol{\xi})\psi_0(\boldsymbol{\xi}') + (1 - \alpha) \sum_{n=0}^{\infty} \psi_n(\boldsymbol{\xi})\psi_n(\boldsymbol{\xi}') \\ &= \alpha[2\pi(RT_0)^2]^{-1}[|\mathbf{c} \cdot \mathbf{n}|\,|\mathbf{c}' \cdot \mathbf{n}|]^{1/2} \exp[-(\boldsymbol{\xi} - \mathbf{u}_0)^2(4RT_0)^{-1} \\ &\quad - (\boldsymbol{\xi}' - \mathbf{u}_0)^2(4RT_0)^{-1}] + (1 - \alpha)\delta(\boldsymbol{\xi} - \boldsymbol{\xi}') \end{aligned} \tag{4.6}$$

Use of Eqs. (4.2) and (4.3) shows that this assumption gives Maxwell's boundary conditions (Eq. 4.8) of Chapter II; a particularly simple choice is $\lambda_0 = 1$, $\lambda_n = 0$ $(n = 1, 2, \ldots)$, which corresponds to $\alpha = 1$ and gives the boundary conditions of pure diffusion according to the wall Maxwellian.

It is clear that in order to introduce more sophisticated boundary conditions, we have either to make some specializing assumptions on the eigenfunctions ψ_n or use a more general relation than Eq. (4.4). If we follow the first procedure, we can assume the ψ_n to have the form

$$\psi_n(\boldsymbol{\xi}) = \psi_0(\boldsymbol{\xi})P_n(\boldsymbol{\xi}) \tag{4.7}$$

where the P_n are half-range polynomials, chosen in such a way that the ψ_n are orthonormal according to the scalar product in the Hilbert space of square summable functions of $\boldsymbol{\xi}$ (for $\mathbf{c} \cdot \mathbf{n} > 0$). We can choose the P_n in two ways, by assuming that they are polynomials in either the three-components of $\boldsymbol{\xi} - \mathbf{u}_0$ or the two tangential components of $(\boldsymbol{\xi} - \mathbf{u}_0)$ and $|\mathbf{c} \cdot \mathbf{n}|^2$. The first procedure yields the half-range polynomials introduced by Gross *et al.* (ref. 14) for different purposes, while the second one has the advantage of yielding classical polynomials (Hermite polynomials in the tangential components, Laguerre polynomials in $|\mathbf{c} \cdot \mathbf{n}|^2$).

The eigenvalues λ_n can also receive a sort of interpretation in terms of accommodation coefficients, in the sense that $1 - \lambda_n$ can be regarded as a generalized accommodation coefficient of a suitable moment of the distribution function. For specular reflection $\lambda_n = 1$ for any n; therefore there is no accommodation. For Maxwell's boundary conditions, all the moments (except density, which never accommodates) have their accommodation coefficient equal to $1 - (1 - \alpha) = \alpha$.

The above procedure is analogous to the models for Maxwell molecules discussed in Section 2 (the main difference is that here we don't know if a dynamical model does exist such that the corresponding ψ_n are exactly given by Eq. (4.7); a more general and justified procedure is analogous to the non-Maxwell modeling for the linearized collision operator (also discussed in Section 2). According to this procedure we take the functions ψ_n given by Eq. (4.7) as a vector base in Hilbert space, even if we do not regard them as eigenfunctions of A; in this case Eq. (4.4) must be written as follows:

$$K(\xi, \xi') = \sum_{m,n=0}^{\infty} \lambda_{nm}\psi_n(\xi)\psi_m(\xi') \tag{4.8}$$

where $\|\lambda_{nm}\|$ is a symmetric matrix which expresses the nonaccommodation properties of the wall [λ_{00} is always equal to 1 because of Eq. (4.11) of Chapter II]. The matrix $\|\delta_{mn} - \lambda_{mn}\|$ can be regarded as an "accommodation matrix;" the diagonal terms can be taken to have the same meaning as the above λ_n, while nondiagonal terms measure how much memory of the moments of the arriving flow is retained by the moments of the emerging flow. It is also clear that the present procedure reduces to the previous one when nondiagonal terms are zero.

As an example of the procedures which have just been described, one can take ψ_1, ψ_2, and ψ_3 to be the tangential components of $\xi - \mathbf{u}_0$, and

$$\begin{aligned} &\lambda_{00} = 1; \quad \lambda_{11} = \lambda_{22} = \lambda_{33} = 1 - \beta; \quad \lambda_{44} = \ldots = \lambda_{nn} = 1 - \alpha; \\ &\lambda_{mn} - 0 \qquad (m \neq n) \end{aligned} \tag{4.9}$$

where $0 < \alpha < 1, 0 < \beta < 1$; then

$$\begin{aligned} K(\xi, \xi') = {} & (1 - \alpha)\delta(\xi - \xi') + [2\pi(RT_0)^2]^{-1} \\ & \times \exp\{ - [(\xi - \mathbf{u}_0)^2 + (\xi' - \mathbf{u}_0)^2](4RT_0)^{-1}\} \\ & \times \left\{\alpha + \frac{\alpha - \beta}{RT_0}[\xi - \mathbf{u}_0]\,[\xi' - \mathbf{u}_0]\right\} [|\mathbf{c} \cdot \mathbf{n}|\, |\mathbf{c}' \cdot \mathbf{n}|]^{1/2} \end{aligned} \tag{4.10}$$

The kernels have to be submitted to the condition of being positive, as first pointed out in an explicit way by Shen; this condition rules out, e.g., the kernel appearing in Eq. (4.10). Kernels having this property (other than Maxwell's) were proposed in the early 1970s. An interesting example, which was pointed out by Cercignani and Lampis and, independently, by Kuščer *et al.*, is the following (where, for simplicity $\mathbf{u}_0$ has been taken to be zero):

$$K(\boldsymbol{\xi}, \boldsymbol{\xi}') = \frac{[\alpha_n \alpha_t(2-\alpha_t)]}{2\pi(RT_0)^2}\,\xi_n \exp\left[-\frac{(2-\alpha_n)(\xi_n^2+\xi_n'^2)}{4RT_0\alpha_n} - \frac{1}{\alpha_t(2-\alpha_t)}\right.$$
$$\left.\times\,\frac{(2-\alpha_t)(\xi_t^2+\xi_t'^2) - 4(1-\alpha_t)\boldsymbol{\xi}_t\cdot\boldsymbol{\xi}_t'}{4RT_0}\right] I_0\left(\frac{\sqrt{1-\alpha_n}\,\xi_n\xi_n'}{\alpha_n RT_0}\right) \tag{4.11}$$

References

Section 1——The BGK model was proposed independently by

1. P. L. Bhatnagar, E. P. Gross, and M. Krook, *Phys. Rev.* **94**, 511 (1954);
2. P. Welander, *Arkiv. Fysik.* **7**, 507 (1954);

The possibility of using a velocity-dependent collision frequency was pointed out by

3. M. Krook, *J. Fluid Mech.* **6**, 523 (1959);

The ES model was proposed by

4. L. H. Holway, Jr., "Approximation Procedures for Kinetic Theory," Ph.D. Thesis, Harvard University (1963)

and rediscovered by

5. C. Cercignani and G. Tironi, in: *Rarefied Gas Dynamics* (C. L. Brundin, ed.), Vol. I, p. 441, Academic Press, New York (1967).

Section 2——The linearized models discussed in this section were proposed by

6. E. P. Gross and E. A. Jackson, *Phys. Fluids* **2**, 432 (1959);
7. L. Sirovich, *Phys. Fluids* **5**, 908 (1962).

The simplicity and usefulness of the linearized ES model was ignored by all the investigators except

8. L. H. Holway, Jr., "Sound Propagation in a Rarefied Gas Calculated from Kinetic Theory by Means of a Statistical Model," Raytheon Co. Report T-578 (1964);
9. C. Cercignani and G. Tironi, *Nuovo Cimento* **43**, 64 (1966).

Section 3——Linearized models with velocity-dependent collision frequency were proposed by

10. C. Cercignani, *Ann. Phys.* (*N.Y.*) **40**, 469 (1966);
11. S. K. Loyalka and J. H. Ferziger, *Phys. Fluids* **10**, 1833 (1967).

Section 4——The theory expounded in this section is taken from

12. C. Cercignani, "Boundary Value Problems in Linearized Kinetic Theory," in: *Transport Theory* (R. Bellman, G. Birkhoff, and I. Abu-Shumays, eds.), SIAM-AMS Proceedings, Vol. I, p. 249, American Mathematical Society, Providence (1968);

but a similar approach was independently considered by

13. S. F. Shen, *Entropie* **18**, 138 (1967).

The half-range Hermite polynomials first appeared in

14. E. P. Gross, E. A. Jackson, and S. Ziering, *Ann. Phys.* (*N.Y.*) **1**, 141 (1957).

The kernel in Eq. (4.11) is known in the literature as the Cercignani–Lampis model. It was proposed independently in ref. 4 of Chapter II and in

15. I. Kuščer, J. Mozina, and F. Krizanic, in: *Rarefied Gas Dynamics* (D. Dini, S. Nocilla, and C. Cercignani, eds.), Vol. I, p. 97, Editrice Tecnico-Scientifica, Pisa (1971).

Other relevant references on this topic are ref. 15 of Chapter I and ref. 5 of Chapter II.

Chapter V

THE HILBERT AND CHAPMAN–ENSKOG THEORIES

1. A Bridge between the Microscopic and Macroscopic Descriptions

It was pointed out in Chapter III, Section 1 that if we want to solve the Boltzmann equation for realistic nonequilibrium situations, we must rely upon approximation methods, in particular, perturbation procedures.

In order to do this, we have to look for a parameter ϵ which can be considered to be small in some situations; accordingly, a first step consists in investigating the order of magnitude of the various terms appearing in the Boltzmann equation. If we denote by T a typical time, by d a typical length, by w a typical molecular velocity, then

$$\partial f/\partial t = O(T^{-1}f); \qquad \boldsymbol{\xi}\cdot\partial f/\partial \mathbf{x} = O(wd^{-1}f); \qquad Q(f,f) = O(nw\sigma^2 f) \tag{1.1}$$

where $n = \rho/m$ is the number density of molecules and σ the molecular diameter.

A well-known elementary argument relates the product $n\sigma^2$ to the mean free path l, i.e., the average length of the free flight of a molecule between two successive collisions, as follows:

$$l \approx (n\sigma^2)^{-1} \tag{1.2}$$

It is to be noted, however, that the mean-free-path concept can be rigorously defined only for rigid-sphere molecules or cutoff potentials (with a cutoff length of the order of the distance at which attraction changes into repulsion); for long-range potentials the molecules are always interacting ($\sigma = \infty$) and, accordingly, the mean free path is, strictly speaking, zero. In spite of this, the mean-free-path concept can be retained as a tool for expressing the order of magnitude of the right-hand side of the Boltzmann equation, which turns out to be $O(wl^{-1}f)$. The combination $(wl^{-1})^{-1}$ can be considered as defining naturally a mean free time θ, and its inverse, wl^{-1}, as a measure of the collision frequency.

This discussion shows the existence of two basic nondimensional numbers in the Boltzmann equation, θ/T and l/d. In a first approach the length and time scales can be taken to be comparable, and $l/d \approx \theta/T$. Accordingly, in a first approach we shall consider the left-hand side of the Boltzmann equation as being a single term of order of magnitude wd^{-1}, in such a way that the simple nondimensional number

$$\mathrm{Kn} = l/d \tag{1.3}$$

expresses the relative magnitudes of left- and right-hand sides of the Boltzmann equation. We shall later find it necessary to consider phenomena for which $d^2\theta/Tl^2 \approx 1$, with T and d having different order of magnitudes.

It is clear that the Knudsen number Kn, defined by Eq. (1.3), ranges from 0 to ∞, Kn $\to 0$ corresponding to a fairly dense gas and Kn $\to \infty$ to a free-molecular flow (i.e., a flow where molecules have negligible interactions with each other).

Two kinds of perturbation methods suggest themselves, one for Kn $\to 0$, the other for Kn $\to \infty$. The latter will be briefly described much later (Chapter VIII, Section 3), but the former will be our first concern; in fact, it is likely that a perturbation expansion for Kn $\to 0$ will help us to complete the task begun in Chapter II, Section 3, i.e., to deduce that a macroscopic description applies in the case of a fairly dense gas and determine its limits of validity. It is clear that such a transition from the microscopic to the macroscopic description must be very singular, since it is based on the transition from an integrodifferential equation in one unknown depending upon seven variables to a set of differential equations in five unknowns depending upon four variables.

2. The Hilbert Expansion

Since we want an expansion valid in the limit Kn $\to 0$ and Kn is the order of magnitude of the ratio of a typical term of the left-hand side to a typical term of the right-hand side, we can formalize our procedure by putting an artificial parameter ϵ (to be treated as small) in front of the left-hand side (or, equivalently, by introducing nondimensional quantities and denoting the Knudsen number by ϵ):

$$\epsilon\left(\frac{\partial f}{\partial t} + \boldsymbol{\xi}\cdot\frac{\partial f}{\partial \mathbf{x}}\right) = Q(f,f) \tag{2.1}$$

The singular nature of a perturbation procedure in the limit $\epsilon \to 0$ is emphasized by the fact that ϵ multiplies all the derivatives which appear in the Boltzmann equation. In spite of this, we try a series expansion in powers of

ϵ (Hilbert's expansion):

$$f = \sum_{n=0}^{\infty} \epsilon^n f_n \tag{2.2}$$

Substituting into Eq. (2.1) gives

$$\sum_{n=1}^{\infty} \epsilon^n \left(\frac{\partial f_{n-1}}{\partial t} + \boldsymbol{\xi} \cdot \frac{\partial f_{n-1}}{\partial \mathbf{x}} \right) = \sum_{n=0}^{\infty} \epsilon^n Q_n \tag{2.3}$$

where Q_n is given by Eq. (1.3) of Chapter III. Accordingly,

$$Q_0 = 0 \tag{2.4}$$

$$\frac{\partial f_{n-1}}{\partial t} + \boldsymbol{\xi} \cdot \frac{\partial f_{n-1}}{\partial \mathbf{x}} = Q_n \qquad (n \geq 1) \tag{2.5}$$

Equation (2.4) ensures that f_0 is a Maxwellian; as a consequence, Eq. (1.6) of Chapter III gives

$$\left(\frac{\partial}{\partial t} + \boldsymbol{\xi} \cdot \frac{\partial}{\partial \mathbf{x}} \right) (f_0 h_{n-1}) = f_0 L h_n + S_n \qquad (n = 1, 2, \ldots) \tag{2.6}$$

$$S_1 = 0; \qquad S_n = \sum_{k=1}^{n-1} Q(f_0 h_k, f_0 h_{n-k}) \qquad (n \geq 2) \tag{2.7}$$

$$f_n = f_0 h_n; \qquad h_0 = 1 \tag{2.8}$$

Accordingly, we have a sequence of equations for the unknowns h_n; we can solve these equations step by step by noting that they have the form

$$Lh = g \tag{2.9}$$

where g is a given source term. Solving this equation amounts to inverting the operator L; this cannot be done in general because the origin belongs to the spectrum of L [$\lambda = 0$ is a fivefold degenerate eigenvalue, and the collision invariants ψ_α ($\alpha = 0, 1, 2, 3, 4$) are the corresponding eigenfunctions]. However, if g belongs to the Hilbert space $\mathscr{H}$ and is orthogonal to the ψ_α,

$$(\psi_\alpha, g) = 0 \tag{2.10}$$

then a solution h of Eq. (2.9) does exist and belongs to $\mathscr{H}$, provided the potential is hard (no cutoff is required). To prove this, let us consider a function g satisfying Eq. (2.10) and solve Eq. (2.9) in the subspace $\mathscr{W} \subset \mathscr{H}$ orthogonal to the subspace spanned by the five collision invariants (note that $\mathscr{W}$ is an invariant subspace for L). In $\mathscr{W}$, L is a self-adjoint operator, and the origin is not in its spectrum, thanks to Eq. (3.13) of Chapter III (which was shown to hold in general for hard potentials in Chapter III,

Section 4); because of the definition of spectrum, L^{-1} exists in $\mathscr{W}$, and a solution $h^{(0)} \in \mathscr{W} \subset \mathscr{H}$ is found. While $h^{(0)}$ is the only solution in $\mathscr{W}$, if we return to $\mathscr{H}$, we can add to $h^{(0)}$ a linear combination of the five collision invariants with arbitrary coefficients and still satisfy Eq. (2.9).

We therefore have this situation: at each step we can find h_n provided the five conditions expressed by Eq. (2.10) are satisfied by the source term, but the solution h_n is determined up to five parameters c_n^α (which can depend upon the time and space variables). Since the source term is constructed by means of the previous approximations, the two circumstances mentioned above combine cyclically in such a way that the five orthogonality conditions on the nth source term "determine" the five parameters left unspecified by the $(n-1)$th step; the start of the cycle is made possible by the fact that the zeroth-order approximation already contains five disposable parameters (the density, the mass velocity components, and the temperature of the Maxwellian f_0).

Let us now see in what sense the orthogonality conditions determine the five disposable parameters. We note that the source term of the nth step is constituted by the left-hand side of Eq. (2.6) minus S_n. Because of Eq. (2.7) and the properties of $Q(f, g)$ (Chapter II, Section 1), S_n automatically satisfies the orthogonality conditions, and we are left with

$$\left(f_0^{-1}\psi_\alpha, \left(\frac{\partial}{\partial t} + \boldsymbol{\xi}\cdot\frac{\partial}{\partial \mathbf{x}}\right) f_0 h_{n-1}\right) = 0 \qquad (n \geq 1) \tag{2.11}$$

or

$$\int \psi_\alpha \left(\frac{\partial}{\partial t} + \boldsymbol{\xi}\cdot\frac{\partial}{\partial \mathbf{x}}\right) f_n \, d\boldsymbol{\xi} = 0 \qquad (n \geq 0) \tag{2.12}$$

Since the collision invariants depend only upon $\boldsymbol{\xi}$, we can exchange the order of the differentiations and integrations under suitable assumptions and write

$$\frac{\partial \rho_n^\alpha}{\partial t} + \operatorname{div} \mathbf{j}_n^\alpha = 0 \qquad (n \geq 0; \alpha = 0, 1, 2, 3, 4) \tag{2.13}$$

where

$$\rho_n^\alpha = \int \psi_\alpha f_n \, d\boldsymbol{\xi}; \qquad \mathbf{j}_n^\alpha = \int \boldsymbol{\xi} \psi_\alpha f_n \, d\boldsymbol{\xi} \tag{2.14}$$

constitute the nth-order terms of the expansion of

$$\rho^\alpha = \int \psi_\alpha f \, d\boldsymbol{\xi}; \qquad \mathbf{j}^\alpha = \int \boldsymbol{\xi} \psi_\alpha f \, d\boldsymbol{\xi} \tag{2.15}$$

into a series of powers of ϵ (any expansion of f into a series implies a related expansion of its moments).

The ρ^α can be visualized as a five-dimensional vector and $\mathbf{j}^\alpha$ as a 5×3 matrix; their components can be expressed in terms of the physical quantities defined in Chapter II, Section 3:

$$\|\rho^\alpha\| = \left\| \begin{matrix} \rho \\ \rho v_j \\ \rho(\frac{1}{2}v^2 + e) \end{matrix} \right\|; \qquad \|\mathbf{j}^\alpha\| = \left\| \begin{matrix} \rho v_i \\ \rho v_i v_j + p_{ij} \\ \rho v_i(\frac{1}{2}v^2 + e) + p_{ij}v_j + q_i \end{matrix} \right\| \tag{2.16}$$

$$(i, j = 1, 2, 3; \qquad \alpha = 0, 1, 2, 3, 4)$$

In terms of ρ^α and $\mathbf{j}^\alpha$ the conservation equations [Eq. (3.20) of Chapter III with $X_i = 0$] can be written formally as follows:

$$\frac{\partial \rho^\alpha}{\partial t} + \operatorname{div} \mathbf{j}^\alpha = 0 \qquad (\alpha = 0, 1, 2, 3, 4) \tag{2.17}$$

Accordingly, the compatibility equations, Eq. (2.13), are nothing other than the nth terms of the expansion of Eqs. (2.17). The expanded equations, however, contain more than Eqs. (2.17). In order to show this, let us observe that the only unknown quantities in f_n are the five coefficients c_n^α of the linear combination of ψ_α to be added to $f_0 h_n^{(0)}$; hence

$$\rho_n^\alpha = \int \psi_\alpha f_0 h_n^{(0)}\, d\boldsymbol{\xi} + \sum_{\beta=0}^{4} c_n^\beta \int \psi_\alpha f_0 \psi_\beta\, d\boldsymbol{\xi} \tag{2.18}$$

This equation can be used to express the c_n^β in terms of the ρ_n^α (the determinant of $\| \int \psi_\alpha f_0 \psi_\beta\, d\boldsymbol{\xi} \|$ is easily shown to be different from zero); since this is true for each n, the coefficients c_n^α are expressed in terms of the ρ_n^α, i.e., the only unknown quantities in f_n can be considered to be the five ρ_n^α. But this implies that the $\mathbf{j}_n^\alpha$ are also expressed in terms of the ρ_n^α, i.e., we have found (under the form of a series expansion) the relations between the stress tensor and the heat-flow vector on the one hand, and the basic macroscopic unknowns (density, velocity, temperature, or internal energy) on the other hand. This means that the present procedure accomplishes (at least formally) the closure of the system of conservation equations, and extracts a macroscopic model based upon the concepts of density, velocity, and temperature from the microscopic description based upon a distribution function!

In order to obtain a better appreciation of the situation, let us consider the conservation equations written as follows:

$$E^\alpha(\rho^\beta) = S^\alpha \tag{2.19}$$

where E^α is the (nonlinear) Euler operator (such that $E^\alpha(\rho^\beta) = 0$ gives the inviscid fluid equations) and

$$S^\alpha = \left\| \begin{array}{l} 0 \\ -\dfrac{\partial}{\partial x_j}(p_{ij} - p\delta_{ij}) \\ -\dfrac{\partial}{\partial x_i}(p_{ij}v_j - pv_i + q_i) \end{array} \right\| \tag{2.20}$$

When ρ^β is expanded into a series of powers of ϵ

$$E^\alpha(\rho^\beta) = \sum_{n=0}^{\infty} \epsilon^n E_0^{\alpha\beta}\rho_n^\beta + \sum_{n=1}^{\infty} \epsilon^n E_n^\alpha(\rho_k^\beta) \tag{2.21}$$

where a sum from 0 to 4 over repeated indices is understood. Here $E_0^{\alpha\beta}$ is an operator which contains the ρ_0^β and is linear if the latter are regarded as known; when applied to the ρ_0^α, however, $E_0^{\alpha\beta}\rho_0^\beta = E^\alpha(\rho_0^\beta)$. The E_n^α contain only the ρ_k^α for $k \leq n - 1$. When we expand the ρ^β Eq. (2.19) gives

$$E_0^{\alpha\beta}\rho_0^\beta = 0 \qquad [E^\alpha(\rho_0^\beta) = 0] \tag{2.22}$$

$$E_0^{\alpha\beta}\rho_n^\beta = S_n^\alpha - E_n^\alpha \tag{2.23}$$

where

$$S_n^\alpha = \left\| \begin{array}{l} 0 \\ -\dfrac{\partial}{\partial x_j}(p_{ij}^{(n)} - p^{(n)}\delta_{ij}) \\ -\dfrac{\partial}{\partial x_j}\left[\displaystyle\sum_{k=1}^{n} (p_{ij}^{(k)} - p^{(k)}\delta_{ij})v_i^{(n-k)} + q_i^{(n)}\right] \end{array} \right\| \tag{2.24}$$

Here $v_i^{(n)}$, $p_{ij}^{(n)}$, $p^{(n)}$, and $q_i^{(n)}$ denote, respectively, the nth terms of the power expansions of v_i, p_{ij}, p, and q_i. The fact that the zeroth-order equation, Eq. (2.22), has zero source term comes from the fact that f_0 is a Maxwellian and hence has an isotropic stress tensor and a zero heat flux vector.

We must now explain why we have treated the terms in S_n^α as source terms. We observe that

$$\mathbf{j}_n^\alpha = \int \psi_\alpha f_n \boldsymbol{\xi}\, d\boldsymbol{\xi} = \int \psi_\alpha f_0 h_n^{(0)} \boldsymbol{\xi}\, d\boldsymbol{\xi} + \sum_{\beta=0}^{4} c_n^\beta \int \boldsymbol{\xi}\psi_\alpha f_0 \psi_\beta\, d\boldsymbol{\xi} \tag{2.25}$$

It can be easily seen [by explicitly performing the integrals not involving $h_n^{(0)}$ in Eq. (2.18) and (2.25) and substituting from Eq. (2.18) into Eq. (2.25)] that the right-hand side of Eq. (2.25) can be written as $E_0^{\alpha\beta}\rho_n^\beta$ plus something depending only on $h_n^{(0)}$ and hence only on the ρ_k^α $(k \leq n - 1)$; accordingly, the nth-order contributions to $p_{ij} - p\delta_{ij}$ and q_i (and, as a consequence, to S^α) are expressed in terms of the $\rho_k^\alpha(k \leq n - 1)$ and can be regarded as known at the nth step.

In conclusion, the situation is the following: we have to solve the ordinary inviscid fluid equations at the zeroth level of approximation, and inhomogeneous linearized inviscid equations at the next steps.

3. The Chapman–Enskog Expansion

The above treatment shows that if we grant the possibility of expanding the distribution function into a power series in the Knudsen number, then we can extract a macroscopic description of the gas in terms of density, mass velocity, and temperature. This description is essentially in terms of the inviscid fluid equations, but contains corrections which can be computed by solving linearized equations. This is a fine result; in fact, it allows us to express f in terms of its five moments (which are also the five basic macroscopic quantities corresponding to f) and hence shows that in some sense a macroscopic description follows. It must be noted, however, that we have proceeded formally, and an obvious question is: when are we allowed to make the Hilbert expansion? We shall try to answer this question later (Section 4). For the moment let us observe that we have some preliminary warnings from our knowledge of macroscopic theories; to be precise, we know that the inviscid fluid equations are unrealistic and incapable of dealing with certain situations; what is worse, we know that regular perturbation methods are not capable of correcting the unsatisfactory features of an inviscid fluid description at subsequent steps. In other words, we know that no matter how small the parameter describing the deviation from the inviscid fluid equations (the viscosity and heat conduction coefficients in continuum theory, the mean free path here), there are situations where the solution cannot be expanded in powers of this parameter; in other words, the dependence of f on ϵ is, in general, not analytic, a circumstance which is emphasized by the fact that ϵ multiplies the derivatives in Eq. (2.1) (compare with nonanalytic character of the solutions of $\epsilon\,\partial f/\partial t + f = 0$ at $\epsilon = 0$).

A possible way of overcoming the difficulties just discussed is offered by the method of inner–outer expansions, well-known from the continuum theory; it will be considered later (Section 5). In this section, instead, we shall consider early and widely known attempts due to Chapman (ref. 3) and Enskog (ref. 4); this method is based on a plausible argument which runs as follows: The solutions of both the continuum conservation equations and the Boltzmann equation are, in general, nonanalytic in a certain parameter ϵ, and, accordingly, series expansions in powers of ϵ fail to give uniformly valid solutions for specific initial and boundary value problems; some of the troublés can be avoided, however, if we do not expand the solutions, and expand the equations instead. To understand this statement, we note that if we multiply Eqs. (2.23) by ϵ^n and sum from 0 to ∞, we have

$$E^{\alpha}(\rho^{\beta}) = \sum_{n=1}^{\infty} \epsilon^n S_n^{\alpha} = S^{\alpha}(\rho^{\beta}) \tag{3.1}$$

where $S^{\alpha}(\rho^{\beta})$ is the same source term as in Eq. (2.19), but is now expressed as a nonlinear operator acting on the ρ^{β}, while E^{α} is the nonlinear Euler operator [Eq. (2.21) has been used]. The operator S^{α} goes to zero with ϵ, and at present is known only when the ρ^{β} are expanded into a power series in ϵ (this is the basic result of the Hilbert expansion). We can, however, assume that S^{α} exists (at least in an asymptotic sense for $\epsilon \to 0$) even when the ρ^{β} are not expanded. In fact, although the Hilbert expansion does not give uniformly valid solutions, it does give valid solutions, if we restrict ourselves to small regions far away from certain singular surfaces (see Section 4); hence the existence of S^{α} almost everywhere follows (at least in an asymptotic sense when $\epsilon \to 0$).

We can think of the S_n^{α} as containing only space derivatives and not time derivatives, because the latter can always be eliminated from Eq. (2.6) by observing that $f_{n-1} = f_0 h_{n-1}$ varies with time and space variables through its dependence upon the ρ_k^{α} $(k \leq n - 1)$, whose time derivative is known in terms of space derivatives according to Eq. (2.13). As a consequence, the operator S^{α} appearing in Eq. (3.1) can be thought of as acting only upon the space dependence of the ρ^{β}.

The idea behind the Chapman–Enskog method is to expand S^{α} while leaving the ρ^{β} unexpanded; this assumes that although the dependence of the ρ^{β} upon ϵ is nonanalytic in many cases, the operator S^{α} is analytic in ϵ (or, at least, possesses an asymptotic expansion in powers of ϵ). This assumption is far from being contradictory, since, e.g., the Navier–Stokes equations depend analytically (actually linearly) upon the viscosity and heat conduction, but have, in general, solutions which cannot be expanded in a series of these parameters. In order to formalize this notion into an algorithm, we observe that Eq. (3.1) can be written as follows:

$$\partial \rho^{\beta}/\partial t = D_{\beta}(\rho^{\alpha}) \tag{3.1'}$$

where D_{β} is a nonlinear operator acting upon the space dependence of the ρ^{α} and is obtained by subtracting the space derivatives in E^{α} from $S^{\alpha}(\rho^{\beta})$. Expanding S^{α} clearly means expanding D_{β} as follows:

$$D_{\beta} = \sum_{n=0}^{\infty} \epsilon^n D_{\beta}^{(n)} \tag{3.2}$$

where the $D_\beta^{(n)}$ are nonlinear operators to be found by the present method; in particular, $D_\beta^{(0)}$ will consist of the space derivatives contained in $E^{\alpha\beta}$ (with a sign change). The expansion of D_β means that we regard the time evolution of ρ^β as influenced by processes of different orders of magnitude in ϵ.

Since our present expansion essentially amounts to a reordering of the Hilbert expansion according to a new criterion, we must also maintain a basic result of the latter expansion, which is also a prerequisite for extracting a self-contained macroscopic theory from the Boltzmann equation, i.e., that the distribution function depends upon the space and time variables only through a functional dependence on the ρ^α. In other words, we have

$$\frac{\partial f}{\partial t} = \sum_{k=0}^{\infty} \frac{\partial f}{\partial(\nabla^k \rho^\alpha)} \frac{\partial(\nabla^k \rho^\alpha)}{\partial t} \tag{3.3}$$

where ∇^k formally denotes the nth-order space derivatives. Since Eq. (3.1′) gives

$$\frac{\partial}{\partial t}\nabla^k \rho^\beta = \nabla^k D_\beta(\rho^\alpha) \tag{3.4}$$

the expansion of D_β, Eq. (3.2), implies an expansion of the time evolution of f as well. We can write this formally as follows:

$$\partial f/\partial t = \sum_{n=0}^{\infty} \epsilon^n \, \partial^{(n)} f/\partial t \tag{3.5}$$

where $\partial^{(n)}/\partial t$ simply denotes the contribution to $\partial f/\partial t$ coming from $D_\beta^{(n)}$ through Eqs. (3.2)–(3.4). The expressions of the operators $\partial^{(n)}/\partial t$ (or, equivalently, $D_\beta^{(n)}$) and the dependence of f upon the ρ^α are the unknowns of the Chapman–Enskog method.

As noted above, the ρ^α are left unexpanded; we must, however, expand f if we want to avoid trivial results. To use these two requirements consistently, we simply have to write

$$f = \sum_{n=0}^{\infty} \epsilon^n f_n \tag{3.6}$$

and

$$\int \psi_\alpha f_n \, d\xi = 0 \qquad (n \geq 1) \tag{3.7}$$

This implies that f is not expanded into a series of powers of ϵ; in fact, the coefficients f_n depend on ϵ in a complex way. However, the dependence of

f_n upon ϵ is only through the ρ^α; this defines the algorithm uniquely. If we now substitute both Eq. (3.5) and Eq. (3.6) into Eq. (2.1), we find

$$Q_0 = 0 \tag{3.8}$$

$$\sum_{k=0}^{n-1} \frac{\partial^{(k)}(f_0 h_{n-k-1})}{\partial t} + \boldsymbol{\xi} \cdot \frac{\partial(f_0 h_{n-1})}{\partial \mathbf{x}} = f_0 L h_n + S_n \qquad (n \geq 1) \tag{3.9}$$

where the notation is the same as in Section 2.

Equation (3.8) shows that f_0 is again a Maxwellian. There is, however, a basic difference from the Hilbert method: the fluid variables (density, velocity, and temperature) which appear in the Maxwellian are now exact (unexpanded), while only the zeroth-order approximation to the fluid variables appeared in the zeroth-order distribution function of the Hilbert method.

Equation (3.9) is again of the general form, Eq. (2.9), discussed in Section 2. The orthogonality conditions, Eq. (2.10), now take the following form:

$$\sum_{k=0}^{n} \frac{\partial^{(k)}}{\partial t} \int \psi_\alpha f_{n-k}\, d\boldsymbol{\xi} + \operatorname{div}\left[\int \boldsymbol{\xi} \psi_\alpha f_n . d\boldsymbol{\xi} \right] = 0 \tag{3.10}$$

Now, however, there are no disposable parameters, since Eq. (3.7) implies that $h_n \in \mathscr{W}$ $(n \geq 1)$, i.e., h_n $(n \geq 1)$ is uniquely determined as a function of $\boldsymbol{\xi}$ and a functional of the ρ^α. Equation (3.7) also implies that all the integrals appearing in the sum in Eq. (3.10) except one are zero, and the nonzero one is simply equal to ρ^α; accordingly,

$$\partial^{(n)} \rho^\alpha / \partial t + \operatorname{div}\left[\int \boldsymbol{\xi} \psi_\alpha f_n\, d\boldsymbol{\xi} \right] = 0 \tag{3.11}$$

Satisfying these compatibility conditions means solving our problem, i.e., finding $\partial^{(n)}\rho^\alpha/\partial t$ (or D_α^n); as a matter of fact, Eq. (3.11) gives D_α^n explicitly in terms of space derivatives of the ρ^α once one has solved the nth equation of the hierarchy (3.9).

The net result of the Chapman–Enskog expansion is that one can write the Nth approximation to the macroscopic equations in the following way:

$$\partial \rho^\alpha / \partial t + \operatorname{div}\left[\int \boldsymbol{\xi} \psi_\alpha \sum_{n=0}^{N} \epsilon^n f_n\, d\boldsymbol{\xi} \right] = 0 \tag{3.12}$$

where the f_n are functionals of the ρ^α and functions of $\boldsymbol{\xi}$, explicitly known from the step-by-step solution. The main difference from the Hilbert procedure, which was based on a nonlinear zeroth-order equation plus linear equations to evaluate higher-order corrections, is that now we choose a fixed N and evaluate the solution at that level; if we want a higher-order solution, we have to solve a completely new, more complicated set of nonlinear equations.

At the zeroth level we have the inviscid fluid equations again, because f_0 is still a Maxwellian. In order to see what happens at the next level, we write Eq. (3.9) for $n = 1$:

$$\frac{1}{\rho}\left(\frac{\partial^{(0)}}{\partial t} + \xi_i \frac{\partial}{\partial x_i}\right)\rho + \left(\frac{c^2}{2RT} - \frac{3}{2}\right)\frac{1}{T}\left(\frac{\partial^{(0)}}{\partial t} + \xi_i \frac{\partial}{\partial x_i}\right)T + \frac{1}{RT}c_k\left(\frac{\partial^{(0)}}{\partial t} + \xi_i \frac{\partial}{\partial x_i}\right)v_k = Lh_1 \tag{3.13}$$

where the explicit form of f_0 has been used and $\mathbf{c} = \boldsymbol{\xi} - \mathbf{v}$, as usual. Equation (3.11) now gives $(n = 0)$

$$\begin{aligned} \frac{\partial^{(0)}\rho}{\partial t} &= -\frac{\partial}{\partial x_i}(\rho v_i) \\ \frac{\partial^{(0)}v_k}{\partial t} &= -v_i\frac{\partial v_k}{\partial x_i} - \frac{1}{\rho}\frac{\partial p}{\partial x_k} \qquad (p = \rho RT) \\ \frac{\partial^{(0)}T}{\partial t} &= -v_i\frac{\partial T}{\partial x_i} - \frac{2}{3}T\frac{\partial v_j}{\partial x_j} \end{aligned} \tag{3.14}$$

These relations, substituted back into Eq. (3.13), give

$$\left(\frac{c^2}{2RT} - \frac{5}{2}\right)c_i\frac{1}{T}\frac{\partial T}{\partial x_i} + \frac{1}{RT}\left(c_i c_k - \frac{1}{3}c^2\delta_{ik}\right)\frac{\partial v_i}{\partial x_k} = Lh_1 \tag{3.15}$$

This equation can be used to evaluate h_1 or, equivalently, f_1, which is required to write Eq. (3.12) for $N = 1$ in an explicit form. We note that the left-hand side of Eq. (3.15) is a polynomial in $\mathbf{c}$; to be precise, the polynomials multiplying the velocity and temperature gradients can be identified with the functions ψ_{nlm} ($n = 0$, $l = 2$, and $n = 1$, $l = 1$, with $-l \le m \le l$) introduced in Chapter III, Section 5. It follows that Eq. (3.15) can be trivially solved in the case of Maxwell's molecules, because in this case the ψ_{nlm} are exactly the eigenfunctions of L; to be precise, we have

$$h_1 = \rho^{-1}A(T)c_i\left(\frac{c^2}{2RT} - \frac{5}{2}\right)\frac{\partial T}{\partial x_i} + \rho^{-1}B(T)\left(c_i c_k - \frac{1}{3}c^2\delta_{ik}\right)\frac{\partial v_i}{\partial x_k} \tag{3.16}$$

where ρ is the density and $A(T)$ and $B(T)$ depend only upon temperature and molecular constants and can be expressed in terms of the eigenvalues λ_{11} and λ_{02} of the collision operator [$A(T) = (T\lambda_{11}/\rho)^{-1}$; $B(T) = (RT\lambda_{02}/\rho)^{-1}$; note that ρ appears as a factor in the λ's]. We observe that

$f_1 = f_0 h_1$, as given by Eq. (3.16), already satisfies Eq. (3.7) and yields the following contributions to the stress tensor and heat flux:

$$p_{ij}^{(1)} = -\mu\left(\frac{\partial v_i}{\partial x_j} + \frac{\partial v_j}{\partial x_i}\right) + \frac{2}{3}\mu\frac{\partial v_k}{\partial x_k}\delta_{ij}; \qquad q_i^{(1)} = -k\frac{\partial T}{\partial x_i} \tag{3.17}$$

where μ and k are proportional to $B(T)$ and $A(T)$, respectively, and depend only upon temperature and molecular constants. It is clear that Eq. (3.12) with $N = 1$ is nothing other than a Navier–Stokes system with suitable values of the viscosity and heat conduction coefficients (the so-called transport coefficients).

In the case of more general interactions with a central force law Eq. (3.15) cannot be solved in closed form; since, however, the angular dependence of the eigenfunctions is given by the spherical harmonics for any central law (this follows from the fact that L commutes with any rotation operator acting upon $\mathbf{c}$), Eq. (3.16) is simply modified as follows:

$$h_1 = A(c; T)c_i\frac{\partial T}{\partial x_i} + B(c; T)\left(c_i c_k - \frac{1}{3}c^2\delta_{ik}\right)\frac{\partial v_i}{\partial x_k} \tag{3.18}$$

where $A(c; T)$ and $B(c; T)$ are now functions of the molecular speed as well. $A(c; T)$ and $B(c; T)$ satisfy the following equations:

$$\begin{gathered} c_i\left(\frac{c^2}{2RT} - \frac{5}{2}\right)\frac{1}{T} = L(c_i A) \\ \frac{1}{RT}\left(c_i c_k - \frac{1}{3}c^2\delta_{ik}\right) = L[(c_i c_k - c^2\delta_{ik})B] \end{gathered} \tag{3.19}$$

with the additional condition

$$\int_0^\infty A(c; T)c^4 \exp[-c^2(2RT)^{-1}]\, dc = 0 \tag{3.20}$$

which is required in order to satisfy Eq. (3.7) (with $n = 1$).

It is easily verified that Eqs. (3.17) still hold. Accordingly, the first two terms ($n \leq 1$) of the Chapman–Enskog expansion provide us with a macroscopic model of the Navier–Stokes type, the transport coefficients depending only upon temperature and molecular constants. We recall that the independence of viscosity from density was one of the first successes of kinetic theory, since it came before any experimental measurement had been made.

We add that the actual computation of μ and k [*via* $A(c; T)$ and $B(c; T)$] from specific molecular models is a thoroughly investigated subject and will not be considered in detail here. We only observe that the basic problem is the solution of the equation

$$Lh = g, \qquad (g, h \in \mathscr{W}) \tag{3.21}$$

for a given polynomial g. One can use expansions in the ψ_{nlm} ($l = 1, 2$; $0 \leq n \leq \infty$) or, more effectively, exploit a variational procedure. The latter is based upon the observation that if h is the solution of Eq. (3.21) and $\tilde{h} = h + \delta h$ is any other function of $\mathscr{W}$, then, if we define

$$T(\tilde{h}) = (\tilde{h}, L\tilde{h}) - 2(g, \tilde{h}) \tag{3.22}$$

we have

$$\begin{aligned} T(\tilde{h}) &= (h, Lh) + (h, L\,\delta h) + (\delta h, Lh) + (\delta h, L\,\delta h) - 2(g, h) - 2(g, \delta h) \\ &= -(g, h) + (\delta h, L\,\delta h) \end{aligned} \tag{3.23}$$

where we have used the symmetry of L and the fact that h satisfies Eq. (3.21). But Eq. (3.13) of Chapter III implies that

$$T(\tilde{h}) \leq -(g, h) - \mu_0 \|\delta h\|^2 \qquad (\mu_0 > 0, \delta h \in \mathscr{W}) \tag{3.24}$$

This means that the functional T attains its (absolute) maximum when $\delta h = 0$, i.e., $\tilde{h} = h$. One can then take a trial function $\tilde{h}$ containing a certain number of constants, and maximize $T(\tilde{h})$ by a suitable choice of the constants; the resulting $\tilde{h}$ gives an approximation to h. The usefulness of the method is further enhanced by the circumstance that one can relate the value assumed by $T(\tilde{h})$ for $\tilde{h} = h$ to the values of the transport coefficients; due to this circumstance a relative error of order, say, δ in approximating h by $\tilde{h}$ implies a relative error of order δ^2 in evaluating μ and k.

We also note that the Chapman–Enskog theory allows us to find the viscosity and heat conduction coefficients in terms of molecular constants, and makes it clear that both of them are proportional to the mean free path (in fact, L^{-1} is of the order of the mean free path, according to the qualitative definition given in Section 1); this explains why the Prandtl number

$$\mathrm{Pr} = c_p(\mu/k) \tag{3.25}$$

where c_p is the specific heat at constant pressure, is always of order unity for a gas. In the case of Maxwell's molecules it is very easy to calculate Pr because if we express μ and k in terms of λ_{02} and λ_{11} and take into account that $c_p = 5R/2$ for a monatomic gas, we obtain

$$\mathrm{Pr} = \lambda_{11}/\lambda_{02} \tag{3.26}$$

and the latter ratio can be shown to be $\frac{2}{3}$ from Eq. (5.26) of Chapter III.

For non-Maxwell molecules the Prandtl number is slightly temperature-dependent, but is always very close to $\frac{2}{3}$. Accordingly, the latter value can be taken as typical for a monatomic gas; these results are in good agreement with experimental data.

The circumstance noted above, that the viscosity coefficient is proportional to a qualitatively defined mean free path, suggests a precise definition of the latter in terms of the former; a definition used frequently is

$$l = \frac{\mu(\pi RT/2)^{1/2}}{p} = \frac{\mu(\pi/2RT)^{1/2}}{\rho} \tag{3.27}$$

This definition is in agreement with the qualitative definition given in Section 1, and allows us to associate a definite number with the concept of mean free path; it also has the advantage that one can easily compare results corresponding to different macroscopic models because the above definition is in terms of macroscopically measurable quantities, and, accordingly, does not depend upon any assumption on the molecular interactions.

4. Advantages and Disadvantages of the Hilbert and Chapman–Enskog Expansions

We have already pointed out that the Hilbert expansion cannot provide uniformly valid solutions; this follows from the fact that the solutions of the inviscid fluid equations cannot be improved to describe the (viscous) boundary layers even by taking into account higher-order corrections, from the singular manner in which the parameter ϵ enters into the Boltzmann equation, from the study of time-dependent problems (where secular terms are introduced by higher approximations), etc. These circumstances, however, do not prevent a truncated Hilbert expansion from representing solutions of the Boltzmann equation with arbitrary accuracy in suitably chosen space-time regions (which can be called the normal regions), provided that we stay at finite distance from certain singular surfaces and ϵ is sufficiently small; we shall now briefly discuss this subject. It is obvious that if one substitutes a truncated Hilbert expansion into the Boltzmann equation, the latter is satisfied except for an error term of order ϵ^n; therefore the Hilbert expansion can be used to approximate certain solutions (normal solutions) of the Boltzmann equation, the error being arbitrarily small for a sufficiently small ϵ [a rigorous proof with estimates is available in the case of the linearized Boltzmann equation (Grad, ref. 6)]. In order to see what kind of solutions these normal solutions are, one can observe that the abovementioned remainder of order ϵ^n contains space derivatives of order n; accordingly, a high degree of smoothness is required for the normal solutions to exist. This suggests that the normal solutions cease to be valid in space-time regions where the density, velocity, and temperature profiles tend to be very steep; such regions are immediately identified as the neighborhoods of boundaries (boundary layers), the initial stage (initial layer), and shock waves (shock layers). The same result is obtained if we observe that the Hilbert expansion treats $\partial f/\partial t$ and $\partial f/\partial \mathbf{x}$ as if they were of the same order of f, while they are of order f/ϵ whenever changes of f are sensible on the scale of the mean free path. An additional region where the Hilbert expansion fails is given by the "final layer," i.e., the evolution at times of order $1/\epsilon$; on such a scale $\partial f/\partial t \sim \epsilon f$ is negligible with respect to f, but a nonuniform expansion arises because $\partial f/\partial \mathbf{x} \sim f$ is forced to be of the same order of magnitude as $\partial f/\partial t$.

According to this discussion, the normal solutions are capable of approximating (for sufficiently small ϵ) the solutions of arbitrary problems provided the above-mentioned layers are excluded. However, in order to solve the differential equations which, according to the Hilbert method, regulate the fluid variables ρ^α, it is necessary to complete them with suitable initial data, boundary conditions, or matching conditions across a shock, i.e., we must go through those regions where the theory does not hold. It is evident that to complete the theory, one has to solve three connection problems across the layers within which the Hilbert expansion fails:

1. To relate a given initial distribution function to the Hilbert solution which takes over after an initial transient.
2. To relate a given boundary condition on the distribution function to the Hilbert solution which holds outside the boundary layer.
3. To find the correct matching conditions for the two Hilbert solutions prevailing on each side of a shock layer.

The present state of these three connection problems will be reviewed in Section 5; briefly, a rather complete theory exists for the initial layer, a qualitative theory exists for the boundary layer, while the matching through shock layers has only been treated from a phenomenological point of view.

The Chapman–Enskog procedure tries to overcome one of the many nonuniformities of the Hilbert expansion; it starts from the macroscopical information that, besides kinetic layers (of order ϵ), viscous layers (of order $\epsilon^{1/2}$) exist in the vicinity of boundaries, and incorporates both the viscous layers and the normal regions into a uniform description. At the same time the Chapman–Enskog procedure eliminates the nonuniformity of the "final layer," since it allows contributions of different orders in ϵ to the time derivative from the space derivatives. Actually, the existences of the viscous layers and the "final layer" are interrelated, and the Chapman–Enskog theory simply takes into account the existence and practical importance of regimes with $d^2(\epsilon T)^{-1} \approx 1$ (where T and d are a typical time and a typical length; T can be replaced by another typical length, different from d). The basic result of the Chapman–Enskog procedure is that one can recover the

Navier–Stokes–Fourier macroscopic description by a suitable expansion of certain solutions of the Boltzmann equation. Thus we may expect the Chapman–Enskog theory to be much more accurate than the Hilbert theory; on the other hand, if we consider higher-order approximations of the Chapman–Enskog method, we obtain differential equations of higher order (the so-called Burnett and super-Burnett equations), about which nothing is known, not even the proper boundary conditions. These higher-order equations have never achieved any noticeable success in describing departures from continuum fluid mechanics; furthermore, a preliminary treatment of the connection problem for boundary layers seems to yield the same number of boundary conditions at any order of approximation (see next section), while the order of differentiation increases. These and other facts seem to suggest that the Chapman–Enskog theory goes too far in the direction of taking into account the contributions of different orders in ϵ to the time derivative; in fact, not only regimes with $d^2(\epsilon T) \approx 1$, but also regimes with $d^{n+1}(\epsilon^n T)^{-1} \approx 1$ $(n \geq 0)$ are allowed. This implies that the "final layer" is described in an ultrarefined (physically irrelevant) fashion, while (possibly nonexistent) boundary layers of order $\epsilon^{n/(n+1)}$ $(n \geq 2)$ are taken into account; the practical result is that we complicate the equations by inserting details which are either irrelevant or nonexistent. The fact that the Chapman–Enskog expansion can bring in solutions which are simply nonexistent is not strange; in fact, the Chapman–Enskog procedure expands the operator $S^\alpha(\rho^\beta)$ (whose existence is proved, at least asymptotically for $\epsilon \to 0$, by the Hilbert expansion) into a series of differential operators, in spite of the fact that nothing is known about the nature of $S^\alpha(\rho^\beta)$. It is obvious that such a procedure can bring in extraneous solutions, as illustrated by the operator Δ_ϵ such that

$$\Delta_\epsilon f = f(x + \epsilon) - f(x) \tag{4.1}$$

and the related equation

$$\Delta_\epsilon f = 0 \tag{4.2}$$

Δ_ϵ possesses a series expansion

$$\Delta_\epsilon = \epsilon \frac{\partial}{\partial x} + \frac{\epsilon^2}{2!} \frac{\partial^2}{\partial x^2} + \cdots + \frac{\epsilon^n}{n!} \frac{\partial^n}{\partial x^n} + \cdots \tag{4.3}$$

which is convergent when applied to sufficiently smooth functions of x (analytic functions). If we truncate the expansion at the first step, Eq. (4.2) becomes

$$\epsilon\, \partial f/\partial x = 0 \tag{4.4}$$

while if we truncate at the second step, we obtain

$$\epsilon\frac{\partial f}{\partial x} + \frac{\epsilon^2}{2}\frac{\partial^2 f}{\partial x^2} = 0 \tag{4.5}$$

Equation (4.4) has the general solution $f = \text{const}$, which satisfies Eq. (4.2) as well, while the general solution of Eq. (4.5), $f = A + B\exp[-2x/\epsilon]$, satisfies Eq. (4.2) only for $B = 0$; this means that the Chapman–Enskog expansion has introduced spurious solutions. We note that Eq. (4.5) has other solutions besides $f = \text{const}$, i.e., all the periodic functions with period ϵ; however, these functions vary so rapidly (for small ϵ) that they cannot be obtained by an expansion into a series of powers in ϵ. A Hilbert-like expansion of Eq. (4.2) (f expanded together with Δ_ϵ in powers of ϵ) again yields $f = \text{const}$, but no additional solutions (at any order). Note also that the first-order Chapman–Enskog solution contains a term (the spurious one) which varies on a scale of order ϵ, i.e., the same scale as the neglected solutions. Similar phenomena are observed when studying higher-order terms of the Chapman–Enskog expansion of the solutions of the Boltzmann equation.

A possible way of avoiding the troubles of higher-order terms, while retaining the advantages of the Chapman–Enskog expansion, can be based upon recognizing the importance of regimes with $d^2(\epsilon T)^{-1} \approx 1$, while discussing regimes with $d^{n+1}(\epsilon^n T) \approx 1$ $(n \geq 2)$ as physically irrelevant. This brings us to a new kind of expansion where the time derivative is split into just two parts,

$$\frac{\partial f}{\partial t} = \frac{\partial^{(0)} f}{\partial t} + \epsilon\frac{\partial^{(1)} f}{\partial t} \tag{4.6}$$

while f is expanded into a series,

$$f = \sum_{n=0}^{\infty} \epsilon^n f_n \tag{4.7}$$

The presence of the second term in Eq. (4.1) requires a restriction on the expansion of f. If we take into account that the splitting in Eq. (4.6) is based on the interplay between terms of zeroth and first order in ϵ, the simplest choice is to assume that there is no contribution to the fluid variables from the odd-order terms:

$$\int \psi_\alpha f_n \, d\xi = 0 \qquad (n = 1, 3, 5, \ldots) \tag{4.8}$$

It is clear that the proposed expansion is only a particular instance of the infinitely many possible expansions of this kind. The most general is based upon a truncated expansion of the time derivatives [Eq. (3.5) truncated at $n = N$] plus certain conditions regulating the contribution of different orders to the fluid variables [Eq. (3.7) for $n \neq (N + 1)k$; $k = 1, 2, 3, \ldots$]. The Hilbert expansion corresponds to $N = 0$ and the Chapman–Enskog expansion to $N = \infty$; the particular procedure which has just been proposed corresponds to $N = 1$. This particular choice is dictated by the available information at a macroscopic level; it is also suggested by the preliminary investigations of the kinetic boundary layer (see next section). In fact, these investigations seem to prove that $N = 1$ is the only choice for which the matching procedure provides boundary conditions in both necessary and sufficient number for yielding a mathematically well-posed problem.

The proposed procedure ($N = 1$) is identical to the Chapman–Enskog method up to the Navier–Stokes level; the subsequent steps, however, are similar to the Hilbert expansion, the only essential changes being the presence of the linearized Navier–Stokes operator $N_0^{\alpha\beta}$ in place of the linearized Euler operator $E_0^{\alpha\beta}$ and the fact that we can write a complete evolution equation every two steps instead of each step.

We end this section with some remarks about a question which is frequently asked: do the Chapman–Enskog and Hilbert expansions converge? This is a difficult question in general, although convergence can be proved or disproved in particular cases for the linearized Boltzmann equation (see Chapter VI). This question, however, has little meaning from the practical point of view; what is more important is that the above-mentioned series provide asymptotic solutions (for $\epsilon \to 0$) of the Boltzmann equation in certain regions, identified by the condition that all the space derivatives of the fluid variables exist and are bounded (not uniformly with respect to the order, in general). The existence and boundedness of all the derivatives is of course a powerful requirement, but is essentially the condition which defines the normal regions.

What is more interesting is to prove rigorously the asymptotic agreement of the solutions of the Boltzmann equation (in the lowest orders in ϵ) with the Euler ($n = 0$) or Navier–Stokes ($n = 1$) equations. Rigorous proofs have been given that the first few terms of the expansions agree asymptotically (for a vanishing mean free path ϵ) with the corresponding solutions of the Boltzmann equation in certain cases. The main results are the following:

1. For the linearized Boltzmann equation the Hilbert or Chapman–Enskog expansions (with modification) are asymptotic to the Boltzmann equation. This was proved by Ellis and Pinsky (refs. 12 and 13), who extended the work of Grad, previously mentioned in this section.
2. For initial data close to a global Maxwellian, several Japanese authors showed that the nonlinear Boltzmann and Navier–Stokes equations agree to leading order as $t \to \infty$, and ϵ is held constant (refs. 14 and 15).

3. For initial data close to a global Maxwellian, Grad (ref. 16) and Nishida (ref. 17) proved that the nonlinear Boltzmann and Euler equations agree for at least a short time as $\epsilon \to 0$.
4. Caflisch showed that, if the nonlinear Euler equations have a smooth solution in some time interval, then there is a solution of the Boltzmann equation which agrees with the Euler solution as $\epsilon \to 0$ (ref. 18).
5. More recently, De Masi *et al.* (ref. 19) and Bardos *et al.* (ref. 20) proved interesting results about an expansion valid in the case when the mean free path goes to zero and the speed of sound goes to infinity in such a way that their product [and hence the transport coefficients, see Eq. (3.27)] remain finite. In this limit the incompressible Navier–Stokes equations hold exactly.

It is perhaps appropriate at this point to remark that the Hilbert and Chapman–Enskog expansions as well as the other expansions discussed in this section by no means exhaust the possible expansions valid for a mean free path going to zero. In addition to the important limit considered in refs. 19 and 20 there is the possibility of giving a different order to the gradients of different physical quantities, so that some of the Burnett terms might turn out to be of the same order as the Navier–Stokes terms (refs. 21 and 22).

5. The Problems of Initial Data, Boundary Conditions, and Shock Layers

As mentioned above, the Hilbert theory is not complete, and in order to complete it, one has to solve three connection problems concerning the initial, boundary, and shock layers; the same connection problems arise for the Chapman–Enskog expansion as well as for the modified expansion proposed in Section 4. We first consider the problem of the initial layer, following a paper by Grad (ref. 6). A complete theory should deal with the connection of the above-mentioned expansions with an arbitrary initial datum; such a theory, however, implies the solution of nonlinear integro-differential equations and is of little practical use. Actually if we take the spirit of the Hilbert and related expansions into account, we can restrict ourselves to matching an initial datum of the same type as the solution, i.e., a datum which reduces to a Maxwellian when $\epsilon \to 0$; as a consequence, the initial datum is arbitrary within the condition that it can be written in the form $f_M + \epsilon f_N$, where f_M is a Maxwellian.

In order to study the initial layer, we rescale the time variable as follows:

$$\tau = t/\epsilon \tag{5.1}$$

This rescaling gives the initial layer a finite duration of the scale τ. We now look for solutions of the Boltzmann equation, Eq. (2.1), in the form

$$f = f_H(\mathbf{x}, \boldsymbol{\xi}, t; \epsilon) + \epsilon f_R(\mathbf{x}, \boldsymbol{\xi}, \tau; \epsilon) \tag{5.2}$$

where $\tau = t/\epsilon$..The leading term f_H is given by a Hilbert or related type of expansion (to be called simply the Hilbert expansion in the following), and the "remainder" f_R is expanded as follows:

$$\epsilon f_R = \sum_{n=0}^{\infty} \epsilon^{n+1} f_{R(n+1)}(\mathbf{x}, \boldsymbol{\xi}, t) \tag{5.3}$$

The additional factor ϵ comes from the above assumption on the allowed initial states. The Hilbert expansion f_H formally satisfies the Boltzmann equation; as a consequence, substituting into Eq. (2.1), we obtain

$$\frac{\partial f_R}{\partial \tau} + \epsilon \boldsymbol{\xi} \cdot \frac{\partial f_R}{\partial \mathbf{x}} = Q(f_H, f_R) + \epsilon Q(f_R, f_R) \tag{5.4}$$

To convert this into an equation in τ alone, we write $f_H(\mathbf{x}, \boldsymbol{\xi}, t; \epsilon) = f_H(\mathbf{x}, \boldsymbol{\xi}, \epsilon\tau; \epsilon)$ and expand the latter in powers of ϵ. This gives a reordering of the power series in ϵ considered in the Hilbert expansion:

$$f_H = \sum_{n=0}^{\infty} \epsilon^n f_{H(n)}(\mathbf{x}, \boldsymbol{\xi}, \tau) \tag{5.5}$$

in which the leading term $f_{H(0)}$ is time-independent and locally Maxwellian. We also note that the expansion in $\epsilon\tau$ does not alter the original series at $t = \tau = 0$.

Now, substituting Eqs. (5.3) and (5.5) into Eq. (5.4), we obtain

$$\partial g_n/\partial \tau = L g_n + G_n \qquad (n \geq 1) \tag{5.6}$$

where

$$f_{R(n)} = f_{H(0)} g_n \qquad (n \geq 1)$$

$$G_n = -\left[\boldsymbol{\xi} \cdot \frac{\partial f_{R(n-1)}}{\partial \mathbf{x}} + \sum_{R=1}^{n-1} Q(f_{H(k)} + f_{R(k)}, f_{R(n-k)})\right] f_0^{-1} \tag{5.7}$$

The linearized collision operator L is based on the Maxwellian $f_{H(0)}$; accordingly, its eigenvalues and eigenfunctions are space- but not time-dependent. Since f as given by Eq. (5.2) must satisfy the initial conditions, $f_{H(0)}$ is nothing other than the Maxwellian f_M appearing in the initial data; this defines L completely. The next step is to expand the initial datum $\bar{f}$ into a power series in ϵ:

$$\bar{f} = \sum_{n=0}^{\infty} \epsilon^n \bar{f}_n \tag{5.8}$$

and satisfy the initial conditions term by term:

$$f_{H(n)} + f_{R(n)} = \bar{f}_n \qquad (n \geq 1) \tag{5.9}$$

Equation (5.6) can be projected onto the space $\mathscr{F}$ spanned by the collision invariants to yield

$$\frac{\partial \rho^\alpha_{(n)}}{\partial_\tau} = \int \psi_\alpha f_0 G_n \, d\xi = -\int \psi_\alpha \xi \cdot \frac{\partial f_{R(n-1)}}{\partial \mathbf{x}} \, d\xi \tag{5.10}$$

where $\rho^\alpha_{(n)}$ is the contribution to ρ^α from the nth term of the expansion of the remainder $f_{R(n)}$. Equation (5.10) gives

$$\rho^\alpha_{(n)}(\mathbf{x}, \tau) = \rho^\alpha_{(n)}(\mathbf{x}, 0) - \int_0^\tau d\tau \int \psi_\alpha \xi \cdot \frac{\partial f_{R(n-1)}}{\partial \mathbf{x}} \, d\xi \tag{5.11}$$

We observe now that we can evaluate the nth-order contribution to $\rho^\alpha(\mathbf{x}, 0)$, $\bar{\rho}^\alpha_n(\mathbf{x})$, from the initial datum $\bar{f}$; however, we have to decide how this datum is to be distributed between $\rho^\alpha_{(n)}$ (i.e., the remainder) and ρ^α_n (i.e., the Hilbert expansion). Once this problem is solved we know the initial data to be matched with the Hilbert expansion; at present we only know that

$$\rho^\alpha_n(\mathbf{x}, 0) + \rho^\alpha_{(n)}(\mathbf{x}, 0) = \bar{\rho}^\alpha_n(\mathbf{x}) \tag{5.12}$$

Assigning the initial datum for the Hilbert expansion means selecting that particular Hilbert expansion which describes a specific problem; accordingly, the ρ's which appear in the Hilbert expansion must be the physical ρ's when the Hilbert expansion takes over, i.e., after several mean free times. We cannot say that the initial ρ^α_n for the Hilbert solution are simply the $\bar{\rho}^\alpha_n$ because this would be like pretending that the asymptote (Hilbert solution) of a curve (certain solution) crosses the curve at $t = 0$, an obviously meaningless requirement. We have to require instead that for large times the complete solution f differs from f_H by a negligible quantity. In particular, $\rho^\alpha_{(n)}(\mathbf{x}, t)$, the contribution of the remainder to the fluid variables, must be zero when $\tau \to \infty$; this is easily accomplished if we use Eq. (5.11) and set

$$\rho^\alpha_{(n)}(\mathbf{x}, 0) = \int_0^\infty d\tau \int \psi_\alpha \xi \cdot \frac{\partial f_{R(n-1)}}{\partial \mathbf{x}} \, d\xi \tag{5.13}$$

Inserting this into Eq. (5.12), we obtain the initial conditions for the Hilbert equations:

$$\rho_n^\alpha(\mathbf{x}, 0) = \bar{\rho}_n^\alpha(\mathbf{x}) - \int_0^\infty d\tau \int \psi_\alpha \boldsymbol{\xi} \cdot \frac{\partial f_{R(n-1)}}{\partial \mathbf{x}} d\boldsymbol{\xi} \tag{5.14}$$

These conditions involve the physical data $\bar{\rho}_n^\alpha$ as well as a contribution from the solution $f_{R(n-1)}$ at the previous step. It can be verified that the integral term gives no contribution to first-order initial data, so that the first correction is of order ϵ^2; to this order there is no correction to the initial density, and the corrected initial conditions for velocity and temperature take on the form (Grad, ref. 6)

$$\begin{aligned} \mathbf{v}(\mathbf{x}, 0) &= \bar{\mathbf{v}} - a[\operatorname{div}(a' \operatorname{grad} \bar{\mathbf{u}}) + \tfrac{1}{3} \operatorname{grad}(a' \operatorname{div} \bar{\mathbf{u}})] \\ T(x, 0) &= \bar{T} - b \operatorname{div}(b' \operatorname{grad} \bar{\mathbf{u}}) \end{aligned} \tag{5.15}$$

where $\bar{\mathbf{v}}$ and $\bar{T}$ denote the physical initial data for velocity and temperature, respectively, and a, a', b, and b' are four coefficients such that aa' and bb' are of order ϵ^2. These coefficients can be computed exactly for Maxwell molecules; in particular, we have

$$aa' = l^2/\pi; \qquad bb' = 15l^2/4\pi \tag{5.16}$$

where l is the mean free path defined by Eq. (3.26).

The above results show that the naive approach based upon putting $\rho_n^\alpha(\mathbf{x}, 0) = \bar{\rho}_n^\alpha(\mathbf{x})$ is essentially correct at the Euler and Navier–Stokes level; it is insufficient at the level of the Burnett equations, whose status, however, is not clear and whose practical importance is negligible. This means that an expansion of the Hilbert type treats the initial layer in a correct way except for (usually negligible) terms of order ϵ^2.

The situation is essentially different for boundary layers. We already know that the Hilbert expansion misses completely not only the kinetic boundary layers but also the viscous boundary layers; the latter are recovered by the Chapman–Enskog method and the method described briefly in Section 4. The kinetic layers of order ϵ, however, are missed by all the expansions in powers of ϵ described so far; in order to recover them, we have to use a magnified variable $X = x/\epsilon$, analogous to the variable τ used before for the initial layer. We shall consider the case of boundaries whose radius of curvature is large with respect to the mean free path, and boundary conditions which do not change appreciably along the boundary (on the scale of the mean free path); if these conditions are not satisfied, the analysis is complicated and becomes two- or three-dimensional instead of one-dimensional in space variables. We also assume that the deviation of the distribution function from a Maxwellian remains of order ϵ in the vicinity of the boundary; this assumption is analogous to the above hypothesis on admissible initial data.

Under our assumptions we can take in the neighborhood of the boundary a non-Cartesian reference frame made as follows: we take a pair of coordinates α_i $(i = 1, 2)$ on the surface; then, through each point $\mathbf{x}$ we draw the straight line normal to the surface; finally, we take as coordinates of $\mathbf{x}$ the distance x along the normal and the coordinates α_i $(i = 1, 2)$ of the point, $\mathbf{x}_0$, where the straight line meets the surface. If $\mathbf{n}$ is the normal unit vector at $\mathbf{x}$, we have

$$x_j = x_{0j} + x n_j \qquad (j = 1, 2, 3)$$

From this relation and the parametric equations of the surface $[x_{0j} = x_{0j}(\alpha_1, \alpha_2)]$ one can obtain $\partial\alpha_i/\partial X_k$ $(i = 1, 2;\ k = 1, 2, 3)$ and $n_k = \partial x/\partial x_k$ as functions of the α's and write

$$f = f_c(x, \alpha_i, \boldsymbol{\xi}, t; \epsilon) + \epsilon f_R(X, \alpha_i, \boldsymbol{\xi}, t; \epsilon) \qquad (X = x/\epsilon) \tag{5.17}$$

The leading term f_c is given by the Chapman–Enskog expansion (or any expansion capable of describing the viscous layer), while f_R satisfies

$$\boldsymbol{\xi} \cdot \mathbf{n} \frac{\partial f_R}{\partial X} + \epsilon \left(\frac{\partial f_R}{\partial t} + \sum_{i=1}^{2} \boldsymbol{\xi} \cdot \frac{\partial \alpha_i}{\partial \mathbf{x}} \frac{\partial f}{\partial \alpha_i} \right) = Q(f_c, f_R) + \epsilon Q(f_R, f_R) \tag{5.18}$$

If we now expand as above (interchanging the roles of X and τ), the basic equation turns out to be:

$$\boldsymbol{\xi} \cdot \mathbf{n} \frac{\partial g_n}{\partial X} = L g_n + G_n \qquad (n \geq 1) \tag{5.19}$$

where G_n is known in terms of the previous steps of the approximation.

The problem of solving Eq. (5.19) is much more difficult than the analogous problem of solving Eq. (5.6); in particular, one cannot project Eq. (5.19) onto $\mathscr{F}$ in order to obtain an equation for the fluid variables $\rho^{\alpha}_{(n)}$ because the factor $\boldsymbol{\xi} \cdot \mathbf{n}$ couples the fluid variables to the whole distribution function. This does not allow us to write equations similar to Eqs. (5.14) until we have constructed a theory for solving Eq. (5.19); this is not an easy task even for Maxwell molecules. Furthermore, this solution depends on the boundary conditions, which are much more complicated than the initial conditions.

The solution of the connection problem for boundary layers is much more important than the solution of the problem for the initial layer, since it can be shown that the influence of the boundary layers is already felt at the first order in ϵ, i.e., at the Navier–Stokes level. One can show that the extrapolated boundary conditions are (to order ϵ)

$$\begin{aligned} \mathbf{v} - \zeta \mathbf{n} \cdot \frac{\partial}{\partial \mathbf{x}}[\mathbf{v} - \mathbf{n}(\mathbf{n} \cdot \mathbf{v})] - \omega (2R/T)^{1/2} \mathbf{n} \times \left[\left(\frac{\partial T}{\partial \mathbf{x}} \right) \times \mathbf{n} \right] &= \mathbf{u}_0 \\ T - \tau \mathbf{n} \cdot \frac{\partial T}{\partial \mathbf{x}} - \chi (2RT)^{-1/2} \left[\mathbf{n} \times \left(\mathbf{n} \times \frac{\partial}{\partial \mathbf{x}} \right) \right] \cdot (\mathbf{v} - \mathbf{n}(\mathbf{n} \cdot \mathbf{v})) &= T_0 \end{aligned} \tag{5.20}$$

where $\mathbf{u}_0$ and T_0 are, respectively, the velocity and temperature of the boundary, and ζ, ω, and τ are coefficients of the order of the mean free path. In particular, ζ measures the tendency of the gas to slip over a solid wall in the presence of velocity gradients, and is called the slip coefficient; τ measures the tendency of the gas to have a temperature different from the wall temperature, and is called the temperature jump coefficient. The coefficients ζ, ω, and τ have been evaluated by means of kinetic models; the procedure to be used for these calculations will be expounded in Chapter VII. We note that when the mean free path is not only small but completely negligible the boundary conditions reduce to $\mathbf{v} = \mathbf{u}_0$, $T = T_0$, i.e., the gas does not slip and accommodates completely to the wall temperature.

The third connection problem (shock layer) has been investigated only by an analogy with boundary layers (Pan and Probstein, ref. 26). The kinetic theory solution of the zeroth-order connection problem for the Hilbert expansion is already a difficult one (the shock structure problem) but the matching relations are trivial (the Rankine–Hugoniot relations); the analogous problem for the Chapman–Enskog theory (or the modified theory proposed in Section 4) presents itself as very interesting and possibly easier than the analogous problem for the Hilbert expansion, but its setting in the frame of the theory of the Boltzmann equation has never been attempted.

In addition to his work on deriving the boundary conditions on a general surface (ref. 24), Sone (ref. 27) discovered that there is another boundary layer due to curvature, particularly significant when the gas flows past a (locally) convex body.

6. Kinetic Models versus Chapman–Enskog Theory

The procedures expounded in this chapter apply not only to the full Boltzmann equations, but also to the Boltzmann-like equations which are obtained when the quadratic collision operator is replaced by a model operator $J(f)$ (Chapter IV). The only changes arise in connection with the expansion of the nonlinear terms in powers of ϵ, because the nonlinearity of the models is, in general, more complicated than quadratic. This circumstance, however, does not come in before the second-order approximation (terms in ϵ^2). As a consequence, the models reproduce correctly the inviscid and Navier–Stokes equations, and even the coefficients of viscosity and heat conduction can be adjusted to agree with the correct ones provided that the models contain at least two adjustable parameters. This is not true for the simplest models, as, e.g., the BGK model, and we have to decide whether to adjust viscosity or heat conduction.

Since the BGK model is frequently used, we briefly describe the first steps of the Chapman–Enskog theory for this model. As noted above, the zeroth- and first-order equations are formally the same as for the Boltzmann equation: f_0 is a Maxwellian and $f_1 = f_0 h_1$ is to be found by solving Eq. (3.15), where now, however, L is the linearized BGK operator. The procedure used in Chapter IV, Section 2 to construct the Gross and Jackson models shows that the linearized BGK collision operator has the same eigenfunctions as the Maxwell collision operator; the distinct eigenvalues are now only two, $\lambda = 0$ (corresponding to the five collision invariants) and $\lambda = -\nu$ (corresponding to the remaining eigenfunctions). Accordingly, the solution of Eq. (3.15) for the BGK model is again given by Eq. (3.16), where $A(T)$ and $B(T)$ are essentially the same as for Maxwell's molecules (the only change being in numerical factors); the main consequence of the infinite-fold degeneracy of the eigenvalue $\lambda = -\nu$ is that the Prandtl number [still given by Eq. (3.25)] is now 1 because $\lambda_{11} = \lambda_{02} = -\nu$. This result implies the above-mentioned result that one cannot adjust μ and k at the same time if the BGK model is being used; it can be done, however, if the ES model or the nonlinear models described in Chapter IV, Section 2 are being used.

References

Section 2——The Hilbert expansion was proposed by

1. D. Hilbert, *Math. Ann.* **72**, 562 (1912).

   See also

2. D. Hilbert, *Grundzüge einer Allgemeinen Theorie der Linearen Integralgleichungen*, Chelsea, New York (1953).

Section 3——The Chapman–Enskog procedure was proposed by

3. S. Chapman, *Phil. Trans. R. Soc.* A **216**, 279 (1916); **217**, 115 (1917);
4. D. Enskog, Dissertation, Uppsala (1917); *Arkiv Mat., Ast. och. Fys.* **16**, 1 (1921).

   Results for different force laws, gas mixtures, etc. are expounded in detail in ref. 7 of Chapter I and in

5. J. O. Hirschfelder, C. F. Curtiss, and R. B. Bird, *Molecular Theory of Gases and Liquids*, John Wiley and Sons, New York (1954).

Section 4——A proof that the Hilbert expansion does provide asymptotic solutions to the linearized *Boltzmann equation is given in*

6. H. Grad, *Phys. Fluid* **6**, 147 (1963),

while the three connection problems were posed in ref. 9 of Chapter I. The nonuniformity of the Hilbert expansion for long times was noticed by

7. S. Boguslawski, *Math. Ann.* **76**, 431 (1915).

The Burnett equations were obtained by

8. D. Burnett, *Proc. London Math. Soc.* **39**, 385 (1935); **40**, 382 (1935).

The modified two-times expansion is proposed here for the first time, although previous attempts at reordering the Chapman–Enskog expansion with a similar intent can be found in

9. R. Schamberg, Ph.D. Thesis, California Institute of Technology (1947);
10. C. Cercignani, "Higher Order Slip According to the Linearized Boltzmann Equation," University of California Report No. AS-64-18 (1964);
11. L. Trilling, *Phys. Fluids* 7, 1681 (1964).

The rigorous results on the Chapman-Enskog and Hilbert solutions, mentioned at the end of the section, can be found in ref. 6 and in

12. R. Ellis and M. Pinsky, *J. Math. Pure Appl.* **54**, 125 (1975);
13. R. Ellis and M. Pinsky, *J. Math. Pure Appl.* **54**, 157 (1975);
14. S. Kawashima, A. Matsumura, and J. Nishida, *Commun. Math. Phys.* **70**, 97 (1979);
15. J. Nishida and K. Imai, *Pub. Res. Inst. Math. Sci. Kyoto* **12**, 229 (1976);
16. H. Grad, *Proc. Sympos. Appl. Math.* **17**, 154 (1965);
17. J. Nishida, *Commun. Math. Phys.* **61**, 119 (1978);
18. R. Caflisch, *Commun. Pure Appl. Math.* **33**, 651 (1980);
19. A. De Masi, R. Esposito, and J. L. Lebowitz, *CARR Reports in Mathematical Physics* No. 2/89 (1989);
20. C. Bardos, F. Golse, and D. Levermore, ACMS Pub. 88-37, submitted to *C. R. Acad. Sci. Paris Ser. I* **309**, 727 (1989).
21. V. S. Galkin, M. N. Kogan, and O. G. Fridlender, *Izv. AN SSSR, Mekhanika Zhidkosti i Gaza*, No. 3, 14 (1970);
22. M. N. Kogan, V. S. Galkin, and O. G. Fridlender, *Uspekhi Fizicheskich Nauk* **119**(1), 111 (1976).

Section 5——The theory for the initial layer is described in ref. 6, while the treatment of the boundary layer is sketched in ref. 10 (with explicit calculations for the BGK model in the plane case). The effect of the wall curvature is studied in the following papers:

23. H. Grad, in: *Transport Theory* (R. Bellman, G. Birkhoff, and I. Abu-Shumays, eds.), *SIAM-AMS Proceedings*, Vol. I, p. 249, American Mathematical Society, Providence (1968).
24. Y. Sone, in: *Rarefied Gas Dynamics* (L. Trilling and H. Wachman, eds.), Vol. I, p. 243, Academic Press, New York (1969);
25. J. S. Darrozès, in: *Rarefied Gas Dynamics* (L. Trilling and H. Wachman, eds.), Vol. I, p. 211, Academic Press, New York (1969).

The boundary conditions (5.20) are well known and were first proposed by Maxwell in ref. 1 of Chapter II. The shock layer was treated in

26. Y. S. Pan and R. F. Probstein, in: *Rarefied Gas Dynamics* (J. A. Laurmann, ed.), Vol. II, Academic Press, New York (1963);
27. Y. Sone, Kyoto University Research Report No. 24 (1972).

Chapter VI

BASIC RESULTS ON THE SOLUTIONS OF THE BOLTZMANN EQUATION

1. The Linearized Boltzmann Equation

The Hilbert and Chapman–Enskog methods are perturbation procedures for solving the Boltzmann equation on the basis of the assumption of a small Knudsen number; other procedures based on the assumption of a large Knudsen number will briefly be described later (Chapter VIII, Section 3). The above two procedures are valid in the so-called near-continuum (or slip) regime (Kn → 0) and in nearly-free regime (Kn → ∞). They are both based upon a specific assumption on the order of magnitude of the Knudsen number. Accordingly, the intermediate regime (the so-called transition region) remains untouched by the above procedures because it cannot be described in terms of either a higher-order continuum theory or of small corrections to a picture of essentially noninteracting particles. A treatment of the transition regime requires the full use of the Boltzmann equation (or, at least, sufficiently accurate models of the latter). As a consequence, if we want to investigate the transition regime, we have either to give up the idea of using perturbation methods, or look for some other parameter, different from the Knudsen number, to be regarded as small under suitable conditions.

According to the discussion in Chapter V, Section 1 the Boltzmann equation contains essentially one nondimensional parameter (the Knudsen number); we therefore have to look for a new "small" parameter in the initial and boundary conditions, and not in the Boltzmann equation itself. Accordingly we consider an expansion of f into a power series in ϵ

$$f = \sum_{n=0}^{\infty} \epsilon^n f_n \tag{1.1}$$

and substitute it into the Boltzmann equation

$$\frac{\partial f}{\partial t} + \boldsymbol{\xi} \cdot \frac{\partial f}{\partial \mathbf{x}} = Q(f, f) \tag{1.2}$$

without specifying the physical meaning of ϵ; we only assume that such a parameter (or parameters) exists and is provided by the initial and boundary conditions (in a manner to be described later). The result of inserting Eq. (1.1) into Eq. (1.2) is

$$\frac{\partial f_0}{\partial t} + \boldsymbol{\xi} \cdot \frac{\partial f_0}{\partial \mathbf{x}} = Q(f_0, f_0) \tag{1.3}$$

$$\frac{\partial f_n}{\partial t} + \boldsymbol{\xi} \cdot \frac{\partial f_n}{\partial \mathbf{x}} = Q_n \qquad (n \geq 1) \tag{1.4}$$

where the results of Chapter III, Section 1 have been used.

Equation (1.3) shows that f_0 must be a solution of the Boltzmann equation. Since we do not know any solutions except Maxwellians, we are practically forced to choose f_0 to be a Maxwellian; otherwise, making the zeroth-order step of the approximation procedure would be as hard as solving the original equation. Although there are Maxwellians with variable density, velocity, and temperature which solve the Boltzmann equation, they constitute a very particular class, useful only in very special situations (special initial and boundary value data); accordingly, we shall choose our Maxwellian f_0 to have constant parameters (in particular, we can always choose it as having zero mass velocity by suitably choosing the reference frame). This choice is sufficiently broad for our purposes. We can now put $f_n = f_0 h_n$ $(n \geq 1)$ and write

$$\frac{\partial h_n}{\partial t} + \boldsymbol{\xi} \cdot \frac{\partial h_n}{\partial \mathbf{x}} = Lh_n + S_n \qquad (n \geq 1) \tag{1.5}$$

$$S_1 = 0; \qquad S_n = f_0^{-1} \sum_{k=1}^{n-1} Q(f_0 h_k, f_{n-k}) \qquad (n \geq 2) \tag{1.6}$$

where again results from Chapter III, Section 1 have been used; in particular, L is the linearized collision operator corresponding to f_0. The equations given by (1.5) describe a successive approximation procedure for solving the Boltzmann equation. What is interesting is that at each step we have to solve the same equation, the only change being in the source term, which has to be evaluated in terms of the previous approximations. In this respect the procedure is similar to the Hilbert and Chapman–Enskog methods; here, however, the operator acting upon the unknown h_n is not simply the operator L, but a more complicated integrodifferential operator. In other words, the equations to be solved are as complicated as the original Boltzmann equation, except for the fact that we have gotten rid of the nonlinearity. The fact that the same operator appears at each step allows us to concentrate on the first step, i.e., to study the following equation:

$$\frac{\partial h}{\partial t} + \boldsymbol{\xi} \cdot \frac{\partial h}{\partial \mathbf{x}} = Lh \tag{1.7}$$

The presence of a source term in the subsequent steps is hardly a complication in solving the equations, since well-known procedures allow us to solve an inhomogeneous linear equation once we are able to master the corresponding homogeneous equation.

As noted above, Eq. (1.7) is very similar to the full Boltzmann equation except for the fact that it is linear; accordingly, it is called the linearized Boltzmann equation.

The study of the linearized Boltzmann equation is important for at least three reasons:

1. There are conditions (to be specified below) under which the results obtained from the linearized Boltzmann equation can be retained to faithfully represent the physical situation.
2. The fact that the linearized equation has the same structure (except for the nonlinearity in the collision term) as the full Boltzmann equation suggests that we can obtain a valuable insight into the features of the solutions of the full Boltzmann equation by studying the linearized one; these features are obviously not those related to nonlinear effects, but, e.g., those related to the behavior in the proximity of boundaries. In the latter, in fact, the nonlinear nature of the collision term probably brings in small changes, the main features resulting from the general form of the equation and the boundary conditions.
3. The equations met in the study of the connection problems for the Hilbert and Chapman–Enskog theories are particular cases of the linearized Boltzmann equation; as a consequence, a study of Eq. (1.7) implies, as a byproduct, a study of the kinetic layers discussed in Chapter V.

We now have to specify the conditions under which one can make use of the linearized Boltzmann equation to obtain physically significant results; as noted before, these conditions must be found in the initial and boundary conditions. Since we look for a solution of the Boltzmann equation in the form $f = f_0(1 + h)$ with the condition that h can be regarded, in some sense, as a small quantity with respect to 1, a necessary condition is that h is small for $t = 0$ and at the boundaries.

As a consequence, a first condition is that the initial datum shows little departure from the basic Maxwellian f_0; this does not necessarily mean that h is small everywhere for $t = 0$, but that, e.g., $\|h\| \ll 1$ (the Maxwellian weighting the norm is f_0/ρ_0) and $f(\mathbf{x}, \boldsymbol{\xi}, 0) = \bar{f}(\mathbf{x}, \boldsymbol{\xi})$ could, e.g., be another Maxwellian, possibly space-dependent, with density, velocity, and temperature parameters slightly different from those contained in f_0; this condition means that the quantities

$$|\bar{\rho} - \rho_0|/\rho_0; \qquad |\bar{T} - T_0|/T_0; \qquad |\bar{\mathbf{v}} - \mathbf{v}_0|/(RT_0) \tag{1.8}$$

(where ρ_0, $\mathbf{v}_0$, and T_0 are the parameters in f_0, and $\bar{\rho}$, $\bar{\mathbf{v}}$, and $\bar{T}$ are those in the initial datum $\bar{f}$) are small with respect to 1. It can be verified that these are actually reasonable conditions for considering two Maxwellians as being close to each other.

The situation is very similar when we examine the boundary conditions; now we have to drop the condition on density, obviously meaningless, but the conditions on temperature and mass velocity are the same as above ($\bar{T}$ and $\bar{v}$ denoting now the wall temperature and velocity). Since this result is not as obvious as for initial conditions, we observe that if we put

$$f = f_0(1 + h) \tag{1.9}$$

then the boundary conditions, Eq. (4.7) of Chapter II, can be rewritten as follows:

$$h(\boldsymbol{\xi}) = h_0(\boldsymbol{\xi}) + \int_{\mathbf{c}'\cdot\mathbf{n}<0} A(\boldsymbol{\xi}' \to \boldsymbol{\xi}; \mathbf{x}) h(\boldsymbol{\xi}')\, d\boldsymbol{\xi}' \tag{1.10}$$

where

$$h_0(\boldsymbol{\xi}) = [f_0(\boldsymbol{\xi})|\mathbf{c}\cdot\mathbf{n}|]^{-1} \int_{\mathbf{c}'\cdot\mathbf{n}<0} B(\boldsymbol{\xi}' \to \boldsymbol{\xi}; \mathbf{x}) f_0(\boldsymbol{\xi}')|\mathbf{c}'\cdot\mathbf{n}|\, d\boldsymbol{\xi}' - 1 \tag{1.11}$$

$$A(\boldsymbol{\xi}' \to \boldsymbol{\xi}; \mathbf{x}) = [f_0(\boldsymbol{\xi})|\mathbf{c}\cdot\mathbf{n}|]^{-1} B(\boldsymbol{\xi}' \to \boldsymbol{\xi}; \mathbf{x}) f_0(\boldsymbol{\xi}')|\mathbf{c}'\cdot\mathbf{n}| \tag{1.12}$$

It is now obvious that h can be small, in some sense, only if h_0 (the source term) is small (in some related sense). This means that if $\bar{f}(\xi; \mathbf{x})(\mathbf{x} \in \partial R)$ is the Maxwellian with the same density as $f_0(\boldsymbol{\xi})$ but velocity and temperature equal to the wall velocity and temperature at $\mathbf{x}$ (in such a way that $\bar{f}$ satisfies the detailed balancing at $\mathbf{x}$), then the difference $1 - (\bar{f}/f_0)$ is small with respect to 1 and determines the order of magnitude of h (at the boundary). This in turn implies that the velocity and temperature of the boundaries (including possible boundaries at infinity) must be considered to be small perturbations, i.e., as anticipated, relative velocities and temperature differences must be small, in the sense that the quantities (1.8) must be small.

Equations (1.10)–(1.12) are exact; it is convenient, however, to take into account a simplification which does not alter the accuracy of a linearized treatment, i.e., to linearize $B(\boldsymbol{\xi}' \to \boldsymbol{\xi}; \mathbf{x})$ with respect to the small parameters $|\bar{T} - T_0|/T_0$ and $|\bar{\mathbf{u}} - \mathbf{u}_0|/(RT_0)$. This linearization procedure can be applied to both Eqs. (1.11) and (1.12); first-order terms are to be retained in the former case (zeroth-order terms cancel because of detailed balancing), while only zeroth-order terms are kept in Eq. (2.20) [because first-order terms in $A(\boldsymbol{\xi}' \to \boldsymbol{\xi}; \mathbf{x})$ become second order when multiplied by h as in Eq. (1.10)]. The zeroth-order approximation to $B(\boldsymbol{\xi}' \to \boldsymbol{\xi}; \mathbf{x})$ is no longer dependent upon $\mathbf{x}$ (for boundary conditions depending only on the local velocity and temperature of the boundary) and satisfies the same relations as the full $B(\boldsymbol{\xi}' \to \boldsymbol{\xi}; \mathbf{x})$ except for the fact that the basic Maxwellian $f_0(\boldsymbol{\xi})$ now replaces the wall Maxwellian; all our previous results about boundary conditions remain true, with obvious modifications, and, in particular, we can construct model boundary conditions for the linearized Boltzmann equation in the same way as we did in Chapter IV for the full equation.

We have found that a necessary condition for linearizing is that the inhomogeneous terms in initial and boundary conditions must be small. In order to find whether this is also a sufficient condition, we have to investigate the initial and boundary value problems for the linearized Boltzmann equation and prove that there is one and only one solution to a given boundary problem and that this solution remains small if the above-mentioned inhomogeneous terms are sufficiently small. We also have to prove that the difference between the nonlinear and the linear equation is small of higher order when the necessary conditions for linearization are met, because this is required for the latter conditions to be sufficient also.

These problems will be our main concern in this chapter; another closely related problem which will also be considered is whether a hierarchy of models capable of approximating the linearized Boltzmann equation with arbitrary accuracy also produces a hierarchy of solutions which approximate the solution of the Boltzmann equation with arbitrary accuracy.

Because of the nature of the material presented in this chapter, we find it convenient to use the "symmetrized" notation $\hat{h}$, $\hat{L}$ mentioned in Chapter III, Section 2 in such a way that the scalar product in $\mathscr{H}$ is not affected by the Maxwellian weight; however, we shall continue to write h, L with the understanding that throughout this chapter h and L mean what before was called $\hat{h}$, $\hat{L}$. No confusion will arise, since we shall not use the old h and L in this chapter.

The assumptions made in Chapter IV, Section 4 on the boundary conditions will be assumed to hold for the boundary value problems to be considered. In particular, Eq. (4.5) of Chapter IV implies

$$\int_{\boldsymbol{\xi}\cdot\mathbf{n}>0} |\boldsymbol{\xi}\cdot\mathbf{n}|\,|Ah|^2\,d\boldsymbol{\xi} \le \int_{\boldsymbol{\xi}\cdot\mathbf{n}<0} |\boldsymbol{\xi}\cdot\mathbf{n}|\,|h|^2\,d\boldsymbol{\xi} \tag{1.13}$$

where

$$Ah = \int_{\boldsymbol{\xi}'\cdot\mathbf{n}<0} [f_0(\boldsymbol{\xi})]^{1/2} A(\boldsymbol{\xi}' \to \boldsymbol{\xi}) [f_0(\boldsymbol{\xi})]^{-1/2} h(\boldsymbol{\xi}')\,d\boldsymbol{\xi}' \tag{1.14}$$

Here $A(\boldsymbol{\xi}' \to \boldsymbol{\xi})$ is the zeroth-order approximation to $A(\boldsymbol{\xi}' \to \boldsymbol{\xi}; \mathbf{x})$ defined above ($\boldsymbol{\xi} \cdot \mathbf{n}$ appears throughout in place of $|\mathbf{c} \cdot \mathbf{n}|$ because of the linearization and the fact that the basic Maxwellian has been assumed to have zero mass velocity). In the notation just introduced the boundary conditions are

$$h = h_0 + Ah \qquad (\boldsymbol{\xi} \cdot \mathbf{n} > 0) \tag{1.15}$$

In the following it will be convenient to use the following norm and scalar product for functions defined along the boundary:

$$\|h\|_B^2 = (h, h)_B \tag{1.16}$$

$$(h, g)_B = \int_{\partial R} dS \int_{\boldsymbol{\xi} \cdot \mathbf{n} > 0} d\boldsymbol{\xi} |\boldsymbol{\xi} \cdot \mathbf{n}| hg \tag{1.17}$$

where dS is a surface element. In this notation Eq. (1.13) becomes (for any function h defined for $\boldsymbol{\xi} \cdot \mathbf{n} < 0$)

$$\|Ah\|_B \le \|Rh\|_B \tag{1.18}$$

where R is the specular reflection operator. As pointed out in Chapter IV, the equality sign in Eq. (1.18) holds only if h is a constant multiple of $f_0^{1/2}$; we shall assume a little more than this, i.e., that if

$$(f_0^{1/2}, h)_B = 0 \tag{1.19}$$

then a constant $\sigma < 1$ exists such that

$$\|(\beta\xi)^{1/2} Ah\|_B \le \sigma \|(\beta\xi)^{1/2} Rh\|_B \qquad (\sigma < 1) \tag{1.20}$$

where $\beta = \beta(\xi)$ is a function of the molecular speed such that $0 < \beta(\xi) < \beta_0(1 + \xi^2)^{-1/2}$. We note that Eq. (1.20) is obviously satisfied by Maxwell's boundary conditions with $\sigma = 1 - \alpha$.

In this chapter we shall restrict our attention to the time-separated linearized Boltzmann equation

$$sh + \boldsymbol{\xi} \cdot \frac{\partial h}{\partial \mathbf{x}} = Lh \qquad (\operatorname{Re} s \ge 0) \tag{1.21}$$

which is obtained from Eq. (1.7) by postulating an exponential dependence in time ($h \propto e^{st}$). Equation (1.21) governs time-independent problems ($s = 0$), problems of steady oscillations ($s = i\omega$, ω real), as well as any initial value problem provided we treat it by means of the Laplace transform with respect to time (the initial datum in general comes in when Laplace-transforming, but since it constitutes an inhomogeneous term, it does not essentially complicate the problem, as noted above). Initial value problems can also be treated directly (without Laplace-transforming) by the methods presented in this chapter, but this treatment will not be considered here for the sake of brevity. (For the case of the initial value problem in infinite domains or finite domains with specularly reflecting boundaries see References.)

2. The Free-Streaming Operator

In this section we shall study the free streaming operator $\boldsymbol{\xi} \cdot \partial/\partial \mathbf{x}$ acting on a function h with the boundary conditions

$$h(\mathbf{x}_0, \boldsymbol{\xi}) = Ah \qquad (\mathbf{x}_0 \in \partial R; \boldsymbol{\xi} \cdot \mathbf{n} > 0) \tag{2.1}$$

where ∂R is the boundary ∂R of a bounded region R, and A is an operator satisfying all the assumptions made in Section 1. We also introduce the auxiliary condition

$$((f_0^{1/2}, h)) = 0 \tag{2.2}$$

where the double brackets denote scalar product in the Hilbert space of square summable functions of both $\mathbf{x}$ and $\boldsymbol{\xi}$, defined for any $\boldsymbol{\xi}$ and $\mathbf{x} \in R$. Equation (2.2) is required to define h uniquely when we invert the free-streaming operator because adding to h a constant multiple of $f_0^{1/2}$ does not alter either $\boldsymbol{\xi} \cdot \partial/\partial \mathbf{x}$ or the fact that h satisfies the boundary conditions.

We want to show that for any function $\beta(\xi)$ of the molecular speed which satisfies $0 < \beta(\xi) < \beta_0(1 + \xi^2)^{-1/2}$ we have

$$\int \left| \boldsymbol{\xi} \cdot \frac{\partial h}{\partial \mathbf{x}} \right|^2 \beta(\xi)\, d\boldsymbol{\xi}\, d\mathbf{x} \geq \frac{\gamma}{d^2} \int \xi^2 \beta(\xi) |h|^2\, d\boldsymbol{\xi}\, d\mathbf{x} \tag{2.3}$$

provided the indicated integrals exist. Here $\boldsymbol{\xi}$ is integrated over the whole velocity space, and $\mathbf{x}$ on the whole region R; d is the chord of maximum length which can be drawn in R and γ is a positive constant (depending upon β and R).

We start with the equation

$$\boldsymbol{\xi} \cdot \partial h/\partial \mathbf{x} = g(\mathbf{x}, \boldsymbol{\xi}) \qquad (\mathbf{x} \in R; \boldsymbol{\xi} \in \Xi) \tag{2.4}$$

where h satisfies the above conditions; in order that h can satisfy the conditions of conservation of the number of molecules at the wall ($\mathbf{v} \cdot \mathbf{n} = 0$ at the wall), g must satisfy

$$\int g(\mathbf{x}, \boldsymbol{\xi}) f_0^{1/2}\, d\boldsymbol{\xi}\, d\mathbf{x} = 0 \tag{2.5}$$

as is seen by integrating Eq. (2.4) multiplied by $f_0^{1/2}$ and using Gauss' lemma.

If we integrate Eq. (2.4) along the characteristic lines of $\boldsymbol{\xi} \cdot \partial/\partial \mathbf{x}$, we have

$$h = g_0(s) + h_0 \tag{2.6}$$

$$g_0(s) = \int_0^s g(\sigma)\, d\sigma \tag{2.7}$$

where s is the parameter varying along a characteristic line; h_0 is the value of h corresponding to $s = 0$ and has to be supplied by the boundary conditions. A shorthand notation has been used, in the sense that $g(s)$ means $g(\mathbf{x}, \boldsymbol{\xi})$ evaluated at a certain point (labeled by s) of a chosen characteristic. If we consider Eq. (2.6) at a point of the boundary for arriving molecules, it reads as follows:

$$h_0(\mathbf{x}_0, \boldsymbol{\xi}) = g_0(\mathbf{x}_0, \boldsymbol{\xi}) + h_0(\tilde{\mathbf{x}}_0, \boldsymbol{\xi}) \qquad (\mathbf{x}_0 \in \partial R;\, \boldsymbol{\xi} \cdot \mathbf{n} < 0) \tag{2.8}$$

where $\tilde{\mathbf{x}}_0$ (conjugate of $\mathbf{x}_0$ with respect to $\boldsymbol{\xi}$) means that $\mathbf{x}_0 - \tilde{\mathbf{x}}_0 = k\boldsymbol{\xi}\ (k > 0)$; if more than one point of the boundary exists satisfying the latter relation, the one corresponding to the minimum value of k is to be chosen (note that $\boldsymbol{\xi} \cdot \mathbf{n} > 0$ at $\tilde{\mathbf{x}}_0$ if $\boldsymbol{\xi} \cdot \mathbf{n} < 0$ for the same $\boldsymbol{\xi}$ at $\mathbf{x}_0$).

Equation (2.8) can now be written, interchanging the roles of $\mathbf{x}_0$ and $\tilde{\mathbf{x}}_0$ (obviously $\tilde{\tilde{\mathbf{x}}}_0 = \mathbf{x}_0$),

$$h_0(\mathbf{x}_0, \boldsymbol{\xi}) = -g_0(\tilde{\mathbf{x}}_0, \boldsymbol{\xi}) + h_0(\tilde{\mathbf{x}}_0, \boldsymbol{\xi}) \qquad (\mathbf{x}_0 \in \partial R;\, \boldsymbol{\xi} \cdot \mathbf{n} > 0) \tag{2.9}$$

i.e.,

$$h_0 = -g_0 + \tilde{A}h_0 \tag{2.10}$$

where

$$\tilde{A}h_0 = A\tilde{h}_0; \qquad \tilde{h}_0 = h_0(\tilde{\mathbf{x}}_0, \boldsymbol{\xi}) \tag{2.11}$$

Equation (2.10) is an equation for h_0; it relates the values of h_0 at different points of ∂R for different values of $\boldsymbol{\xi}(\boldsymbol{\xi} \cdot \mathbf{n} > 0)$. Once we have solved Eq. (2.10) we can put the result into Eq. (2.7), thus obtaining the solution h of Eq. (2.4) in terms of g. In order to discuss the solution of Eq. (2.10), we split h_0 and g_0 as follows:

$$\begin{aligned} h_0(\mathbf{x}_0, \boldsymbol{\xi}) &= \mu(\mathbf{x}_0) f_0^{1/2} + \bar{h}(\mathbf{x}_0, \boldsymbol{\xi}) \\ g_0(\mathbf{x}_0, \boldsymbol{\xi}) &= \nu(\mathbf{x}_0) f_0^{1/2} - \bar{g}(\mathbf{x}_0, \boldsymbol{\xi}) \end{aligned} \tag{2.12}$$

where

$$\int_{\boldsymbol{\xi}\cdot\mathbf{n}>0} Y(\mathbf{x}_0, \boldsymbol{\xi}) |\boldsymbol{\xi} \cdot \mathbf{n}| [f_0(\boldsymbol{\xi})]^{1/2}\, d\boldsymbol{\xi} = 0 \qquad (\forall x_0 \in \partial R) \tag{2.13}$$

is satisfied by both $Y = \bar{h}$ and $Y = \bar{g}$. As a consequence,

$$(\bar{h}, f_0^{1/2})_B = (\bar{g}, f_0^{1/2})_B = 0 \tag{2.14}$$

It is obvious that the splittings (2.12) are always possible by a proper choice of μ and ν. If $\bar{h}$ satisfies Eq. (2.13), then $\tilde{A}\bar{h}$ also does; accordingly, Eq. (2.10) splits as follows:

$$\mu(\mathbf{x}_0) = -\nu(\mathbf{x}_0) + (1/\pi) \int_{\mathbf{\Omega}'\cdot\mathbf{n}<0} |\mathbf{\Omega}'\cdot\mathbf{n}|\mu(\tilde{\mathbf{x}}_0)\, d\mathbf{\Omega}' \tag{2.15}$$

$$\bar{h} = -\bar{g} + \tilde{A}\bar{h} \tag{2.16}$$

where $\mathbf{\Omega} = \boldsymbol{\xi}/\xi = (\mathbf{x}_0 - \tilde{\mathbf{x}}_0)/(|\mathbf{x}_0 - \tilde{\mathbf{x}}_0|)$.

According to our assumptions, Eq. (1.20) holds whenever Eq. (1.19) is satisfied; this implies

$$\|(\beta\xi)^{1/2}\tilde{A}\bar{h}\|_B < \sigma\|(\beta\xi)^{1/2}\bar{h}\|_B \qquad (\sigma < 1) \tag{2.17}$$

provided we take into account that $|\boldsymbol{\xi}\cdot\mathbf{n}|\, d\boldsymbol{\xi}\, ds$ is invariant with respect to specular reflection or conjugation of boundary points with respect to $\boldsymbol{\xi}$. The contracting mapping theorem then ensures that a unique solution $\bar{h}$ of Eq. (2.16) exists which satisfies

$$\|(\beta\xi)^{1/2}h\|_B < \frac{1}{1-\sigma}\|(\beta\xi)^{1/2}\bar{g}\|_B \tag{2.18}$$

On the other hand, Eq. (2.15) can be written

$$\mu(\mathbf{x}_0) = -\nu(\mathbf{x}_0) + \frac{1}{\pi}\int_{\partial R(x_0)} \frac{|(\mathbf{x}_0 - \mathbf{x}_0')\cdot\mathbf{n}|\,|(\mathbf{x}_0 - \mathbf{x}_0')\cdot\mathbf{n}'|}{|\mathbf{x}_0 - \mathbf{x}_0'|^4}\mu(x_0')\, dS' \tag{2.19}$$

where we have written $\mathbf{x}_0'$ in place of $\tilde{\mathbf{x}}_0$ (the fact that $\tilde{\mathbf{x}}_0$ is the conjugate of $\mathbf{x}_0$ is no longer necessary) and used $\mathbf{x}_0 - \mathbf{x}_0' = \mathbf{\Omega}|\mathbf{x}_0 - \mathbf{x}_0'|$, $d\mathbf{\Omega} = \cos(\mathbf{x}_0' - \mathbf{x}_0, n')\, dS'$, n' and dS' being the normal and the surface element, respectively at $\mathbf{x}_0'$. In (2.19) $\partial R(\mathbf{x}_0)$ is the part of ∂R which is seen from $\mathbf{x}_0'$; therefore it coincides with the whole boundary ∂R for a simply connected region with nonconvex boundary.

We note that the integral equation (2.19) has a symmetric square integrable kernel for any domain with bonded curvature. This is easily seen provided that we take into account that both $\mathbf{x}_0'$ and $\mathbf{x}_0$ lie on the boundary ∂R, and therefore when $\mathbf{x}_0' \to \mathbf{x}_0$ the singularity is much milder than would appear at a first glance. We observe now that μ = const is the only solution of Eq. (2.19) for $\nu = 0$ (this follows from the fact that $\int_{\mathbf{\Omega}'\cdot\mathbf{n}<0} |\mathbf{\Omega}'\cdot\mathbf{n}|\, d\mathbf{\Omega}' = \pi$ and Schwarz's inequality); therefore a necessary condition in order to have a square integrable solution for a given square integrable source term is that $\nu(\mathbf{x}_0)$ must be orthogonal to 1, i.e., $\int \nu(\mathbf{x}_0)\, dS = 0$. Because of the definition of ν, Eq. (2.12), the definition of g_0, Eq. (2.7), and Eq. (2.5), $\nu(\mathbf{x}_0)$ is automatically orthogonal to 1 (note that $|\boldsymbol{\xi}\cdot\mathbf{n}|\, ds\, dS = d^3\mathbf{x}$). The solution $\mu(\mathbf{x}_0)$ is now determined up to an additive constant, but the latter is fixed by Eq. (2.2). In conclusion, we have proved that both $\mu(\mathbf{x}_0)$ and $\bar{h}(\mathbf{x}_0, \boldsymbol{\xi})$ exist, and are uniquely determined; hence $h_0(\mathbf{x}_0, \boldsymbol{\xi})$ also exists and is uniquely determined. We have also proved that

$$\|(\xi\beta)^{1/2}h_0\|_B \leq k\|(\xi\beta)^{1/2}g_0\|_B \tag{2.20}$$

In fact, this results from Eq. (2.18) and the complete continuity of the kernel of Eq. (2.19) (which implies $\int|\mu^2|\,dS < k_1\int|\nu|^2\,dS$).

Equation (2.6) now gives

$$\|\beta^{1/2}\xi h\|^2 \leq \|\beta^{1/2}\xi g_0\|^2 + \|\beta^{1/2}\xi h_0\|^2 \tag{2.21}$$

where the norms refer to the Hilbert space of the square summable functions of both $\mathbf{x}$ and $\boldsymbol{\xi}$ ($\mathbf{x} \in R$). Using the definition of g_0, Eq. (2.7), and Schwarz's inequality, we obtain

$$\begin{aligned}\|\beta^{1/2}\xi h\|^2 &= \int \beta(\xi)\xi^2\left|\int_0^{s(\mathbf{x},\boldsymbol{\xi})} g(\sigma)\,d\sigma\right|^2 d\boldsymbol{\xi}\,d\mathbf{x}\\ &\leq \int \beta(\boldsymbol{\xi})\xi^2|s(\mathbf{x},\boldsymbol{\xi})| \int_0^{s(\mathbf{x},\boldsymbol{\xi})} |g(\sigma)|^2\,d\sigma\,d\boldsymbol{\xi}\,d\mathbf{x}\end{aligned} \tag{2.22}$$

Now $|s(\mathbf{x},\boldsymbol{\xi})| \leq d/\xi$, where d is the maximum chord in R; if we use this fact, put $t = \xi\sigma$, and interchange the order of integrations, we have

$$\begin{aligned}\|\beta^{1/2}\xi h\|^2 &\leq d\int_0^d dt \int \beta(\xi)\,|g(\mathbf{x},\boldsymbol{\xi})|^2\,d\boldsymbol{\xi}\,d\mathbf{x}\\ &= d^2\int \beta(\xi)|g(\mathbf{x},\xi)|^2\,d\boldsymbol{\xi}\,d\mathbf{x} = d^2\|\beta^{1/2}g\|^2\end{aligned} \tag{2.23}$$

Analogously, we have (using $d^3\mathbf{x} = |\boldsymbol{\xi}\cdot\mathbf{n})\,dS\,d\sigma$ and $|s(\mathbf{x},\boldsymbol{\xi})| \leq d/\xi$)

$$\begin{aligned}\|\beta^{1/2}\xi h_0\|^2 &= \int_{\mathbf{n}\cdot\boldsymbol{\xi}>0} \beta(\xi)\xi^2|h_0(\mathbf{x}_0,\boldsymbol{\xi})|^2|\boldsymbol{\xi}\cdot\mathbf{n}|\,d\boldsymbol{\xi}\,dS\,d\sigma\\ &\leq d\int_{\boldsymbol{\xi}\cdot\mathbf{n}>0} \beta(\xi)\xi|h_0(\mathbf{x}_0,\boldsymbol{\xi})|^2|\boldsymbol{\xi}\cdot\mathbf{n}|\,d\boldsymbol{\xi}\,dS'\\ &= d\|(\beta\xi)^{1/2}h_0\|_B^2 \leq kd\|(\beta\xi)^{1/2}g_0\|_B^2\\ &= kd\int \beta(\xi)\xi\left|\int_0^{s(\mathbf{x}_0,\boldsymbol{\xi})} g(\sigma)\,d\sigma\right|^2|\boldsymbol{\xi}\cdot\mathbf{n}|\,d\boldsymbol{\xi}\,dS\\ &\leq kd^2\int \beta(\xi)|g(\mathbf{x},\boldsymbol{\xi})|^2|\boldsymbol{\xi}\cdot\mathbf{n}|\,d\boldsymbol{\xi}\,dS\,d\sigma\\ &= kd^2\|\beta^{1/2}g\|^2\end{aligned} \tag{2.24}$$

Using Eqs. (2.23) and (2.24) in Eq. (2.21), we have

$$\|\beta^{1/2}\xi h\|^2 \leq d^2(1 + k)\|\beta^{1/2}g\|^2 \tag{2.25}$$

and this is nothing other than Eq. (2.3) provided we recall Eq. (2.4) and put $\gamma = (1 + k)^{-1}$.

3. The Integral Version of the Boltzmann Equation and Its Properties

In order to construct a sound mathematical theory of the boundary value problem, as well as to set up solution procedures in the limit of large Knudsen numbers (see Chapter VIII), it is useful to transform the linearized Boltzmann equation from an integrodifferential to a purely integral form. This can be achieved in many ways, each of which can be convenient for specific purposes. The simplest procedure is to consider Eq. (1.21) and integrate both sides along the characteristics of the differential operator $D = \boldsymbol{\xi} \cdot \partial/\partial\mathbf{x}$, while taking into account the proper boundary conditions; this essentially amounts to constructing the inverse of D under given homogeneous boundary conditions. The Boltzmann equation takes on the form

$$h = h_1 + D^{-1}[Lh - sh] \tag{3.1}$$

where h_1 (the value of h at the boundary) is related to the inhomogeneous term in Eq. (1.15) and is obtained by solving $Dh_1 = 0$ with the boundary conditions given by Eq. (1.15); the discussion is analogous to the one given in Section 2. The result proved in Section 2 shows that ξD^{-1} is bounded in a norm weighted with $\beta(\xi)$; D^{-1} itself, however, is not bounded. This fact has undesirable consequences when one tries to solve Eq. (3.1) by iteration (Knudsen iteration, Chapter VIII); in particular, disastrous results are obtained in one-dimensional problems.

The procedure just sketched, however, is by no means the only one capable of eliminating the space derivatives. One can add, e.g., a term $\mu(\xi)h$ [$\mu(\xi) > 0$, a function of the molecular speed] to both sides of Eq. (1.21) and then construct the inverse $\hat{U}$ of the operator $s + \boldsymbol{\xi} \cdot \partial/\partial\mathbf{x} + \mu(\xi)$ appearing on the left-hand side; again, one has to integrate along the characteristics of $\boldsymbol{\xi} \cdot \partial/\partial\mathbf{x}$ and use the boundary conditions, the only influence of the additional terms being that of introducing certain exponential factors. There is a slight difficulty, however, because we want to apply the inequality (2.3), which applies only to functions satisfying the restriction (2.2). Thus we should consider a slightly modified operator U which is the inverse of $s + \xi \cdot \partial/\partial\mathbf{x} + \mu(\xi) - P_\mu$, where P_μ is a projector defined by $P_\mu h = f_0^{1/2}(\mu + s)[(f_0^{1/2}(\mu + s), h)]/C$, where $C = (f_0^{1/2}(\mu + s), 1)$. The operator U acts on functions satisfying Eq. (2.2) and is given by $Uh = \hat{U}h - f_0^{1/2}(f_0^{1/2}, h)_B/\|f_0^{1/2}\|_B^2$. In order to avoid tedious estimates we shall henceforth assume that the boundary conditions assign h explicitly, i.e., that the operator A in Eq. (1.10) is zero. In this case there is no conservation of the molecules at the boundary, but Eq. (2.3) applies with no restriction. With this understanding, we use the above-mentioned operator U to act on both sides of Eq. (1.21) to obtain

$$h = \bar{h}_0 + UHh \tag{3.2}$$

where

$$H = L + \mu(\xi) \tag{3.3}$$

and $\bar{h}_0$ is constructed by solving

$$\boldsymbol{\xi} \cdot \partial \bar{h}_0/\partial \mathbf{x} + \mu(\xi)\bar{h}_0 = 0 \tag{3.4}$$

with $\bar{h}_0$ satisfying the inhomogeneous boundary conditions, Eq. (1.15).

The essentially arbitrary function $\mu(\xi)$ can be chosen in many ways; its positivity guarantees that the exponentials built up in U control the growth induced by the factor $1/\xi$ and its powers in the neighborhood of $\xi = 0$, as is required for iterative methods to be meaningful. Since we shall consider only the cases when L can be split according to

$$Lh = Kh - \nu(\xi)h \tag{3.5}$$

a reasonable choice is $\mu(\xi) = \nu(\xi)$; in such a case $H = K$. This is in fact the choice usually made. We shall, however, find it useful to choose μ in such a way that H is a positive operator:

$$(h, Hh) > 0 \tag{3.6}$$

This is always possible according to a theorem proved in Chapter III, Section 4 provided that we take $H = K + \lambda\nu$, λ being the constant in Eq. (4.6). It is also probable that one can take $H = K(\lambda = 0)$ and still have Eq. (3.6) satisfied, since the positivity of K has been proved for rigid spheres (Finkelstein, ref. 3); in the absence of more detailed information, however, we stick to the more general description based on H rather than on K. We note that in any case Eq. (3.3) and the nonpositivity of L imply

$$0 < (h, Hh) \leq (h, \mu h) \tag{3.7}$$

Now, we want to prove the following:

Lemma I. A positive numerical constant η (depending upon the shape of the boundary) can be found such that, if

$$\rho(\xi) = \{[\mu(\xi) + \operatorname{Re} s]^2 + \eta^2\xi^2/d^2\}^{1/2} \tag{3.8}$$

(where d is the maximum chord in R, as in Section 2), then

$$((\rho U g, U g)) \leq ((\rho^{-1} g, g)) \tag{3.9}$$

where the double brackets have the same meaning as in Section 2.

In order to show that inequality (3.9) holds, we note that if we put

$$U g = h \tag{3.10}$$

then Eq. (3.9) becomes

$$((\rho h, h)) \leq \left(\left(\rho^{-1}\left\{\boldsymbol{\xi} \cdot \frac{\partial h}{\partial \mathbf{x}} + [\mu(\xi) + s]h\right\}, \left\{\boldsymbol{\xi} \cdot \frac{\partial h}{\partial \mathbf{x}} + [\mu(\xi) + s]h\right\}\right)\right)$$
$$\equiv \text{RHS} \tag{3.11}$$

where h satisfies the homogeneous boundary conditions, Eq. (2.1). The right-hand side of Eq. (3.11) can now be written as follows:

$$\text{RHS} = \int \rho^{-1}\left|\boldsymbol{\xi} \cdot \frac{\partial h}{\partial \mathbf{x}}\right|^2 d\boldsymbol{\xi}\, d\mathbf{x} + 2\int \rho^{-1}[\mu(\xi) + \operatorname{Re} s] h \boldsymbol{\xi} \cdot \frac{\partial h}{\partial \mathbf{x}} d\boldsymbol{\xi}\, d\mathbf{x}$$
$$+ \int |\mu(\xi) + s|^2 \rho^{-1} |h|^2 \, d\boldsymbol{\xi}\, d\mathbf{x} \tag{3.12}$$

$$\int \rho^{-1}[\mu(\xi) + \operatorname{Re} s] h \boldsymbol{\xi} \cdot \frac{\partial h}{\partial \mathbf{x}} d\boldsymbol{\xi}\, d\mathbf{x}$$
$$= \int h \boldsymbol{\xi} \cdot \frac{\partial h}{\partial \mathbf{x}} d\boldsymbol{\xi}\, d\mathbf{x} + \int [\mu(\xi) + \operatorname{Re} s - \rho]\rho^{-1} h \boldsymbol{\xi} \cdot \frac{\partial h}{\partial \mathbf{x}} d\boldsymbol{\xi}\, d\mathbf{x} \tag{3.13}$$

The last term is, in absolute value, smaller than

$$\frac{\eta}{d}((\xi^2 \rho^{-1} h, h))^{1/2}\left(\left(\rho^{-1}\boldsymbol{\xi} \cdot \frac{\partial h}{\partial \mathbf{x}}, \boldsymbol{\xi} \cdot \frac{\partial h}{\partial \mathbf{x}}\right)\right) \tag{3.14}$$

where η is the constant which appears in ρ and has to be suitably determined. Therefore if we apply Gauss's lemma to the first term on the right-hand side of Eq. (3.13) and use Eqs. (2.1) and (1.13), we obtain

$$\text{RHS} \geq \left\{\left(\left(\rho^{-1}\boldsymbol{\xi} \cdot \frac{\partial h}{\partial \mathbf{x}}, \boldsymbol{\xi} \cdot \frac{\partial h}{\partial \mathbf{x}}\right)\right)^{1/2} - 2\eta d^{-1}((\xi^2 \rho^{-1} h, h))^{1/2}\right\}$$
$$\times \left(\left(\rho^{-1}\boldsymbol{\xi} \cdot \frac{\partial h}{\partial \mathbf{x}}, \boldsymbol{\xi} \cdot \frac{\partial h}{\partial \mathbf{x}}\right)\right)^{1/2} + ((\rho^{-1}(\mu + \operatorname{Re} s)^2 h, h)) \tag{3.15}$$

Consider now a closed neighborhood $0 \leq \eta \leq \eta_0$, where η_0 is some arbitrarily fixed number. Then for any η in this neighborhood the corresponding ρ satisfies Eq. (2.3) with $\beta = \rho^{-1}$ and some $\gamma = \gamma(\eta) > 0$. Since γ

depends continuously on η, it will reach a minimum value γ_0 which is also positive. We take for η the smallest of the two numbers $\gamma_0/4$ and η_0. Then we can apply Eq. (2.1) twice and obtain

$$\begin{aligned}\text{RHS} &\geq \gamma_0(2d)^{-1}((\xi^2\rho^{-1}h, h)) + ([\rho^{-1}(\mu + \operatorname{Re} s)^2 h, h]) \\ &= ((\{\mu + \operatorname{Re} s)^2 + \gamma_0(2d)^{-1}\xi^2\}\rho^{-1}h, h)) \geq ((\rho h, h))\end{aligned} \tag{3.16}$$

as was to be shown.

We prove also the following:

Lemma II. For hard potentials with either a radial or an angular cutoff a constant $\alpha < 1$ exists such that

$$(\rho^{-1}Hh, Hh) \leq \alpha^2(\rho h, h) \qquad (\rho^{1/2}h \in \mathscr{H};\ \alpha^2 < 1) \tag{3.17}$$

where $H = K + \lambda\nu\,(0 \leq \lambda \leq 3)$ is the operator defined by Eq. (3.3) and ρ the function defined by Eq. (3.8).

To prove this lemma, we first observe that

$$\begin{aligned}\frac{(\rho^{-1}Hh, Hh)}{(\rho h, h)} &= \frac{\|J(\rho^{1/2}h)\|^2}{\|\rho^{1/2}h\|^2} \leq \max_{\rho^{1/2}h \in \mathscr{H}} \left|\frac{(\rho^{1/2}h, J(\rho^{1/2}h))}{\|\rho^{1/2}h\|^2}\right|^2 \\ &= \max_{\rho^{1/2}h \in \mathscr{H}} \left|\frac{(h, Hh)}{(h, \rho h)}\right|^2\end{aligned} \tag{3.18}$$

where $J = \rho^{-1/2}H\rho^{-1/2}$ is a self-adjoint operator in $\mathscr{H}$. Due to Eq. (3.18) it is sufficient to show that a constant α exists such that

$$(h, Kh) + \lambda(h, \nu h) \geq \alpha(h, \rho h) \qquad (\alpha < 1;\ \rho^{1/2}h \in \mathscr{H}) \tag{3.19}$$

We have

$$\begin{aligned}(h, Kh) + \lambda(h, \nu h) &\leq (h, Lh) + (\lambda + 1)(h, \nu h) \\ &\leq -\mu_0\|(I - P)h\|^2 + (\lambda + 1)(h, \nu h)\end{aligned} \tag{3.20}$$

where P is the projector onto the five-dimensional subspace $\mathscr{F}$, and μ_0 is the constant appearing in Eq. (3.13) of Chapter III (the latter equation has been proved to also hold for operators with a radial cutoff; see Chapter III, Section 4). In order to prove Eq. (3.19) we simply have to show that the least upper bound of the spectrum of the self-adjoint operator

$$A = \rho^{-1}(\lambda + 1)\nu - \mu_0(\rho^{-1} - \rho^{-1/2}P\rho^{-1/2}) \tag{3.21}$$

is smaller than 1. Let us find an upper bound for the essential spectrum first.

This is the same as the essential spectrum of the operator B consisting of the multiplication by $b(\xi) = [(\lambda + 1)\nu - \mu_0]\rho^{-1}$, since $\rho^{-1/2}P\rho^{-1/2}$ is completely continuous. The operator B has a purely continuous spectrum composed of the values of $b(\xi)$. These values are clearly smaller than 1 for any finite ξ $[(\lambda + 1)\nu/\rho < 1$ and $\mu_0 > 0]$. Also, as $\xi \to 0$, $b(\xi) \to 1 - \mu_0[(\lambda + 1)\nu(0)]^{-1} < 1$, and as $\xi \to \infty$, $b(\xi) \to \{1 + \eta^2 k^2/d^2\}^{-1} < 1$, where $k^2 = \lim_{\xi\to\infty}[\xi/\nu(\xi)]^2(\lambda + 1)^{-1}$ [for angular cutoff, $k = \infty$ and $b(\xi) \to 0$]. The essential spectrum of A is therefore bounded by a constant smaller than 1. In order to show that this is also true for the discrete spectrum, let us consider the equation satisfied by the eigenfunctions of A corresponding to the general eigenvalue α_i:

$$\rho^{-1}(\lambda + 1)\nu\varphi_i - \mu_0(\rho^{-1} - \rho^{-1/2}P\rho^{-1/2})\varphi_i = \alpha_i\varphi_i \tag{3.22}$$

If we take the scalar products of φ_i times each side of Eq. (3.22) and take into account the positive nature of $I - P$, we obtain

$$\begin{aligned}\alpha_i(\varphi_i, \varphi_i) &= (\lambda + 1)(\varphi_i, \rho^{-1}\nu\varphi_i) - \mu_0(\rho^{-1/2}\varphi_i, (I - P)\rho^{-1/2}\varphi_i)\\ &\leq (\lambda + 1)(\varphi_i, \rho^{-1}\nu\varphi_i)\end{aligned} \tag{3.23}$$

i.e.,

$$\alpha_i \leq (\varphi_i, (\lambda + 1)\nu\rho^{-1}\varphi)/\|\varphi_i\|^2 \tag{3.24}$$

Now, in any set of nonzero measure $(\lambda + 1)\nu\rho^{-1} < 1$; accordingly,

$$\alpha_i < 1 \tag{3.25}$$

and the equality sign does not apply in this case because φ_i must be different from zero in a nonzero measure set. Thus Eq. (3.19) is proved and the lemma follows.

4. Existence and Uniqueness of the Solution for Linearized and Weakly Nonlinear Boundary Value Problems

We now present a proof of the following:

Theorem I. The integral equation form of the Boltzmann equation, Eq. (3.2), has for a bounded domain R, a unique solution in the Hilbert space $\mathscr{H}$ of the functions which are square integrable with weight $\rho(\xi)$ (with respect to both the space and velocity variables), provided the source term $\bar{h}_0$ also belongs to $\mathscr{H}$. This unique solution can be obtained in principle by a convergent iteration procedure.

This theorem is an immediate and obvious consequence of the contraction mapping theorem and the results of Section 3. Using Lemma I and

Lemma II, we have

$$((\rho UHh, UHh)) \leq ((\rho^{-1}Hh, Hh)) \leq \alpha^2((\rho h, h)) \tag{4.1}$$

i.e., introducing the norm of $\mathscr{H}$, $\|\quad\|_{\mathscr{H}}$,

$$\|UHh\|_{\mathscr{H}}^2 \leq \alpha^2 \|h\|_{\mathscr{H}}^2 \tag{4.2}$$

This means that UH is a contracting operator in $\mathscr{H}$, and the theorem follows.

Theorem I can be considered as a starting point for the construction of a rigorous theory for the boundary value problem, since it allows us to talk about a solution which has been shown to exist and to be unique. We note, however, that h has been shown to exist as a square integrable function of both $\mathbf{x}$ and $\boldsymbol{\xi}$, and, accordingly, we do not know anything about the smoothness properties of the solution; in particular, we do not know whether h possesses space derivatives almost everywhere, in such a way that not only the integral form, Eq. (3.2), but also the original integrodifferential form, Eq. (1.21), of the linearized Boltzmann equation is satisfied. A result which is rather easily shown is that the directional derivative $\boldsymbol{\xi} \cdot \partial h/\partial \mathbf{x}$ exists, and this implies that the original integrodifferential equation is satisfied, at least in a generalized sense.

In order to obtain stronger results, it is necessary to show that new assumptions on $\bar{h}_0$ imply new properties for h; in particular, that $\bar{h}_0 \in \mathscr{H}'(\subset \mathscr{H})$ implies $h \in \mathscr{H}'$, where $\mathscr{H}'$ is (in general) a Banach space whose elements form a subset of $\mathscr{H}$. Research along this line was begun by Pao (ref. 4) by using the function spaces $\mathscr{V}$ and $\mathscr{E}$ based on the following norms:

$$\|h\|_{\mathscr{V}}^2 = \max_{\mathbf{x}} \int d\boldsymbol{\xi}\, \rho(\boldsymbol{\xi}) h^2(\mathbf{x}, \boldsymbol{\xi}) \tag{4.3}$$

$$\|h\|_{\mathscr{E}}^2 = \max_{\mathbf{x},\boldsymbol{\xi}} (1 + \xi^2)|h(\mathbf{x}, \boldsymbol{\xi})| \tag{4.4}$$

Pao considered only the steady, one-dimensional case with special boundary conditions; he started from an early version (Cercignani, ref. 2) of Theorem I valid in the case just mentioned, and considered only collision operators with an angular cutoff. With these restrictions he was able to prove the following results [where $\bar{h}_1$ denotes the solution of Eq. (3.4) for $\mu(c) = \nu(c)$ (or $\lambda = 0$) and $\bar{h}_0$ the source term in Eq. (1.15)]:

Theorem II. If $\bar{h}_1 \in \mathscr{V}$, then the linearized Boltzmann equation has a unique solution $h \in \mathscr{V}$, and $\|h\|_{\mathscr{V}} < \beta \|\bar{h}_1\|_{\mathscr{V}}$, where β is a constant.

Theorem III. If $\bar{h}_1 \in \mathscr{E}$, then the linearized Boltzmann equation has a unique solution $h \in \mathscr{E}$, and $\|h\|_{\mathscr{E}} > m \|\bar{h}_1\|_{\mathscr{E}}$.

Theorem IV. If $h_0 \in \mathscr{E}$ (or $h_0 \in \mathscr{V}$), then all the moments of $f_0^{1/2}$ are continuous in x.

Theorem V. There exists a $\tau > 0$ such that the boundary value problem for the nonlinear Boltzmann equation has a unique solution $f = f_0 + f_0^{1/2}h^*$, with $h^* \in \mathscr{E}$, for all $\bar{h}_1$ which satisfy $\|\bar{h}_1\|_{\mathscr{E}} < \tau$. Moreover, if h is the solution of the linearized Boltzmann equation with the same $\bar{h}_1$, then, as $\|\bar{h}_1\|_{\mathscr{E}} \to 0$, we have

$$\|h - h^*\|_{\mathscr{E}}/\|h\| \to 0 \tag{4.5}$$

The latter theorem gives a rigorous justification for the use of the linearized Boltzmann equation.

Further results on the boundary value problem were considered by Guiraud, who studied both steady and unsteady problems in a compact domain.

5. Convergence of the Solutions of Kinetic Models

The existence and uniqueness theorem given in Section 4 (Theorem I) shows that a solution does exist for any (nonzero) Knudsen number based on the chord of maximum length. The theorem is also constructive in nature, because one can, in principle, write the solution in the form of a series. The complicated form of the operators U and H, however, makes this procedure hopeless from a practical standpoint (unless the Knudsen number is very large). From the point of view of finding a solution, one has to rely upon the use of model equations which approximate the collision operator L by a simpler operator L_N sucht that the equations can be satisfactorily handled by analytical or numerical procedures.

In order to establish a sound connection between the present theory and the practical procedures based on kinetic models, it is important to show that if $L_N \to L$ ($N \to \infty$) in some sense, then in some related sense the sequence of the solutions h_N of the model equations converges to the solution h of the actual (linearized) Boltzmann equation (the existence and uniqueness of h_N is ensured by the fact that the theory developed in the previous sections also applies to model equations). The above-mentioned property ($h_N \to h$ if $L_N \to L$) is made precise by the following:

Theorem VI. Let L be a collision operator which can be split as in Eq. (3.3), L_N be a sequence of operators which can be similarly split with the same $\mu(\xi)$ as L, and

$$\|\rho^{-1/2}(L_N - L)\rho^{-1/2}\| \to 0 \quad \text{as} \quad N \to \infty \tag{5.1}$$

where ρ is given by Eq. (3.8) with appropriate μ and η (the norm appearing in Eq. (5.1) is the operator norm in $\mathscr{H}$]. Then the solution h_N of the integral version of the model equation corresponding to L_N tends (in the $\mathscr{H}$ norm) to the solution h of the analogous equation corresponding to L and the same boundary conditions.

In order to prove this theorem, we consider the integral versions of both the model and the actual equation:

$$h_N = \bar{h}_0 + UH_N h_N \tag{5.2}$$

$$h = \bar{h}_0 + UHh \tag{5.3}$$

We note that $\bar{h}_0$ and U are the same in both equations, and both h and h_N exist and are uniquely determined because of the existence and uniqueness theorem proved in Section 3. Now, if we set

$$h = h_N + r_N \tag{5.4}$$

we obtain from Eqs. (5.2) and (5.3)

$$r_N = U(H - H_N)h + UH_N r_N \tag{5.5}$$

Applying Eq. (4.2) to UH_N, we have

$$\|UH_N r_N\|^2 \leq \alpha_N^2 \|r_N\|^2 \tag{5.6}$$

Because of Eq. (5.1), which implies

$$\epsilon_N \equiv \|\rho^{-1/2}(H_N - H)\rho^{-1/2}\| \to 0 \qquad (N \to \infty) \tag{5.7}$$

We can choose α_N in such a way that $\alpha_N \to \alpha$ as $N \to \infty$, where α is the constant in Eq. (4.2). Since infinity is the only accumulation point of the sequence of the integers, an $\alpha_0 < 1$ exists such that $\alpha_N \leq \alpha_0$ for any N. Accordingly,

$$\|UH_N r_N\|^2 \leq \alpha_0^2 \|r_N\|^2 \qquad (\alpha_0 < 1) \tag{5.8}$$

and one can apply the contraction mapping theorem to Eq. (5.5) and deduce

$$\begin{aligned} \|r_N\|^2 &\leq \frac{1}{1 - \alpha_0^2}((\rho U(H - H_N)h, U(H - H_N)h)) \\ &\leq \frac{1}{1 - \alpha_0^2}([\rho^{-1}(H - H_N)h, (H - H_N)h]) \\ &\leq \frac{\epsilon_N^2}{1 - \alpha_0^2}((\rho h, h)) = \frac{\epsilon_N^2}{1 - \alpha_0^2}\|h\|^2 \end{aligned} \tag{5.9}$$

where Eq. (3.9) has been used. But $\epsilon_N \to 0$ according to Eq. (5.7); hence $\|r_N\|^2 \to 0$ as $N \to \infty$, as was to be shown.

6. Unbounded Regions and External Flows

The theory developed in the previous sections holds when the region R is bounded. It can be extended, however, to unbounded regions with a certain degree of symmetry, as, e.g., the region between parallel plates or within a cylinder having a finite cross section; in these cases there are chords of arbitrary length within the region, but the solution depends on one or two space variables alone, and the chords parallel to the relevant axis or the relevant plane have bounded length. In order to make this extension, it is necessary to introduce a slight modification of the above treatment. Lemma I remains true provided that in the term $\eta^2\xi^2/d^2$ of Eq. (3.8) and related equations, ξ is replaced by the absolute value of the projection of $\boldsymbol{\xi}$ onto the relevant axis or plane. Lemma II then remains true for rigid spheres and angular cutoff potentials, but the proof must be modified and the complete continuity of $\nu^{-1/2}K\nu^{-1/2}$ used explicitly. Since we do not know if the latter property holds for potentials with finite range (in the absence of angular cutoff), further changes are required for such potentials; to be precise, we can replace $\mu(\xi)$ by some multiple of a power of $\nu(\xi)$ in such a way that $\mu^{-1/2}K\mu^{-1/2}$ is completely continuous (see Chapter III, Section 4) and the remaining properties are retained. With these modifications the two basic lemmas remain true and the basic theorems I and VI follow.

The situation is basically different, however, for external flows, i.e., when the region R extends to infinity outside its inner boundary ∂R; the latter coincides with the surface of a solid body (or surfaces of a certain number of bodies) and can be partly concave and partly convex. We shall prove a lemma which shows that the property of square integrability can be too strong a requirement for the solution of the linearized Boltzmann equation for external problems.

Lemma III. Let h satisfy the steady, linearized Boltzmann equation [Eq. (1.21) with $s = 0$] in a region R, outside a certain boundary ∂R, along which boundary conditions of the usual type, Eq. (1.15), hold. Let the body have a fixed temperature T_0, and let h be the perturbation of a Maxwellian f_0 with zero mass velocity, temperature T_0, and density ρ_∞ equal to the density of the gas at infinity. If we define

$$\boldsymbol{\sigma} = \tfrac{1}{2}\int \boldsymbol{\xi} h^2\, d\boldsymbol{\xi} \tag{6.1}$$

and assume that

$$\int_\Sigma \boldsymbol{\sigma}\cdot\mathbf{n}\, dS \to 0 \tag{6.2}$$

as the distance of the surface Σ from ∂R tends to infinity, then h is a constant multiple of $f_0^{1/2}$.

In order to prove this lemma, we first observe that because of the choice of the Maxwellian, h_0 [given by Eq. (1.11)] is zero thanks to detailed balancing, and the boundary conditions are homogeneous. If we multiply Eq. (1.21) (with $s = 0$) by h and integrate with respect to $\boldsymbol{\xi}$, Eq. (6.1) gives

$$\frac{\partial}{\partial \mathbf{x}} \cdot \boldsymbol{\sigma} = \int hLh\, d\boldsymbol{\xi} \tag{6.3}$$

Because of the nonpositive character of L, it follows that

$$\frac{\partial}{\partial \mathbf{x}} \cdot \boldsymbol{\sigma} \leq 0 \tag{6.4}$$

Let us now integrate Eq. (6.4) over the region between Σ_1 and Σ_2 such that Σ_2 is interior to Σ_1 and exterior to ∂R and does not intersect Σ_1; Σ_2 is also required to be simply connected. We obtain

$$\int_{\Sigma_1} \boldsymbol{\sigma} \cdot \mathbf{n}\, dS \leq \int_{\Sigma_2} \boldsymbol{\sigma} \cdot \mathbf{n}\, dS \tag{6.5}$$

Using this equation twice and taking into account Eq. (6.2) gives

$$0 \leq \int_{\Sigma} \boldsymbol{\sigma} \cdot \mathbf{n}\, dS \leq \int_{\partial R} \boldsymbol{\sigma} \cdot \mathbf{n}\, dS \tag{6.6}$$

where Σ is now any simply connected surface without points inside the volume bounded by ∂R. But Eq. (6.1) and the boundary conditions, Eq. (1.15) with $h_0 = 0$, together with Eq. (1.13), give

$$\boldsymbol{\sigma} \cdot \mathbf{n} = \tfrac{1}{2} \int_{\boldsymbol{\xi}\cdot\mathbf{n}>0} |\boldsymbol{\xi} \cdot \mathbf{n}||h|^2\, d\boldsymbol{\xi} - \tfrac{1}{2} \int_{\boldsymbol{\xi}\cdot\mathbf{n}<0} |\boldsymbol{\xi} \cdot \mathbf{n}||h|^2\, d\boldsymbol{\xi} \leq 0 \qquad (\mathbf{x} \in \partial R) \tag{6.7}$$

and

$$\int_{\partial R} \boldsymbol{\sigma} \cdot \mathbf{n}\, dS \leq 0 \tag{6.8}$$

Therefore

$$\int_{\Sigma} \boldsymbol{\sigma} \cdot \mathbf{n}\, dS = \int_{\partial R} \boldsymbol{\sigma} \cdot \mathbf{n}\, dS = 0 \tag{6.9}$$

Because of the arbitrariness of Σ, Eq. (6.9) also implies

$$\frac{\partial}{\partial \mathbf{x}} \cdot \boldsymbol{\sigma} = 0 \qquad (\mathbf{x} \in R) \tag{6.10}$$

and, because of Eq. (6.3),

$$\int hLh\, d\boldsymbol{\xi} = 0 \tag{6.11}$$

i.e.,

$$h = (A + \mathbf{B} \cdot \boldsymbol{\xi} + C\xi^2) f_0^{1/2} \tag{6.12}$$

where A, $\mathbf{B}$, and C can depend upon $\mathbf{x}$.

Equation (6.12) can satisfy both the Boltzmann equation and the boundary conditions only if $\mathbf{B} = 0$ and $C = 0$. As a matter of fact, Eq. (6.9) implies that the equality sign must hold in Eq. (6.7). Because of the properties of the operator A appearing in Eq. (1.15), the equality sign in Eq. (6.7) implies $\mathbf{B} = C = 0$ at the boundary. Substituting Eq. (6.12) into the linearized Boltzmann equation gives A = const, C = const, and $\mathbf{B} = \mathbf{a} + \mathbf{b} \times \mathbf{x}$, where $\mathbf{a}$ and $\mathbf{b}$ are constant vectors; this general form of $\mathbf{B}$ and C together with their vanishing at the boundary implies their vanishing everywhere (note that $\mathbf{a} + \mathbf{b} \times \mathbf{x}$ cannot be zero on a whole surface unless $\mathbf{a} = \mathbf{b} = 0$). Hence h is a constant multiple of $f_0^{1/2}$, as was to be shown.

If all the assumptions made in Lemma III were to be satisfied in actual flows, a very unpleasant consequence would result: the only possible situation described by the linearized Boltzmann equation would be a state of rest for gas surrounding a solid body. We must now discover whether this result can be avoided if we drop the only nonobvious assumption, i.e., Eq. (6.2). The physical meaning of this equation is that there is not a source or a sink of entropy at infinity in space ($-\boldsymbol{\sigma}$ is the relevant entropy flux vector, see Chapter III, Section 2 and Chapter II, Section 2) and Lemma III obviously holds even if we allow a positive source of entropy at infinity. If, however, we allow an entropy sink, Lemma III does not apply. The relaxing of the assumption embodied in Eq. (6.2) is somewhat disturbing, because we expect that the region at infinity has nothing to do with the irreversible processes connected with the flow of a gas past a solid body. However, if the linearization is not uniformly valid, the "region at infinity" in the linearized treatment is not the same thing as the physical "region at infinity." We come therefore to the basic question of whether or not the linearization about a Maxwellian corresponding to a zero mass velocity with respect to the body demands strict validity everywhere. The answer is obviously negative because if we go sufficiently far from the body, the space derivatives in the Boltzmann equation are no longer larger than the neglected quadratic terms. The situation is the same as in classical Stokes flows because the only modification introduced by a kinetic treatment is the presence of a kinetic layer (which can be very large, but is always finite). This kinetic layer or Knudsen layer is approximately a mean free path thick in the vicinity of the body ($r < r_1$, where $r_1 \approx l$; l denotes the mean free path and r the distance from the body). Then we have a large layer [$r_1 < r < r_2$, where $r_2 \approx l/M$; M denotes the Mach number, i.e., the ratio of the velocity of the free stream to the thermal speed $(RT)^{1/2}$]; in this layer the linearization assumption still holds (for $M \ll 1$) and the flow can be essentially described in terms of Stokes's solution. Finally, for $r > r_2$ the linearization is not valid, the flow is strongly influenced by the conditions at infinity and can be essentially described in terms of the Oseen solution (for a discussion of the analogous problems in continuum fluid dynamics and the concepts of Stokes and Oseen flow, see, e.g., Van Dyke, ref. 11). Accordingly, the conditions for a linearization about a Maxwellian distribution having zero mass velocity with respect to the body are certainly satisfied in a large part of the flow region, but not everywhere for small but finite M.

The situation for three-dimensional flows is satisfactory, in a way, since the Stokes solution is known to be well behaved at infinity, in the sense that the velocity profiles approach arbitrarily close (for sufficiently small Ma/l, a being a typical dimension of the body) to the conditions of a uniform stream before the theory breaks down. Accordingly, the Oseen solution is not required unless we are interested in the far field ($r > l/M$), or wish to iterate the solution to obtain terms in M^2. In this second case the procedure breaks down because the linearized solution does not give a uniform approximation to the whole distribution function. If we restrict ourselves to the first approximation in the vicinity of the body, then we have only to drop Eq. (6.2) and use the assumption that when $r \to \infty$ the solution describes a uniform flow.

The situation is less simple for two-dimensional flows. As a matter of fact, if we consider, e.g., the flow past an axisymmetric body, we can prove that the conclusion of Lemma III applies even if we drop Eq. (6.2) and simply assume that the mass velocity is single-valued and bounded as $r \to \infty$. This follows from an asymptotic analysis (Cercignani, ref. 10) of the solutions of the linearized Boltzmann equation for two-dimensional flows, proving that Eq. (6.2) is satisfied if the mass velocity is single-valued and bounded as $r \to \infty$. In order to obtain a nontrivial solution for two-dimensional flows, we are forced to allow a logarithmic behavior of the mass velocity for $r \to \infty$; accordingly, no uniform approximation to the velocity distribution is possible by means of the linearized Boltzmann equation, and we have to resort to a matching procedure of the inner solution (given by the linearized Boltzmann equation) and an outer solution valid for $r \gtrsim l/M$. The latter can be obtained by a stretching of the space coordinates before expanding in the Mach number; a formal expansion in powers of M shows that the solution in the outer region is similar to a Hilbert expansion where the lowest-order approximation to the fluid variables is given by an incompressible Oseen flow (Cercignani, ref. 10).

Although the problem of external flows at low Mach numbers needs further investigation, the analysis just sketched shows that, at least for two-dimensional flows, it is necessary to consider an approach more sophisticated than a naive use of the linearized Boltzmann equation; otherwise we may be looking for solutions which do not exist.

The difficulties which have been just discussed show up in the proofs of existence theorems for external problems. They were circumvented in the case of one space dimension by Rigolot-Turbat, Maslova, Bardos *et al.*, and Cercignani, who proved existence theorems for the half-space problem in the linearized and weakly nonlinear case. Ukai and Asano (ref. 18) proved existence for the weakly nonlinear flow past a body in three dimensions, but not in two dimensions.

Interesting problems also appear when one linearizes about a drifting Maxwellian. These problems occur in connection with the study of strong evaporation and condensation in a half space. If v_∞ and T_∞ are the velocity and temperature of the vapor at infinity, it was conjectured (ref. 19) that the Mach number at infinity

$$\mathrm{M}_\infty = \frac{|v_\infty|}{\sqrt{5RT_\infty/3}} \tag{6.13}$$

should play an important role in determining the existence and uniqueness of the solution; in particular, one is led to conjecture that, in the evaporation case ($v_\infty > 0$), for any given $\mathrm{M}_\infty < 1$ there should be a solution, but no solution should be possible in the supersonic case. These results should show up in a change in the completeness properties of the elementary solutions (see Chapter 7) for the Boltzmann equation linearized about the Maxwellian at infinity. This equation can be written as follows:

$$(\xi_1 + v_\infty)\frac{\partial h}{\partial x} = Lh \tag{6.14}$$

where L is the usual collision operator linearized with respect to a nondrifting Maxwellian. A notable feature of Eq. (6.14) is that, at variance with the usual linearized Boltzmann equation for one-dimensional steady problems, it does not admit the parity transformation $(x, \xi_1) \to (-x, -\xi_1)$ as a symmetry. This requires more careful arguments. If L is replaced by the one-dimensional BGK model, the problem can be discussed with the apparatus discussed in the next chapter, as was done by Arthur and Cercignani (ref. 20). Results on models and abstract equations (refs. 21–24) have proved the conjecture as far as the linearized equation is concerned. In particular, Greenberg and Van der Mee (ref. 24) have proved the following general result: If we denote by $\hat{\psi}_\alpha$ the basic collision invariants chosen in such a way that

$$(\hat{\psi}_\alpha(\xi_1 + v_\infty), \hat{\psi}_\beta) = 0 \qquad (\alpha, \beta = 0, 1, 2, 3, 4; \alpha \neq \beta) \tag{6.15}$$

then the numbers

$$N_\alpha = (\hat{\psi}_\alpha(\xi_1 + v_\infty), \hat{\psi}_\alpha) \qquad (\alpha = 0, 1, 2, 3, 4) \tag{6.16}$$

determine the possibility of solving the problem. In fact, the number of negative values among the N_α gives the number of additional conditions which can be imposed on a solution h bounded at infinity. A simple calculation indicates that we can take

$$\begin{gathered}\psi_0 = 1, \quad \psi_1 = |\boldsymbol{\xi}|^2 - 3v_\infty\xi_1, \quad \psi_2 = \xi_2, \\ \psi_3 = \xi_3, \quad \psi_4 = |\boldsymbol{\xi}|^2 3RT_\infty\end{gathered} \tag{6.17}$$

and

$$\begin{gathered}N_0 = v_\infty, \quad N_1 = 9v_\infty RT_\infty(v_\infty^2 - \tfrac{5}{3}RT_\infty), \quad N_2 = N_3 = v_\infty RT_\infty, \\ N_4 = 6v_\infty(RT_\infty)^2\end{gathered} \tag{6.18}$$

Obviously, if $v_\infty > 0$, then there is one negative value for $v_\infty < (\frac{5}{3}RT_\infty)^{1/2}$, i.e., $M_\infty < 1$, and none if $v_\infty > (\frac{5}{3}RT_\infty)^{1/2}$, or $M_\infty > 1$. Thus in the subsonic case one can obtain solutions with one free parameter (which can be the Mach number M_∞ itself) by imposing the condition that h does not modify the bulk velocity v_∞.

The case $v_\infty < 0$ was not mentioned in the quoted papers, but is briefly discussed in ref. 17. It is clear that if $v_\infty < 0$, the number of additional conditions is four for $M_\infty < 1$ and five for $M_\infty > 1$. (Two of these can be disposed of by letting the motion in the y- and z-directions vanish at infinity.)

7. Influence of Various Spectra

A notable advantage in using the linearized Boltzmann equation instead of the nonlinear one is that one can use the superposition principle to write the general solution as a linear combination of a "complete set" of "elementary solutions" with separated variables. This method will be investigated in detail and used to solve specific problems in the next chapter; here we shall state only a few general considerations. When we separate the variables we find in general that the space and time dependence is exponential, say, $\exp[i\mathbf{k}\cdot\mathbf{x} + i\omega t]$ (although some polynomial solutions are sometimes required to have a complete set, see next chapter), while the function describing the dependence upon molecular velocity satisfies

$$(i\omega + i\mathbf{k}\cdot\boldsymbol{\xi})g = Lg \tag{7.1}$$

In general, ω and $\mathbf{k}$ will be complex, and this equation will have solutions only when particular relations between ω and $\mathbf{k}$ are satisfied; corresponding to the allowed values of ω and $\mathbf{k}$ one can find g's which are either square summable ("eigensolutions") or not ("generalized eigensolutions"). In the former case the relation between ω and $\mathbf{k}$ is usually called a dispersion relation and the solution $g \exp[i\mathbf{k} \cdot \mathbf{x} + i\omega t]$ a normal mode. Both the "eigensolutions" and the "generalized eigensolutions" combine to form the general solution of the linearized Boltzmann equation:

$$h = \int_{D(\omega,\mathbf{k})} A(\omega, \mathbf{k}) \exp[i\mathbf{k} \cdot \mathbf{x} + i\omega t] g(\boldsymbol{\xi}; \omega, \mathbf{k}) \, dD \tag{7.2}$$

where g is a solution of Eq. (7.1) properly normalized and the integral is a set integral over a complete set $D(\omega, \mathbf{k})$ of allowed values of ω, $\mathbf{k}$ (accordingly, it will contain summations over discrete values of ω and $\mathbf{k}$ and integrals over continuous sets). In order to solve specific problems, we must be able to solve the following two problems:

1. To find the solutions $g(\boldsymbol{\xi}; \omega, \mathbf{k})$ corresponding to a set $D(\omega, \mathbf{k})$ which is complete; this allows us to write the most general solution of the Boltzmann equation in the form shown in Eq. (7.2).
2. To find a way of calculating the coefficient $A(\omega, \mathbf{k})$ from the initial and boundary conditions.

It is clear that these two problems (especially the second one) are far from being trivial; they can, however, be solved in particular cases, as we shall see in the next chapter. In many situations even the bare knowledge of the set $D(\omega, \mathbf{k})$ is capable of providing a qualitative understanding of the features of the solutions and estimates of asymptotic behaviors. In particular, the distinction between a continuous and a discrete spectrum, although basically mathematical, can imply very important physical consequences.

There are two possible sources of continuous spectra in the linearized Boltzmann equation; one is the free-streaming operator $\boldsymbol{\xi} \cdot \partial/\partial\mathbf{x}$ and the other the collision operator L. As we noted above, we have to find a complete set $D(\omega, \mathbf{k})$, but in general the total set of values of ω and $\mathbf{k}$ for which Eq. (6.1) has a solution is too large, in the sense that not all the solutions of the set are linearly independent. In order to simplify the present sketch, we shall restrict ourselves to the cases when the complete set $D(\omega, \mathbf{k})$ can be found by assuming $\mathbf{k} = k\mathbf{e}$, where $\mathbf{e}$ is a real unit vector, and restricting either ω or $\mathbf{k}$ to a suitable line or contour C in the complex plane (the choice being dictated by the nature of a specific problem); then we have to find the set T of admissible ω or k (the spectrum), and the set D is given by $C \otimes T$. Accordingly, Eq. (7.1) must be approached with the idea that $\mathbf{e}$ is a given real unit vector and either ω or k is a given complex number; we have to find the allowed ω or k and the corresponding solution g. After this explanation we can investigate the influence of the two operators $\boldsymbol{\xi} \cdot \partial/\partial \mathbf{x}$ and L on the spectrum (of either ω or k).

We start from particular cases obtained by neglecting certain terms in order to isolate the influence of the various operators. We first neglect Lh (free-molecular flow); then Eq. (7.1) becomes

$$(\omega + k\mathbf{e} \cdot \boldsymbol{\xi})g = 0 \qquad (\mathbf{k} = k\mathbf{e}; \mathbf{e} \cdot \mathbf{e} = 1) \tag{7.3}$$

If we regard ω as given, then k can take all the values taken by $-\omega/(\mathbf{e} \cdot \boldsymbol{\xi})$, i.e., we have a continuous spectrum which fills a straight line of the complex plane (or, rather, half a straight line; we can restrict k to half a straight line because the presence of $-\mathbf{k}$ together with $\mathbf{k}$ is already taken into account by the unit vector $\mathbf{e}$). The slope of the straight line is $\operatorname{Im} \omega/\operatorname{Re} \omega$ (in particular, the straight line is the real axis for real values of ω). If, conversely, we regard k as given, then ω can have all the values taken by $-k(\mathbf{e} \cdot \boldsymbol{\xi})$, i.e., again the continuous spectrum fills a straight line. In both cases the generalized eigensolution (apart from an arbitrary normalization factor depending on $\boldsymbol{\xi}$) is

$$g = \delta(\omega + k\mathbf{e} \cdot \boldsymbol{\xi}) \tag{7.4}$$

[see Eq. (3.7) of Chapter I]. It follows that the general solution can be written

$$h = \int A(\boldsymbol{\xi}, \mathbf{k}, \omega) \exp[i\omega t + i\mathbf{k} \cdot \mathbf{x}]\, \delta(\omega + \mathbf{k} \cdot \boldsymbol{\xi})\, d\omega\, d\mathbf{k} \tag{7.5}$$

the integration being extended to, say, all the real values of ω and $\mathbf{k}$. The presence of the delta function allows us to write

$$h = \int A(\boldsymbol{\xi}; \mathbf{k}) \exp[i\mathbf{k} \cdot (\mathbf{x} - \boldsymbol{\xi}t)]\, d\mathbf{k} \tag{7.6}$$

or, because this is a Fourier integral,

$$h = \tilde{A}(\boldsymbol{\xi}; \mathbf{x} - \boldsymbol{\xi}t) \tag{7.7}$$

i.e., essentially, h is an arbitrary function of $\boldsymbol{\xi}$ and $\mathbf{x} - \boldsymbol{\xi}t$, a result which is trivially recovered by integrating the equation

$$\frac{\partial h}{\partial t} + \boldsymbol{\xi} \cdot \frac{\partial h}{\partial \mathbf{x}} = 0 \tag{7.8}$$

along its characteristic lines.

As stated above, the second possible source of a continuous spectrum is the collision operator; this is already clear in the simplest case of a spatially uniform problem ($\partial h/\partial x = 0$), for which we have

$$i\omega g = Lg \tag{7.9}$$

Since we have fixed **k** (**k** = 0), we cannot arbitrarily fix ω; in fact, ω must be such that $i\omega$ is in the spectrum of L. We already know (Chapter III) that in general there is both a discrete and a continuous spectrum, a notable exception being provided by Maxwell's molecules. If we denote by g_i the eigenfunctions of L, by g_λ the generalized eigenfunctions, and by $-\lambda_i$ and $-\lambda$ ($\lambda_i, \lambda > 0$) the corresponding eigenvalues, we can write the general solution

$$h(\boldsymbol{\xi}; t) = \sum_i c_i g_i(\boldsymbol{\xi}) e^{-\lambda_i t} + \int_{\mu_0}^{\mu_1} A(\lambda) g_\lambda(\boldsymbol{\xi}) e^{-\lambda t} \, d\lambda \tag{7.10}$$

In particular, the first five discrete eigenfunctions are the collision invariants and correspond to $\lambda_i = 0$ ($0 \leq i \leq 4$). Equation (7.10) shows that $h(\boldsymbol{\xi}; t)$ relaxes to a linear combination of the five collision invariants (i.e., to the linearized version of a Maxwellian); only for hard potentials ($\mu_0 > 0$), however, do all solutions decay faster than a certain exponential at a rate which does not depend upon the initial data. The presence of a continuous spectrum extending to the origin (as in the case of soft potentials with angular cutoff, not realistic for a neutral gas) would have the physical consequence that the decay to equilibrium may not be exponential and depends upon the degree of smoothness of the initial data.

The spectrum becomes more complicated when we allow the streaming operator to interact with the collision operator. The simplest case is given by steady problems ($\partial h/\partial t = 0$), for which we have

$$ik\mathbf{e} \cdot \boldsymbol{\xi} g = Lg \tag{7.11}$$

the spectrum now being composed of the admissible values of k.

The values of ik are now real, as were the values of $i\omega$ in Eq. (7.9), although now the proof is less trivial (and the result dependent upon the assumption that **e** is a real unit vector); roughly speaking, however, the realness of ik follows from the fact that both $(g, \mathbf{e} \cdot \boldsymbol{\xi} g)$ and (g, Lg) are real [the scalar product here is the usual one for complex functions, $(f, g) = \int \bar{f}(\boldsymbol{\xi}) g(\boldsymbol{\xi}) \, d\boldsymbol{\xi}$, the bar denoting complex conjugate]. For the case of rigid spheres it is also possible to show that the origin $k = 0$ is an isolated point, and the same result seems reasonable for potentials with a strictly finite range; $k = 0$, however, is not isolated for power-law potentials with angular cutoff and kinetic models with constant collision frequency. [It can be shown that the continuous spectrum contains at least the set of values taken by $-\nu(\xi)/(\boldsymbol{\xi} \cdot \mathbf{e})$.] The spectrum of admissible k's has been investigated in some detail for model equations, and in some cases the above-mentioned problems (1) and (2) have been solved (Cercignani, refs. 25, 28–30); the pertinent theory will be expounded in the next chapter.

The spectral theory of Eq. (7.11) can be used to give a more explicit form (for $\omega = 0$ and $\mathbf{k} = k\mathbf{e}$, $\mathbf{e}$ a fixed real unit vector) to the representation shown in Eq. (7.2). The fact that $k = 0$ is a degenerated point of the spectrum leads to discrete terms similar to those in Eq. (7.10), but more complex; in fact, one has to add particular solutions [to be discussed in the next section (Eq. (8.1)] to the linear combination of the collision invariants present in Eq. (7.2).

The more general situation is the one in which we retain all the terms in Eq. (7.1). We have two possibilities because we can fix either ω or k; the second choice is useful for the initial value problem in the absence of boundaries, the first has greater flexibility. The general case has again been investigated with model equations, and in some cases the problems (1) and (2) mentioned above have been solved (Cercignani and Sernagiotto, ref. 26; Cercignani, refs. 28–30; Buckner and Ferziger, ref. 31). In general, for both fixed ω and fixed k the continuous spectrum fills two-dimensional regions, the exceptions being offered by Maxwell's molecules and models with constant collision frequency; in fact, the continuous spectrum contains at least the set of values taken by $-[\nu(\xi) + i\omega]/(\boldsymbol{\xi} \cdot \mathbf{e})$ in one case and by $-[\nu(\xi) + ik(\mathbf{e} \cdot \boldsymbol{\xi})]$ in the other case. A detailed and rigorous treatment of the various spectra was given by Nicolaenko (ref. 33).

In both steady and time-dependent problems the fact that the origin $k = 0$ belongs to the discrete spectrum for certain collision models and does not for others seems to introduce a basic qualitative difference; in principle, this difference should be experimentally detectable by measuring the phase speed and the attenuation of sound waves, but the actual measurements seem to be outside the range of the present experimental techniques.

8. The Linearized Boltzmann Equation and the Chapman–Enskog Theory

We have seen that both the linearization of the Boltzmann equation and the Chapman–Enskog and related expansions are the result of suitable perturbation procedures applied to the Boltzmann equation. The two procedures are basically different because the expansion parameters are completely different: the "small" deviation of initial and boundary data from a Maxwellian distribution in the case of linearization, the smallness of the mean free path or mean free time with respect to other typical lengths or times in the case of the Chapman–Enskog expansion. It is clear that when both circumstances are realized one can apply the Chapman–Enskog method to the linearized Boltzmann equation. One of the main reasons for doing this comes from the possibility of answering certain questions concerning the Chapman–Enskog expansion in the case of the linearized equation, thanks to the simplifications of the algorithm, and thus gaining a heuristic insight into the more complicated nonlinear case.

As already mentioned (Chapter V, Section 4), if we consider the initial value problem, we can obtain a rigorous proof that the Hilbert expansion provides asymptotic solutions (for $\epsilon \to 0$) to the Boltzmann equation, and the same is true for a Chapman–Enskog theory truncated at the Navier–Stokes level. These results make it clear that the expansions considered do provide legitimate approximations (for suitable ranges of the parameters), but do not settle the question of convergence of the expansions, and, consequently, of the very existence of the normal solutions. In view of the importance that is sometimes attached to convergence (although the basic feature of a series with regard to usual applications is its asymptotic rather than convergent nature), we shall briefly discuss the question of the convergence of the Chapman–Enskog expansion for the linearized Boltzmann equation.

As we know, the Chapman–Enskog series essentially consists in the expansion of an operator; accordingly, its convergence has a meaning only if referred to a certain class of functions upon which the operator acts. In the case of the linearized Boltzmann equation it is easy to exhibit, for any kind of collision operator, normal solutions whose expansions are trivially convergent because they contain just a finite number of terms. These solutions are polynomials in the space and time variables; just to give an example, it can easily be checked that the time-independent function

$$H = \beta_{ij}\xi_j x_i + \gamma_i x_i(\xi^2 - 5RT_0) + L^{-1}[\beta_{ij}\xi_i\xi_j + \gamma_i\xi_i(\xi^2 - 5RT_0)] \quad (8.1)$$

where γ_i is an arbitrary (constant) vector, β_{ij} an arbitrary (constant) traceless tensor, and T_0 the temperature of the basic Maxwellian, is an exact solution of the linearized Boltzmann equation, in the form of a power series in the mean free path (containing just the zeroth- and first-order terms).

When we pass to the nonlinear theory, these solutions have certain counterparts, the homoenergetic affine solutions, which will be discussed in the next chapter. For these solutions the Chapman–Enskog method does not lead to finite sums, but for small values of the gradients turns out to give convergent series.

In order to investigate less trivial examples of normal solutions, we can assume a space- and time-dependence in exponential form; then the velocity dependence is given by a factor $g(\boldsymbol{\xi})$ which satisfies Eq. (6.1). If we recall that the Chapman–Enskog procedure expands the time derivative into a series of space derivatives, it will be clear that the Chapman–Enskog series for the functions $g(\boldsymbol{\xi})\exp[i\mathbf{k}\cdot\boldsymbol{\xi} + i\omega t]$ will converge or not depending on whether an expansion of $\omega = \omega(\mathbf{k})$ $[\omega(0) = 0]$ in a series of powers of $\mathbf{k}$

converges or not. It has been shown by McLennan (ref. 14) that convergence for sufficiently small k's holds for the case of rigid spheres, but does not hold, in general, for power-law potentials with angular cutoff. The latter statement is only a very plausible conjecture, because no rigorous proof is available, the conjecture is a consequence of the fact that the spectrum $\omega = i\nu(c) - \mathbf{k}\cdot\mathbf{c}$ extends to the origin if we let k vary in the complex plane within any circle centered at $k = 0$ (see Section 7).

The result for rigid spheres (which is likely to hold for any potential with a strictly finite range) follows rigorously from certain results of perturbation theory. In fact, if we write Eq. (7.1) as follows:

$$i\omega g = Lg + \epsilon(\mathbf{e}\cdot\boldsymbol{\xi})g \qquad (\epsilon = -ik) \tag{8.2}$$

we can consider the self-adjoint operator $\mathbf{e}\cdot\boldsymbol{\xi} = L'$ as a perturbation and the self-adjoint operator L as unperturbed operator. Well-known results in the theory of the perturbation of spectra can be applied provided that

$$\|L'h\| \leq M(\|h\| + \|Lh\|) \tag{8.3}$$

(M independent of h, $\|h\|$ norm in $\mathscr{H}$) for all h which are in the domains of both L and L'. In particular, if λ is a multiple eigenvalue of L with multiplicity m, and if Eq. (8.3) holds, then $L + \epsilon L'$ has, for sufficiently small values of the complex parameter ϵ, m isolated point eigenvalues (not necessarily distinct) $\lambda_i(\epsilon)$ $(i = 1, \ldots, m)$, which reduce to λ for $\epsilon = 0$ and are analytic in ϵ. In order to apply this result to the fivefold eigenvalue $\lambda = 0$ of L and hence prove the above statement on the Chapman–Enskog expansion for rigid spheres, we have only to show that Eq. (8.3) applies. But this is a very simple consequence of the following properties:

1. $L = K - \nu$.
2. $\nu(\xi)$ grows linearly for large ξ in the case of rigid spheres [hence an M_1 exists such that $|\mathbf{e}\cdot\boldsymbol{\xi}| < M_1\nu(\xi)$].
3. K is completely continuous for rigid spheres (hence an M_2 exists such that $\|Kh\| < M_2\|h\|$).
4. The triangle inequality.

Using these properties, we have

$$\begin{aligned}\|(\mathbf{e}\cdot\boldsymbol{\xi})h\| \leq M_1\|\nu h\| &= M_1\|(K - L)h\| \leq M_1(\|Kh\| + \|Lh\|)\\ &\leq M_1(M_2\|h\| + \|Lh\|)\\ &\leq M(\|h\| + \|Lh\|) \qquad [M = \max(M_1, M_1M_2)]\end{aligned} \tag{8.4}$$

as was to be shown.

It is clear that the convergence of the expansion of ω in powers of $\mathbf{k}$ for sufficiently small values of $|k|$, which has just been proved, implies convergence of the Chapman–Enskog expansion for a very restricted type of space dependence: all the derivatives of the fluid variables must be bounded uniformly with respect to the order, and this means that they are not only analytic, but are also entire functions.

9. The Initial Value Problem for the Nonlinear Boltzmann Equation

In this chaper devoted to the existence and uniqueness theory for the Boltzmann equation we have paid a great deal of attention to boundary value problems. This is in agreement with the scope of this book; we cannot, however, conclude this chapter without mentioning the important and elegant results on the purely initial value problem obtained by different authors. The main problem is, of course, to prove (globally in time) the existence and uniqueness of the solution of the initial value problem.

The simplest kind of problem which can be considered is the initial value problem for an infinite expanse of gas with initial data independent of the space variables (the spatially homogeneous problem). The first results were proved by Carleman, but important results were later proved by Maslova and Tchubenko, Gluck, Wild, Morgenstern, Arkeryd, Elmroth, Gustafsson, and Di Blasio (refs. 35–47). This part of the theory is now virtually complete.

The more difficult space-inhomogeneous case was initially considered in the weakly nonlinear case by Ukai, Nishida and Imai, Shizuta and Asano, and Caflisch (in the case of the perturbation of equilibrium; refs. 48–52)) and by Illner and Shinbrot, Bellomo and Toscani, and Hamdache (in the case of the perturbation of a free-molecular solution; refs. 53-55). Other perturbation results (about a general homogeneous solution and an inhomogeneous Maxwellian) are due to Arkeryd *et al.* (ref. 56) and Toscani (ref. 57). A class of special solutions with large deviations from equilibrium were discussed by Cercignani (ref. 58).

The main result on the space-inhomogeneous case is, however, due to DiPerna and Lions (ref. 59), who provided an ingenious proof of existence without a corresponding uniqueness result. Previously, a general existence theorem, due to Arkeryd, was available only in the frame of the so-called nonstandard analysis (ref. 60).

We remark that the theory of existence and uniqueness has an important relation to the problem of a rigorous justification of the Boltzmann equation, discussed in Chapter I.

References

Section 2——The property of the free-streaming operator investigated in this section was first pointed out in

1. C. Cercignani, *J. Math. Phys.* **9**, 633 (1968).

   For the initial-value problem in infinite domains or finite domains with specularly reflecting boundaries see ref. 2 of Chapter III.

Section 3——The properties of the integral version of the Boltzmann equation were first investigated in ref. 1

2. C. Cercignani, *J. Math. Phys.* **8**, 1653 (1967),

   and in ref. 5 of Chapter III. The positive nature of the operator K for rigid spheres was proved in

3. L. Finkelstein, "Transport Phenomena in Rarefied Gases," Ph.D. thesis, Hebrew University, Jerusalem (1962).

Section 4——The existence and uniqueness theory in $\mathscr{H}$ was given in refs. 1 and 2. Theorems II–V were proved by

4. Y. P. Pao, *J. Math. Phys.* **8**, 1893 (1967),

   using the results of ref. 2 as starting point.

   Other results were proved by

5. J. P. Guiraud, *J. Méc. Théor. Appl.* **7**, 171 (1968);
6. J. P. Guiraud, *J. Méc. Théor. Appl.* **9**, 443 (1970);
7. J. P. Guiraud, *J. Méc. Théor. Appl.* **11**, 2 (1972);
8. J. P. Guiraud, *C. R. Acad. Sci. Paris* **274**, 417 (1972);
9. J. P. Guiraud, *C. R. Acad. Sci. Paris* **275**, 171 (1968).

Section 5——The content of this section follows ref. 1.

Section 6——The results for bounded one-dimensional and two-dimensional domains were given in refs. 1 and 2. The remainder of the section summarizes the results of

10. C. Cercignani, *Phys. Fluids* **11**, 303 (1967).

    Details on the Stokes and Oseen flow in continuum theory can be found in

11. M. Van Dyke, *Perturbation Methods in Fluid Mechanics*, Academic Press, New York (1964).

    Results for the half-space problem were proved by

12. C. Rigolot-Turbat, *C. R. Acad. Sci. Paris* **272**, 617 (1971);
13. C. Rigolot-Turbat, *C. R. Acad. Sci. Paris* **272**, 763 (1971);
14. C. Rigolot-Turbat, *C. R. Acad. Sci. Paris* **273**, 58 (1971);
15. N. B. Maslova, *USSR Comput. Math. Math. Phys.* **22**, 208 (1982);
16. C. Bardos, R. E. Caflisch, and B. Nicolaenko, *Commun. Pure Appl. Math.* **39**, 323 (1986);
17. C. Cercignani, in: *Trends in Applications of Pure Mathematics to Mechanics* (E. Kröner and K. Kirchgässner, eds.), Lecture Notes in Physics, Volume 249, p. 35, Springer, Berlin (1986);

    while the weakly nonlinear theory for the flow past a body is discussed in

18. S. Ukai and K. Asano, *Arch. Rat. Mech. Anal.* **84**, 249 (1983).

    The half-space problems linearized about a drifting Maxwellian were discussed in ref. 17 and in

19. C. Cercignani, in: *Mathematical Problems in the Kinetic Theory of Gases* (D. C. Pack and H. Neunzert, eds.), p. 129, Lang, Frankfurt (1980);
20. M. D. Arthur and C. Cercignani, *Z. Angew. Math. Phys.* **31**, 634 (1980);
21. C. E. Siewert and J. R. Thomas, *Z. Angew. Math. Phys.* **32**, 421 (1981);
22. C. E. Siewert and J. R. Thomas, *Z. Angew. Math. Phys.* **33**, 202 (1982);
23. C. E. Siewert and J. R. Thomas, *Z. Angew. Math. Phys.* **33**, 626 (1982);
24. W. Greenberg and C. V. M. Van der Mee, *Z. Angew. Math. Phys.* **35**, 156 (1984).

Section 7——The continuous spectrum for space transients was first pointed out and explicitly treated by

25. C. Cercignani, *Ann. Phys.* (*N.Y.*) **20**, 219 (1962);

in the case of BGK model. The continuous spectrum for spatially uniform problems was pointed out by Grad in ref. 2 of Chapter III. More general situations were investigated in

26. C. Cercignani and F. Sernagiotto, *Ann. Phys.* (*N.Y.*) **30**, 154 (1964);
27. H. Grad, "Theory of the Boltzmann Equation," New York University Report MF 40, AFOSR 64-1377 (1964);
28. C. Cercignani, "Elementary Solutions of Linearized Kinetic Models and Boundary Value Problems in the Kinetic Theory of Gases," Brown University Report (1965);
29. C. Cercignani, *Ann. Phys.* (*N.Y.*) **40**, 469 (1966);
30. C. Cercignani, *Ann. Phys.* (*N.Y.*) **40**, 454 (1966);
31. J. K. Buckner and J. H. Ferziger, *Phys. Fluids* **9**, 2309 (1966).

A general discussion of the various spectra is contained in the fourth chapter of ref. 15 of Chapter I. The general form of the solution for the steady one-dimensional equation based on the spectral properties was given in

32. C. Cercignani, in: *Rarefied Gas Dynamics* (M. Becker and M. Fiebig, eds.), Vol. I, p. A6, DFVLR Press, Porz-Wahn (1974);

while a rigorous treatment of the general spectral problem was given by

33. B. Nicolaenko, in: *The Boltzmann Equation* (F. A. Grünbaum, ed.), p. 125, Courant Institute, New York University, New York (1972).

Section 8——The convergence of the Chapman–Enskog series for the linearized Boltzmann equation was investigated by

34. J. A. McLennan, *Phys. Fluids* **8**, 1580 (1965).

Section 9——The results on existence and uniqueness mentioned in this section were presented by

35. T. Carleman, *Acta Math.* **60**, 91 (1933);
36. T. Carleman, *Problèmes Mathématiques dans la Théorie Cinétique des Gaz*, Almqvist and Wiksells, Uppsala (1957);
37. H. B. Maslova and R. P. Tchubenko, *Dokl. Akad. Nauk SSSR* **202**(4), 800 (1972);
38. H. B. Maslova and R. P. Tchubenko, *Vestnik Leningrad Univ.*, No. 7, 109 (1976);
39. P. Gluck, *Transport Theory Stat. Phys.* **9**, 43 (1980);
40. E. Wild, *Proc. Camb. Phil. Soc.* **47**, 602 (1951);
41. D. Morgenstern, *Proc. Natl. Acad. Sci. U.S.A.* **40**, 719 (1954);
42. L. Arkeryd, *Arch. Rat. Mech. Anal.* **45**, 1 (1972);
43. L. Arkeryd, *Arch. Rat. Mech. Anal.* **45**, 17 (1972);
44. L. Arkeryd, *Arch. Rat. Mech. Anal.* **77**, 11 (1981);
45. T. Elmroth, *Arch. Rat. Mech. Anal.* **82**, 1 (1983);
46. T. Gustafsson, *Arch. Rat. Mech. Anal.* **92**, 23 (1986);
47. G. Di Blasio, *Commun. Math. Phys.* **38**, 331 (1974);
48. S. Ukai, *Proc. Japan. Acad. Ser. A, Math. Sci.* **50**, 179 (1974);

49. T. Nishida and K. Imai, *Publ. Res. Inst. Math. Sci. Kyoto Univ.* **12**, 229 (1977);
50. Y. Shizuta and K. Asano, *Proc. Japan. Acad. Ser. A, Math. Sci.* **53**, 3 (1977);
51. R. E. Caflisch, *Commun. Math. Phys.* **74**, 71 (1980);
52. R. E. Caflisch, *Commun. Math. Phys.* **74**, 97 (1980);
53. R. Illner and M. Shinbrot, *Commun. Math. Phys.* **95**, 217 (1984);
54. N. Bellomo and G. Toscani, *J. Math. Phys.* **26**, 334 (1985);
55. K. Hamdache, *C. R. Acad. Sci. Paris* **299**, 431 (1984);
56. L. Arkeryd, R. Esposito, and M. Pulvirenti, *Commun. Math. Phys.* **111**, 393 (1987);
57. G. Toscani, *Arch. Rat. Mech. Anal.* (to appear);
58. C. Cercignani, *Arch. Rat. Mech. Anal.* (to appear);
59. R. DiPerna and P. L. Lions, *Ann. Math.* (to appear);
60. L. Arkeryd, *Arch. Rat. Mech. Anal.* **86**, 85 (1984).

For a more detailed survey of the results of existence and uniqueness before 1987, one should consult Chapter VIII of ref. 15 of Chapter I. A more recent survey, particularly useful for the details of rigorous perturbation solutions, is

61. N. Bellomo, A. Palczewski, and G. Toscani, *Mathematical Topics in Nonlinear Kinetic Theory*, World Scientific, Singapore (1988).

Chapter VII

ANALYTICAL METHODS OF SOLUTION

1. Introduction

The theory developed in the previous chapters shows that the study of the linearized Boltzmann equation is a worthwhile undertaking and that many of the features of its solutions can be retained by using model equations. We can say more, that practically all the features are retained by a properly chosen model. The advantages offered by the models consist essentially in simplifying both the analytical and numerical procedures for solving boundary value problems of special interest. In particular, the use of models is invaluable in those cases when the solution of the latter is explicit (in terms of quadratures or functions whose qualitative behavior can be studied by analytical means). Accordingly, we shall devote this chapter to the analytical manipulations which can be used to obtain interesting information from the model equations. The method used throughout is the method of separation of variables already sketched in Chapter VI, Section 7. The first step is to construct a complete set of separated-variable solutions ("elementary solutions") and then represent the general solution as a superposition of the elementary solutions; the second step is to use the boundary and initial conditions to determine the coefficients of the superposition. While the first problem can be solved for the model equations discussed in Chapter IV, the second problem can be solved exactly in only a few cases. The method retains its usefulness, however, even when the second problem is not solvable, or is only approximately solvable, because it is capable of providing an analytical representation of the solution and hence a picture of its qualitative behavior.

It must be stated that the method of separation of variables is not the only one capable of solving these problems; transform techniques of the Wiener–Hopf type are completely equivalent to the method of elementary solutions. It is this author's opinion, however, that the method to be described in the next few sections is more concise and easier to handle, especially when one cannot obtain an analytical solution (by either method) and one must resort to an approximate or qualitative description.

2. Splitting of the One-Dimensional BGK Equation

We begin by considering the simplest kind of problems, steady problems in one-dimensional geometry, and the simplest model, the BGK model.

Accordingly, we consider the equation

$$\xi_1\, \partial h/\partial x = Lh \tag{2.1}$$

where ξ_1 is the x component of the molecular velocity $\boldsymbol{\xi}$ and

$$Lh = \nu\left[\sum_{\alpha=0}^{4} (\psi_\alpha, h)\psi_\alpha - h\right]; \qquad (\psi_\alpha, \psi_\beta) = \delta_{\alpha\beta} \tag{2.2}$$

Here the ψ_α denote the collision invariants, ν is a constant, and (,) is the usual scalar product in $\mathscr{H}$. We assume $(2RT_0)^{1/2}$ as velocity unit and $\theta(2RT_0)^{1/2} = \nu^{-1}(2RT_0)^{1/2} = 2\pi^{-1/2}l$ as length unit (T_0 is the unperturbed temperature, l the mean free path given by Eq. (3.27) of Chapter V, and $\theta = \nu^{-1}$).

Integrating Eq. (2.1) over the whole velocity space gives the following relation (the macroscopic continuity equation):

$$\frac{\partial}{\partial x}(\xi_1, h) = 0 \tag{2.3}$$

i.e.,

$$(\xi_1, h) = C \tag{2.4}$$

where C is a constant. In usual problems C is equal to zero; if it is not zero, we define a new unknown

$$\bar{h} = h - 2C\xi_1 \tag{2.5}$$

which is easily verified to satisfy Eq. (2.1) and

$$(\xi_1, \bar{h}) = 0 \tag{2.6}$$

Accordingly, without any loss of generality we can assume $C = 0$ in Eq. (2.4). Consequently, the right-hand side of Eq. (2.1) can be written as follows:

$$Lh = (P - I)h \qquad (\nu = 1) \tag{2.7}$$

where I is the identity operator and P the orthogonal projector onto the subspace of $\mathscr{H}$ spanned by the functions, $1, \xi_2, \xi_3, \xi^2 = \xi_1^2 + \xi_2^2 + \xi_3^2$. Let us now consider, for a moment, x and ξ_1 as parameters and h as a vector of the Hilbert space of the functions of ξ_2 and ξ_3, defined by the inner product

$$(h, g)_2 = \iint_{-\infty}^{\infty} h(\xi_2, \xi_3)g(\xi_2, \xi_3)\exp[-\xi_2^2 - \xi_3^2]\, d\xi_2\, d\xi_3 \qquad (2RT_0 = 1) \tag{2.8}$$

Then h can be split into five components, one along the (Hilbert space) vector $\varphi_0 = 1$, one along $\varphi_1 = \xi_2^2 + \xi_3^2 - 1$, one along $\varphi_2 = \xi_2$, one along $\varphi_3 = \xi_3$, and one belonging to the subspace orthogonal to the finite dimensional one spanned by $\varphi_0, \varphi_1, \varphi_2, \varphi_3$:

$$h(x, \xi_1, \xi_2, \xi_3) = Y_0(x, \xi_1) + (\xi_2^2 + \xi_3^2 - 1)Y_1(x, \xi_1) + 2\xi_2 Y_2(x, \xi_1) + 2\xi_3 Y_3(x, \xi_1) + Y_4(x, \xi_1, \xi_2, \xi_3) \quad (2.9)$$

Because of the definition of Y_0, Y_1, Y_2, Y_3, and Y_4 and the structure of the collision operator, Eq. (2.1) when projected onto φ_0, φ_1, φ_2, φ_3 and the orthogonal subspace gives

$$\xi_1 \frac{\partial Y_0}{\partial x} + Y_0 = \pi^{-1/2} \int_{-\infty}^{\infty} Y_0(x, \xi_1') \exp(-\xi_1'^2)\, d\xi_1' + \frac{2}{3}\left(\xi_1^2 - \frac{1}{2}\right)\pi^{-1/2} \times \left[\int_{-\infty}^{\infty} \left(\xi_1'^2 - \frac{1}{2}\right) \exp(-\xi_1'^2) Y_0(x, \xi_1')\, d\xi_1' + \int_{-\infty}^{\infty} \exp(-\xi_1'^2) Y_1(x, \xi_1)\, d\xi_1\right] \quad (2.10)$$

$$\xi_1 \frac{\partial Y_1}{\partial x} + Y_1 = \frac{2}{3}\pi^{-1/2} \int_{-\infty}^{\infty} \left(\xi_1'^2 - \frac{1}{2}\right) \exp(-\xi_1'^2) Y_0(x, \xi_1')\, d\xi_1' + \frac{2}{3}\pi^{-1/2} \int_{-\infty}^{\infty} \exp(-\xi_1'^2) Y_1(x, \xi_1')\, d\xi_1' \quad (2.11)$$

$$\xi_1 \frac{\partial Y_i}{\partial x} + Y_i = \pi^{-3/2} \int_{-\infty}^{\infty} \exp(-\xi_1'^2) Y_i(x, \xi_1')\, d\xi_1' \qquad (i = 2, 3) \quad (2.12)$$

$$\xi_1\, \partial Y_4/\partial x + Y_4 = 0 \quad (2.13)$$

Equation (2.13) is easily solved, and the general solution is

$$Y_4(x, \xi_1, \xi_2, \xi_3) = A(\boldsymbol{\xi}) \exp(-x/\xi_1) \quad (2.14)$$

where $A(\boldsymbol{\xi})$ is an arbitrary function of $\boldsymbol{\xi}$, provided it is orthogonal to φ_0, φ_1, φ_2, φ_3 with respect to the scalar product in Eq. (2.8).

Equations (2.10) and (2.11) form a system of two coupled equations which describe the heat transfer effects; Eq. (2.12) gives, for $i = 2, 3$, two uncoupled equations which describe the shear effects due, respectively, to motions in the y and z directions. We shall begin by considering the simplest case, one of the two (identical) equations (2.12).

3. Elementary Solutions of the Shear Flow Equation

Writing ξ, ξ_1 and Y in place of ξ_1, ξ_1' and Y_i, Eq. (2.12) can be rewritten as follows:

$$\xi\frac{\partial Y}{\partial x} + Y(x,\xi) = \pi^{-1/2}\int_{-\infty}^{\infty} \exp(-\xi_1^2)Y(x,\xi_1)\,d\xi_1 \tag{3.1}$$

Let us begin by separating the variables. Putting

$$Y(x,\xi) = g(\xi)X(x) \tag{3.2}$$

it is easily seen that either $Y = A_0$ (arbitrary constant) or

$$Y_u(x,\xi) = e^{-x/u}g_u(\xi) \tag{3.3}$$

where $g_u(\xi)$ satisfies

$$[(-\xi/u) + 1]g_u(\xi) = \pi^{-1/2}\int_{-\infty}^{\infty} g_u(\xi_1)\exp(-\xi_1^2)\,d\xi_1 \tag{3.4}$$

and u, the separation parameter, has been used to label the elementary solutions.

Though, *a priori*, u may assume any complex value, it is easily seen that u is a real number. This follows from a direct reasoning (ref. 25, Chapter VI) or from the general reasoning sketched in Chapter VI, Section 3. Thus the values of u must be real. This requires some care, because one cannot divide by $u - \xi$ in Eq. (3.4). This difficulty is overcome by letting $g_u(\xi)$ be a generalized function (see Chapter I, Sections 2 and 3). Then, if we disregard a multiplicative constant [i.e., normalizing g_u in such a way that the right-hand side of Eq. (3.4) is equal to 1], $g_u(\xi)$ will be a generalized function of the type

$$g_u(\xi) = P\frac{u}{u-\xi} + p(u)\,\delta(u-\xi) \tag{3.5}$$

where the constant C appearing in Eq. (3.18) of Chapter I can now depend upon u, and has been called $p(u)$. In order that Eq. (3.4) be satisfied by Eq. (3.5), the normalization condition for $g_u(\xi)$ must be satisfied, i.e., the right-hand side of Eq. (3.4) must be equal to 1. This condition can be satisfied for any real u and serves for determining $p(u)$:

$$p(u) = e^{u^2} P \int_{-\infty}^{\infty} \frac{\xi e^{-\xi^2}}{\xi - u} d\xi$$

$$= \pi^{1/2}\left(e^{u^2} - 2u \int_0^u e^{t^2}\, dt\right) \tag{3.6}$$

The first of these expressions follows directly, the second by manipulation of the first; $p(u)$ can be expressed in terms of tabulated functions (see References).

The generalized eigenfunctions $g_u(\xi)$ have many properties of orthogonality and completeness. Some of the orthogonality properties in the full range $(-\infty < \xi < \infty)$ can easily be proved by means of the usual procedures. Other properties of orthogonality in partial ranges (notably $0 < \xi < \infty$) and completeness are far from trivial to prove, since they require solving singular integral equations. However, standard techniques are available for treating such problems (Muskhelishvili, ref. 4; see also Appendix) and the following results can be obtained (Cercignani—refs. 25 and 28, Chapter VI):

Theorem I. The generalized functions $g_u(\xi)$ $(-\infty < u < +\infty)$ and $g_\infty = 1$, complemented with $g_* = \xi$, form a complete set for the functions $g(\xi)$ defined on the real axis, satisfying a Hölder condition in any open interval of the real axis, and such that

$$\int_{-\infty}^{\infty} \xi^2 e^{-\xi^2} g(\xi)\, d\xi < \infty \tag{3.7}$$

Furthermore, the coefficients of the generalized expansion

$$g(\xi) = A_0 + A_1\xi + \int_{-\infty}^{\infty} A(u) g_u(\xi)\, du \tag{3.8}$$

are uniquely and explicitly determined by

$$A_0 = 2\pi^{-1/2} \int_{-\infty}^{\infty} \xi^2 e^{-\xi^2} g(\xi)\, d\xi \tag{3.9}$$

$$A_1 = 2\pi^{-1/2} \int_{-\infty}^{\infty} \xi e^{-\xi^2} g(\xi)\, d\xi \tag{3.10}$$

$$A(u) = [C(u)]^{-1} \int_{-\infty}^{\infty} \xi e^{-\xi^2} g(\xi) g_u(\xi)\, d\xi \tag{3.11}$$

where

$$C(u) = u e^{-u^2} \{[p(u)]^2 + \pi^2 u^2\} \tag{3.12}$$

Theorem II. The generalized eigenfunctions $g_u(\xi)$ $(0 < u < \infty)$ and $g_\infty = 1$ form a complete set for the functions $g(\xi)$ defined on the positive real semiaxis, satisfying a Hölder condition in any open interval of this semiaxis, bounded by $A|\xi|^{-\gamma}$ with $\gamma < 2$ in the neighborhood of $\xi = 0$, and integrable with respect to the weight $\xi^2 e^{-\xi^2}$. Also, the coefficients of the generalized expansion

$$g(\xi) = A_0 + \int_0^{\infty} A(u) g_u(\xi)\, du \tag{3.13}$$

are uniquely and explicitly determined by

$$A_0 = 2\pi^{-1/2} \int_0^{\infty} P(\xi) \xi e^{-\xi^2} g(\xi)\, d\xi \tag{3.14}$$

$$A(u) = [C(u)P(u)]^{-1} \int_0^{\infty} \xi e^{-\xi^2} P(\xi) g_u(\xi) g(\xi)\, d\xi \tag{3.15}$$

Here we have set

$$P(\xi) = \xi \exp\left\{-\frac{1}{\pi}\int_0^{\infty} \tan^{-1}[\pi t/p(t)] \frac{dt}{t+\xi}\right\} \qquad (\xi > 0) \tag{3.16}$$

where the arctan varies from $-\pi$ to 0 when t varies from 0 to ∞. Theorem I shows that the generalized eigenfunctions are orthogonal with respect to the weight $\xi e^{-\xi^2}$ on $(-\infty, +\infty)$. Since, however, this weight is not everywhere positive, some irregularities are noticed. First, we note that the expressions for A_0 and A_1 [Eqs. (3.9) and (3.10)] are exactly interchanged with respect to what one would expect by analogy with familiar expansions. Second, the appearance of $g_* = \xi$ is completely unexpected, because g_* does not solve Eq. (3.4) for any value of u (g_∞ does for $u = \infty$). There is, however, a particular solution of Eq. (3.1), $Y_* = \xi - x$, which does not have separated variables (in the usual sense) and yet cannot be represented as a superposition of the elementary solutions $Y_u = e^{-x/u} g_u(\xi)$ and $Y_\infty = 1$. The latter statement follows from Theorem I itself, because if we could write

$$\xi - x = A_0 + \int_{-\infty}^{\infty} A(u)e^{-x/u}g_u(\xi)\,du \tag{3.17}$$

for some x (we can take $x = 0$ without loss of generality) and $-\infty < \xi < \infty$, then

$$0 = A_0 - \xi + \int_{-\infty}^{\infty} A(u)g_u(\xi)\,du \qquad (-\infty < \xi < \infty) \tag{3.18}$$

for suitable A_0 and $A(u)$. But this implies that the function which is identically zero on the real axis has two different expansions of the form (3.8): the trivial one with coefficients $A_0 = A_1 = A(u) = 0$ and another given by Eq. (3.18). This is in contrast to the uniqueness of the expansion (3.8) asserted by Theorem I; hence Eq. (3.17) is wrong and $\xi - x$ cannot be represented as a superposition of separated-variable solutions. Accordingly, the elementary solutions $Y_u(-\infty < u < \infty)$ and Y_∞ are not sufficient to construct the general solution. We want to prove now that if we add $Y_* = \xi - x$ to the set, we obtain a complete set, i.e., the general solution of Eq. (3.1) is given by

$$Y(x, \xi) = A_0 + A_1(x - \xi) + \int_{-\infty}^{\infty} A(u)e^{-x/u}g_u(\xi)\,du \tag{3.19}$$

In fact, given any solution $Y(x, \xi)$ of Eq. (3.1), Theorem I proves that for any fixed x we can find A_0, A_1, $A(u)$ such that Eq. (3.19) holds. We have to show now that A_0, A_1, and $A(u)$ do not depend on x, and our result will be proved. If we substitute Eq. (3.19) into Eq. (3.1) and use the fact that Y_0, Y_*, Y_u satisfy Eq. (3.1), we obtain

$$\frac{\partial A_0}{\partial x} + \frac{\partial A_1}{\partial x}(x - \xi) + \int_{-\infty}^{\infty} \frac{\partial A}{\partial x} e^{-x/u}g_u(\xi)\,du = 0 \tag{3.20}$$

But this equation gives for any fixed x the expansion of the function which is identically zero on the real axis; by the uniqueness of the expansion it follows that $\partial A_0/\partial x = \partial A_1/\partial x = \partial A/\partial x = 0$, as was to be shown.

Theorem I therefore ensures that Eq. (3.19) gives the most general solution of Eq. (3.1). Theorem II is equally, or, perhaps, more important, because it allows us to solve boundary value problems. This theorem shows that the generalized eigenfunctions are orthogonal on $(0, +\infty)$ with respect to the weight $\xi e^{-\xi^2}P(\xi)$. This orthogonality property is more classical than the full-range orthogonality because the weight function is positive. The only trouble now is the complicated expression of $P(\xi)$; it is to be noted that $P(\xi)$, though far from being an elementary function, satisfies two important identities which make the manipulation of integrals involving $P(\xi)$ much easier than would be expected. These identities are (Cercignani, refs. 25 and 28, Chapter VI)

$$2\pi^{-1/2}\int_0^\infty \frac{te^{-t^2}P(t)}{t+u}\,dt = [P(u)]^{-1} \tag{3.21}$$

$$u\frac{1}{\pi}\int_0^\infty \tan^{-1}\{\pi t/[p(t)]\}\,dt - \frac{\pi^{1/2}}{2}\int_0^\infty \frac{te^{t^2}[P(t)]^{-1}}{[p(t)]^2+\pi^2t^2}\frac{dt}{t+u} = P(u) \tag{3.22}$$

In addition,

$$P(0) = 2^{-1/2} \tag{3.23}$$

4. Application of the General Method to the Kramers Problem

In this section we shall apply the above results to a typical problem in kinetic theory. This problem has been investigated by many authors, and is usually referred to as the Kramers problem. It consists in finding the molecular distribution function of a gas in the following situation (Fig. 2): the gas fills the half-space $x > 0$ bounded by a physical wall in the plane $x = 0$, and is nonuniform because of a gradient along the x axis of the z component of the mass velocity; this gradient tends to a constant when x goes to infinity. It is seen that this problem can be considered as the limiting case of plane Couette flow, when one of the plates is pushed to infinity. More generally, the Kramers problem can be interpreted as a connection problem through the kinetic boundary layer (see Chapter V, Section 5); in this case "infinity" is simply the region where the Hilbert solution holds, and the velocity gradient "at infinity" can be regarded as constant because it does not vary appreciably on the scale of the mean free path (this is rigorously correct up to terms of order ϵ^2).

Both of these interpretations of the Kramers problem suggest that a convenient linearization is about a Maxwellian endowed with a translational velocity kx in the z direction. Because of the nonuniformity of this Maxwellian distribution linearization gives an inhomogeneous linearized Boltzmann equation:

$$2kc_1c_3 + c_1\,\partial h/\partial x = Lh \tag{4.1}$$

where $\mathbf{c} = (c_1, c_2, c_3) = (\xi_1, \xi_2, \xi_3 - kx)$. Equation (4.1) can be reduced to the homogeneous Boltzmann equation by subtracting a particular solution.

One particular solution independent of x is suggested by the Hilbert theory; this solution, $L^{-1}(2kc_1c_3)$, is given by $-2kc_1c_3\theta$ for the BGK model (θ being the mean free time, previously taken to be unity). Therefore we have

$$h = -2kc_1c_3\theta + 2c_3Y(x, c_1) \tag{4.2}$$

where $Y(x, \xi)$ satisfies Eq. (3.1) if x is measured in θ units. Again we write ξ in place of ξ_1, c_1, since no confusion arises. The mass velocity is given by

$$v_3 = kx + \pi^{-1/2} \int_{-\infty}^{\infty} e^{-\xi^2} Y(x, \xi)\, d\xi \tag{4.3}$$

the first term being the contribution from the Maxwellian.

Concerning the boundary conditions, we shall assume that the molecules are reemitted from the wall according to a Maxwellian distribution completely accommodated to the state of the wall (a more general assumption will be considered later). Therefore the boundary condition for h reads as follows:

$$h(0, \mathbf{c}) = 0 \qquad (c_1 > 0) \tag{4.4}$$

and this, in terms of Y, becomes

$$Y(0, \xi) = k\theta\xi \tag{4.5}$$

In addition, Y must satisfy the condition of boundedness at infinity.

According to the discussion in Section 3, the general solution of Eq. (3.1) which also satisfies the condition of boundedness at infinity is given by

$$Y(x, \xi) = A_0 + \int_0^{\infty} A(u) e^{-x/u} g_u(\xi)\, du \tag{4.6}$$

and the condition to be satisfied at the plate gives

$$k\theta\xi = A_0 + \int_0^{\infty} A(u) g_u(\xi)\, du \tag{4.7}$$

But solving this equation means expanding $k\theta\xi$ according to Theorem II of Section 3; therefore A_0 and $A(u)$ are given by Eqs. (3.14) and (3.15) if $k\theta\xi$ is substituted for $g(\xi)$. The result is as follows:

$$A_0 = -k\theta\pi^{-1/2} \int_0^{\infty} \tan^{-1}\left[\frac{\pi_\xi}{p(\xi)}\right] d\xi$$

$$= k\theta\pi^{1/2} \int_0^{\infty} \frac{\xi e^{\xi^2}}{[p(\xi)]^2 + \pi^2\xi^2}\, d\xi \tag{4.8}$$

$$A(u) = -k\theta\pi^{1/2}e^{u^2}[P(u)]^{-1}\{[p(u)]^2 + \pi^2u^2\}^{-1} \tag{4.9}$$

where use has been made of Eq. (3.21), which yields Eq. (4.9) directly and the following identity by asymptotically expanding for large values of u and comparing with Eq. (3.22) or Eq. (3.16):

$$2\pi^{-1/2}\int_0^\infty \xi^2 e^{-\xi^2}P(\xi)\,d\xi = (1/\pi)\int_0^\infty \tan^{-1}\{\pi t/[p(t)]\}\,dt \tag{4.10}$$

Equation (4.10) yields the first expression for A_0, while the second one is obtained by partial integration.

Substituting Eqs. (4.8) and (4.9) into Eq. (4.6) gives the solution of the Kramers problem. The mass velocity is readily obtained from Eqs. (4.3) and (4.6):

$$v_3(x) = kx + A_0 + \int_0^\infty A(u)e^{-x/u}\,du \tag{4.11}$$

where A_0 and $A(u)$ are given by Eqs. (4.8) and (4.9). From Eq. (4.11) we recognize that A_0 is the macroscopic slip of the gas on the plate (see Chapter V, Section 5): it has the form ζk, where ζ is the slip coefficient:

$$\zeta = \theta\pi^{1/2}\int_0^\infty \frac{\xi e^{\xi^2}\,d\xi}{[p(\xi)]^2 + \pi^2\xi^2} = 2l\int_0^\infty \frac{\xi e^{\xi^2}\,d\xi}{[p(\xi)]^2 + \pi^2\xi^2} \tag{4.12}$$

Here l is the mean free path defined by Eq. (3.27) of Chapter V and hence is related to θ by $\theta = 2\pi^{-1/2}l$ $(2RT_0 = 1)$. The integral appearing in Eq. (4.12) has been evaluated numerically (Albertoni *et al.*, ref. 5) with the following result:

$$\zeta = (1.01615)\theta = (1.1466)l \tag{4.13}$$

Restoring general units, Eq. (4.11) can be written as follows:

$$v_3(x) = k[x + \zeta - (\pi^{1/2}\theta/2)I(x/\theta)] \tag{4.11'}$$

where $I(x/\theta)$ is practically zero outside the kinetic layer; the velocity profile is sketched in Fig. 2. The quantity $I(x/\theta)$ can be easily evaluated (ref. 17); a plot is given in Fig. 3.

A direct evaluation of the microscopic slip, i.e., the velocity of the gas at the wall, results without any numerical calculation. We have

$$v_3(0) = k\left(\zeta - l\int_0^\infty \frac{e^{\xi^2}[P(\xi)]^{-1}}{[p(\xi)]^2 + \pi^2\xi^2}\,d\xi\right) = (2/\pi)^{1/2}kl \tag{4.14}$$

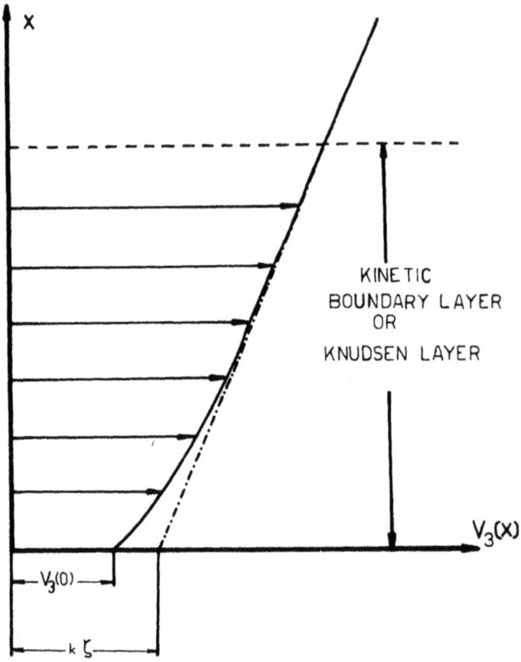

Fig. 2. Sketch of the velocity profile in the neighborhood of a wall in the presence of a velocity gradient k in the main body of the flow. Here $k\zeta$ is the macroscopic slip, $v_3(0)$ the microscopic one.

where the last result is obtained by letting $u \to 0$ in Eq. (3.22) and taking into account Eqs. (3.23) and (4.12).

Analogously, we can evaluate the distribution function of the molecules arriving at the plate. We obtain

$$Y(0, \xi) = 2\pi^{-1/2}kl\xi + 2\pi^{-1/2}klP(-\xi) \qquad (\xi < 0) \tag{4.15}$$

where Eq. (3.22) has been used. Then $h(0, \mathbf{c})$ (the perturbation of the Maxwellian distribution at the plate) is given by

$$h(0, \mathbf{c}) = 4\pi^{-1/2}klc_3P(|c_1|) \qquad (c_1 < 0) \tag{4.16}$$

and the function $P(\xi)$ $(\xi > 0)$ receives a physical interpretation in terms of the distribution function of the molecules arriving at the wall. From Eq. (3.22) it is easily inferred that

$$|c_1| + 0.7071 < P(|c_1|) < |c_1| + 1.01615 \tag{4.17}$$

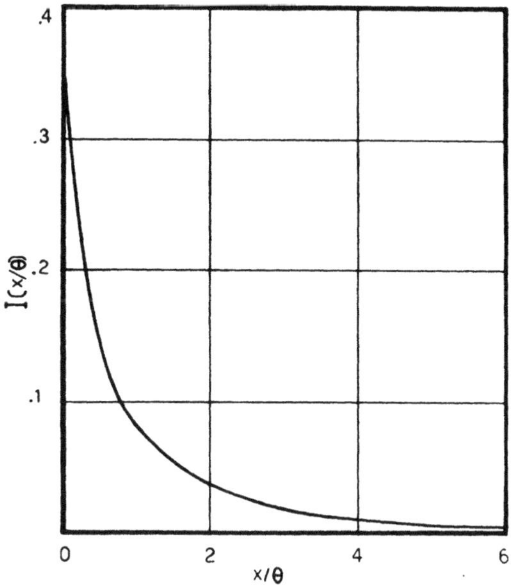

Fig. 3. The structure of the kinetic boundary layer in a shear flow according to the BGK model. The plot shows the correction to the Navier–Stokes profile with slip boundary conditions [see Eq. (4.11′)].

Hence the distribution function of the arriving molecules is rather close to a Hilbert distribution; in fact, a Hilbert expansion would predict Eq. (4.16) with $P(|c_1|)$ linear in $|c_1|$ [this is the distribution holding outside the kinetic layer; see Eqs. (4.2) and (4.6)]. The fact that the distribution of the molecules arriving at the plate is close to the one prevailing outside the kinetic boundary layer is not surprising; in fact, each molecule has the velocity acquired after its last collision, which, on the average, happened a mean free path from the wall, i.e., in a region where the distribution function is of the Hilbert type. It is interesting to note that Maxwell (ref. 1, Chapter II) assumed that the distribution function of the arriving molecules was exactly the one prevailing far from the wall; by using this assumption and conservation of momentum he was able to evaluate the slip coefficient without solving the Kramers problem. He found $\zeta = l$ (with an error of 15%) and

$$h(0, \mathbf{c}) = 4\pi^{-1/2}klc_3(|c_1| + 0.8863) \qquad (c_1 < 0) \tag{4.18}$$

i.e., a good approximation to the correct result [see Eqs. (4.16) and (4.17)].

Let us now briefly consider the case of a more general boundary condition. We shall assume Eq. (4.10) of Chapter IV with $\alpha = 1$:

$$h(0, \mathbf{c}) = \frac{4}{\pi}(1 - \beta)c_3 \int_{c_1' < 0} c_3' |c_1'| h(0, \mathbf{c}') \exp(-c'^2)\, d\mathbf{c}' \qquad (c_1 > 0) \quad (4.19)$$

In terms of $Y(x, \xi)$ this boundary condition becomes

$$Y(0, \xi) = k\theta[\xi + \tfrac{1}{2}(1 - \beta)\pi^{1/2}] - 2(1 - \beta) \int_{-\infty}^{0} \xi_1 Y(0, \xi_1) \exp(-\xi_1^2)\, d\xi_1 \qquad (\xi > 0) \quad (4.20)$$

or

$$Y(0, \xi) = k\theta[\xi + \tfrac{1}{2}(1 - \beta)\pi^{1/2}] + 2(1 - \beta) \int_{0}^{\infty} \xi_1 Y(0, \xi_1) \exp(-\xi_1^2)\, d\xi_1 \qquad (\xi > 0) \quad (4.21)$$

since Eq. (4.6) and Theorem I show that

$$\int_{-\infty}^{\infty} Y(x, \xi)\xi e^{-\xi^2}\, d\xi = 0 \qquad (4.22)$$

If we now multiply Eq. (4.21) by $\xi e^{-\xi^2}$ and integrate from 0 to ∞, we obtain

$$\beta \int_{0}^{\infty} \xi Y(0, \xi) e^{-\xi^2}\, d\xi = \tfrac{1}{4}k\theta(2 - \beta)\pi^{1/2} \qquad (4.23)$$

which inserted back into Eq. (4.21) gives

$$Y(0, \xi) = k\theta\xi + [k\theta\pi^{1/2}(1 - \beta)/\beta] \qquad (4.24)$$

Equation (4.24) shows that $\bar{Y} = Y - [k\theta\pi^{1/2}(1 - \beta)/\beta]$ satisfies the same equation and boundary conditions as the Y corresponding to $\beta = 1$. Therefore

$$Y(x, \xi; \beta) = [k\theta\pi^{1/2}(1 - \beta)/\beta] + Y(x, \xi; 1) \qquad (4.25)$$

and $Y(x, \xi; 1)$ has been found above. In particular, the slip coefficient is now

$$\zeta(\beta) = [\theta\pi^{1/2}(1 - \beta)/\beta] + \zeta(1) \qquad (4.26)$$

or

$$\zeta(\beta) = \frac{2 - (0.8534)\beta}{\beta} l \tag{4.27}$$

5. Application to the Flow between Parallel Plates

We have just seen that half-space problems connected with Eq. (3.1) can be solved by analytical means. This is not true for flows between parallel plates, such as Couette flow and Poiseuille flow. However, the method of elementary solutions can be used to obtain series solutions and gain insight into the qualitative behavior of the solution. Let us rewrite the general solution, Eq. (3.19), of Eq. (3.1) as follows:

$$Y(x, \xi) = A_0 + A_1(x - \xi) + \int_{-\infty}^{\infty} A(u) \exp\left[-\frac{x}{u} - \frac{\delta}{2|u|}\right] g_u(\xi)\, du \tag{5.1}$$

where δ is the distance between the plates in θ units, $A(u)$ has been redefined by inserting a factor $\exp[-\delta/(2|u|)]$ for convenience, and the plates are assumed to be located at $x = \pm\delta/2$.

The general form of Eq. (5.1) shows that any solution is split into two parts, one of which,

$$A_0 + A_1(x - \xi) \tag{5.2}$$

gives just the Navier–Stokes solution for the geometry considered (constant stress, straight profile of velocity) and the other part,

$$\int_{-\infty}^{\infty} A(u) f_u(\xi) \exp\left[-\frac{\delta}{2|u|} - \frac{x}{u}\right] du \tag{5.3}$$

is expected to be important only near boundaries, because of the fast decay of the exponential factor. Therefore for sufficiently large δ the picture is the following: a core, where a continuum description (based on the Navier–Stokes equations) prevails, surrounded by kinetic boundary layers produced by the interaction of the molecules with the walls. As δ becomes smaller, however, the exponentials are never negligible, i.e., the kinetic layers merge with the core to form a flow field which cannot be described in simple terms. Finally, when δ is negligibly small, $Y(x, \xi)$ does not depend appreciably on x, and the molecules retain the distribution they had just after their last interaction with a boundary. In the case of Couette flow (i.e., when there are two plates at $x = \pm\delta/2$ moving with velocities $\pm U/2$ in the z direction) the situation is well described by the above short discussion, although it is possible to obtain a more detailed picture by finding approximate expressions for A_1 and $A(u)$ [A_0 is zero and $A(u)$ is odd in u because of the antisymmetry inherent in the problem]. Accordingly, we shall consider in more detail the case of Poiseuille flow between parallel plates, which lends itself to more interesting considerations.

Plane Poiseuille flow (Fig. 4), is the flow of a fluid between two parallel plates induced by a pressure gradient parallel to the plates. In the continuum case no distinction is made between a pressure gradient arising from a density gradient and one arising from a temperature gradient. This distinction, on the contrary, is to be taken into account when a kinetic theory description is considered. We shall restrict ourselves to the former case (for the case of a temperature gradient see Cercignani, ref. 8).

The basic linearized Boltzmann equation for Poiseuille flow in a channel of arbitrary cross section (including the slab as particular case) will now be derived. We assume that the walls reemit the molecules with a Maxwellian distribution f_0 with constant temperature and an unknown density $\rho = \rho(z)$ (z being the coordinate parallel to the flow). If the length of the channel is much larger than any other typical length (mean free path, distance between the walls), then we can linearize about the Maxwellian f_0; in fact $\rho(z)$ is slowly varying and f_0 would be the solution in the case of a rigorously constant ρ. Accordingly, we have

$$\xi_1 \frac{\partial h}{\partial x} + \xi_2 \frac{\partial h}{\partial y} + \xi_3 \frac{\partial h}{\partial z} + \frac{1}{\rho} \frac{\partial \rho}{\partial z} \xi_3 = Lh \tag{5.4}$$

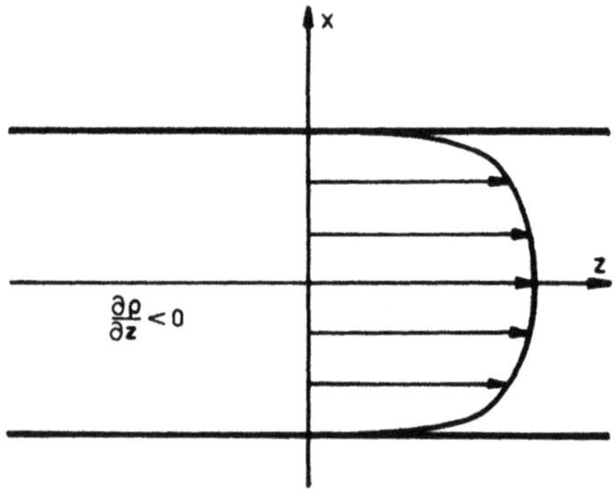

Fig. 4. Sketch of the velocity profile in plane Poiseuille flow.

Because of the assumption of a slowly varying ρ (long tube), we can regard $(1/\rho)\ \partial\rho/\partial z$ as constant (i.e., we disregard higher-order derivatives of ρ as well as powers of first-order derivatives). If $(1/\rho)\ \partial\rho/\partial z$ is regarded as a constant, it follows that $\partial h/\partial z = 0$, since z does not appear explicitly in the equation nor in the boundary conditions. The latter can be written

$$h(x, y, z, \boldsymbol{\xi}) = 0 \qquad [(x, y) \in \partial\Sigma; xn_1 + yn_2 > 0] \tag{5.5}$$

where $\partial\Sigma$ is the contour of the cross section and $\mathbf{n} = (n_1, n_2)$ the normal pointing into the channel. Therefore we can write

$$\xi_1 \frac{\partial h}{\partial x} + \xi_2 \frac{\partial h}{\partial y} + k\xi_3 = Lh \tag{5.6}$$

where $k = (1/\rho)\, d\rho/dz = (1/p)\, dp/dz$. Equation (5.6) governs linearized Poiseuille flow in a very long tube of arbitrary cross section. If we specialize to the case of a slab and use the BGK model, we have

$$h = 2\xi_3 Z(x, \xi_1) \tag{5.7}$$

where $Z(x, \xi)$ satisfies

$$\xi \frac{\partial Z}{\partial x} + \frac{1}{2}k = \pi^{-1/2} \int_{-\infty}^{\infty} \exp(-\xi_1^2) Z(x, \xi_1)\, d\xi_1 - Z(x, \xi) \tag{5.8}$$

$$Z\left(-\frac{\delta}{2}\operatorname{sgn}\xi, \xi\right) = 0 \tag{5.9}$$

provided x is measured in θ units. The above equations follow from a splitting analogous to the one considered in Section 2.

Equation (5.8) differs from Eq. (3.1) because of the inhomogeneous term $k/2$. If we find a particular solution of Eq. (5.8), then we can add it to the general solution of Eq. (3.1) in order to have the general solution of Eq. (5.8). By differentiation of the latter equation we deduce that $\partial Z/\partial x$ satisfies Eq. (3.1); since the general solution of the latter contains exponentials (which reproduce themselves by integration and differentiation) and a linear function of x, we try a particular solution of Eq. (5.8) in the form of a quadratic function of x (with coefficients depending upon ξ). It is verified that solutions of this form exist, and one of them is

$$Z_0(x, \xi) = \tfrac{1}{2}k[x^2 - (\delta^2/4) - 2x\xi - (1 - 2\xi^2)] \tag{5.10}$$

Therefore

$$Z(x, \xi) = Z_0(x, \xi) + Y(x, \xi) \tag{5.11}$$

where $Y(x, \xi)$ is given by Eq. (5.1). Equation (5.9) gives the following boundary condition for $Y(x, \xi)$:

$$Y\left(-\frac{\delta}{2}\operatorname{sgn}\xi, \xi\right) = -\left[|\xi| - (1 - 2\xi^2)\frac{1}{\delta}\right]\frac{k\delta}{2} \tag{5.12}$$

Since the symmetry inherent in this problem implies that

$$Y(x, \xi) = Y(-x, -\xi) \tag{5.13}$$

we have $A_1 = 0$ and $A(u) = A(-u)$. If we take this into account, Eq. (5.12) becomes

$$A_0 + \int_0^\infty A(u)g_u(\xi)\,du = -\frac{k\delta}{2}\left[\xi - (1 - 2\xi^2)\frac{1}{\delta}\right] - \int_0^\infty \frac{uA(u)}{u+\xi}e^{-\delta/u}\,du$$

$$(\xi > 0) \tag{5.14}$$

and the equation for $\xi < 0$ is not required, because $A(u) = A(-u)$. If we call the right-hand side of this equation $g(\xi)$, Eq. (5.14) becomes Eq. (3.13), and we can apply Eqs. (3.14) and (3.15), thus obtaining

$$A_0 = -\left\{\sigma + \frac{1}{\delta}\left(\frac{1}{2} + \sigma^2\right) - \int_0^\infty ue^{-\delta/u}[P(u)]^{-1}A(u)\,du\right\}\frac{k\delta}{2} \tag{5.15}$$

$$A(u) = \frac{k}{2}\pi^{1/2}\left(u + \frac{\delta}{2} + \sigma\right)e^{u^2}[P(u)]^{-1}\{[p(u)]^2 + \pi^2u^2\}^{-1}$$

$$+ \frac{\pi^{1/2}}{2}e^{u^2}[P(u)]^{-1}\{[p(u)]^2 + \pi^2u^2\}^{-1}\int_0^\infty \frac{\xi[P(\xi)]^{-1}}{u+\xi}e^{-\delta/\xi}A(\xi)\,d\xi \tag{5.16}$$

$$\xi > 0$$

where permissible inversions of the orders of integration have been performed and Eq. (3.21) used. Here $\sigma = \zeta/\theta$ [see Eqs. (4.12) and (4.13)]. Thus the problem has been reduced to the task of solving an integral equation in the unknown $A(u)$, Eq. (5.16). This equation is a classical Fredholm equation of the second kind with symmetrizable kernel. The corresponding Neumann–Liouville series can be shown to converge for any given positive value of δ (Cercignani, ref. 7).

It is also obvious that the larger δ, the more rapid is the convergence. This allows the ascertainment of certain results in the near-continuum regime. In particular, if terms of order $\exp[-3(\delta/2)^{3/2}]$ are negligible, only the zero-order term of the series need be retained:

$$A(u) = \frac{k}{2}\pi^{1/2}\left\{\left(\frac{\delta}{2} + \sigma\right) + u\right\}e^{u^2}[P(u)]^{-1}\{[p(u)]^2 + \pi^2u^2\}^{-1} \quad (5.17)$$

Within the same limits of accuracy A_0 is given by

$$A_0 = -\frac{k\delta}{2}\left[\sigma + \frac{1}{\delta}\left(\frac{1}{2} + \sigma^2\right)\right] \quad (5.18)$$

We note that this zero-order approximation is by far more accurate than a continuum treatment (even if slip boundary conditions are used in the latter). In fact, even in the zero-order approximation:

1. Kinetic boundary layers are present near the walls.
2. In the main body of the flow the mass velocity satisfies the Navier–Stokes momentum equation; however, the corresponding extrapolated boundary conditions at the walls are not the usual slip boundary conditions, but show the presence of a second-order slip:

$$v_3\left(\pm\frac{\delta}{2}\right) = \mp\sigma\left(\frac{\partial v_3}{\partial x}\right)_{x=\pm\delta/2} - \frac{1}{2}\left(\frac{1}{2} + \sigma^2\right)\left(\frac{\partial^2 v_3}{\partial x^2}\right)_{x=\pm\delta/2} \quad (5.19)$$

In order to obtain these results, we observe that the mass velocity is given by

$$v_3(x) = \pi^{-1/2}\int_{-\infty}^{\infty} Z(x,\xi)e^{-\xi^2}\,d\xi$$

$$= \frac{k}{2}\left(x^2 - \frac{\delta^2}{4}\right) + A_0 + \int_{-\infty}^{\infty} A(u)\exp\left[-\frac{x}{u} - \frac{\delta}{2|u|}\right]du \quad (5.20)$$

where Eqs. (5.10) and (5.11) have been taken into account. Equation (5.20) is exact; if terms of order $\exp[-3(\delta/2)^{3/2}]$ can be neglected, A_0 and $A(u)$ are given by Eqs. (5.18) and (5.17). In particular, the integral term in Eq. (5.20) describes the space transients in the kinetic boundary layers; in the main body of the flow the integral term is negligible, and we have

$$v_3(x) = (k/2)[x^2 - (\delta^2/4) - \sigma\delta - (\tfrac{1}{2} + \sigma^2)] \quad (5.21)$$

It is easily checked that this expression solves the Navier–Stokes momentum equation for plane Poiseuille flow and satisfies the boundary conditions (5.19).

We can also easily write the distribution function in the main body of the flow. As a matter of fact, $Y(x, \xi)$ reduces here to A_0, so that Eqs. (5.10) and (5.11) give

$$Z(x, \xi) = (k/2)[x^2 - (\delta^2/4) - 2x\xi - (1 - 2\xi^2) - \sigma\delta - (\tfrac{1}{2} + \sigma^2)] \qquad (5.22)$$

By taking into account Eq. (5.21), Eq. (5.22) can be rewritten as

$$Z(x, \xi) = v_3(x) - \theta\frac{\partial v_3}{\partial x}\xi + \left(\xi^2 - \frac{1}{2}\right)\theta^2\frac{\partial^2 v_3}{\partial x^2} \qquad (5.23)$$

where general units for x have been restored (i.e., we have written x/θ in place of x). Equation (5.23) clearly shows that in the main body of the flow the distribution function is of the Hilbert–Chapman–Enskog type (power series in θ), as was to be expected. However, it is not the usual distribution corresponding to the Navier–Stokes level of description. Equation (5.23) gives a Burnett distribution function, and this explains, from a formal point of view, the appearance of a second-order slip. From an intuitive standpoint the second-order slip can be attributed to the fact that molecules with nonzero velocity in the z direction move into a region with different density before having any collisions, and there is a net transport of mass because of the density gradient; i.e., molecules move preferentially toward smaller densities even before suffering any collision, and therefore at a mean free path from the wall an effect of additional macroscopic slip appears.

The presence of an additional slip means that for a given pressure gradient and a given geometry more molecules pass through a cross section than predicted by Navier–Stokes equations with first-order slip. This is easily shown for sufficiently large δ by using Eq. (5.20), which gives for the flow rate

$$F = \int_{-d/2}^{d/2} \rho v_3(x)\, dx = -\frac{1}{2}\frac{dp}{dx}d^2\left\{\frac{1}{6}\delta + \sigma + \frac{2\sigma^2 - 1}{\delta}\right\} \qquad (\delta \gg 1) \qquad (5.24)$$

provided terms of higher order in $1/\delta$ are neglected [Eqs. (5.17) and (5.18) have been used]. In Eq. (5.24) x and z are in general units, and $d = \delta\theta$ is the distance between the plates in the same units. Therefore for given geometry and pressure gradient the nondimensional flow rate is

$$Q(\delta) = \tfrac{1}{6}\delta + \sigma + [(2\sigma^2 - 1)/\delta] \qquad (\delta \gg 1) \qquad (5.25)$$

The last term is the correction to the Navier–Stokes result; it arises in part from the second-order slip and in part from the kinetic boundary layers. In fact, the gas near the walls moves more slowly than predicted by an extrapolation of Eq. (5.21); this brings in a correction to $Q(\delta)$ of the same order as the second-order slip. This correction reduces the effect of the second-order slip, but does not eliminate it completely, at least for sufficiently large values of δ. However, it is clear that the effect is not completely canceled even for smaller values of δ, because the molecules with velocity almost parallel to the wall give an appreciable contribution to the motion by traveling downstream for a mean free path. This qualitative reasoning is confirmed by a study of the nearly-free regime ($\delta \to 0$). This study can be based either on the iteration procedures to be described in Chapter VIII (Cercignani, ref. 9) or on a different use of the method of elementary solutions (Cercignani, ref. 7). In both cases the conclusion is

$$Q(\delta) \approx -\pi^{-1/2} \log \delta \qquad (\delta \to 0) \tag{5.26}$$

This means that higher-order contributions from kinetic layers destroy the $1/\delta$ term in Eq. (5.25) but leave a weaker divergence for $\delta \to 0$ (essentially related to the above-mentioned molecules traveling parallel to the plates). The behavior for large values of δ [Eq. (5.25)] and for small values of δ imply the existence of at least one minimum in the flow rate. This minimum was experimentally found in 1909 by Knudsen (ref. 10) for circular tubes and then by different authors for long tubes of various cross sections. The above discussion gives a qualitative explanation of the presence of the minimum, although its precise location for slabs and more complicated geometries must be found by approximate techniques (see Chapter VIII, Section 5).

6. Elementary Solutions for Time-Dependent Shear Flows

If one considers the time-dependent BGK equation in one-dimensional plane geometry, shear effects can be separated from effects related to normal stresses and heat transfer in the same way as for steady situations. The relevant equation for shear flow problems is

$$\frac{\partial Y}{\partial t} + \xi \frac{\partial Y}{\partial x} + Y(x, \xi) = \pi^{-1/2} \int_{-\infty}^{\infty} \exp(-\xi_1^2) Y(x, \xi_1)\, d\xi_1 \tag{6.1}$$

i.e., the time-dependent analog of Eq. (3.1). Here both x and t are expressed in θ units.

The procedure we shall use to deal with Eq. (6.1) can be summarized as follows: A Laplace transform is taken with respect to time, and, accordingly, the time-dependent problem is reduced to a steady one. The solution of the problem now depends on a complex parameter s. After separating the space and velocity variables the spectrum of values of the separation parameter v must be studied in its dependence on s. This study is essential in order to treat the problem of inversion.

Let us take the Laplace transform of Eq. (6.1). Without any loss of generality a zero initial value for Y will be assumed. In fact, a particular solution of the inhomogeneous transformed equation, which would result from a nonzero initial condition, can be constructed by using the Green function, and the latter can be easily obtained when the general solution of the homogeneous equation is known. Accordingly, we shall restrict ourselves to the homogeneous transformed equation

$$(s + 1)\tilde{Y} + \xi\, \partial\tilde{Y}/\partial x = \pi^{-1/2} \int_{-\infty}^{\infty} \exp(-\xi_1^2)\tilde{Y}(x, \xi_1)\, d\xi_1 \tag{6.2}$$

where $\tilde{Y}$ is the Laplace transform of Y. The same equation (with $s = i\omega$) governs the state of a gas forced to undergo steady transverse oscillations with frequency ω.

Separating the variables in Eq. (6.2) gives

$$\tilde{Y}_u(x, \xi; s) = g_u(\xi) \exp[-(s + 1)x/u] \tag{6.3}$$

where u is the separation parameter and $g_u(\xi)$ satisfies

$$(s + 1)\left(1 - \frac{\xi}{u}\right) g_u(\xi) = \pi^{-1/2} \int_{-\infty}^{\infty} g_u(\xi_1) \exp(-\xi_1^2)\, d\xi_1 \tag{6.4}$$

The right-hand side does not depend on ξ and can be normalized to unity. Accordingly, we are led to a typical division problem, in complete analogy with the steady case. If the factor $(u - \xi)$ cannot be zero, i.e., u is not a real number, $g_u(\xi)$ is an ordinary function given by

$$g_u(\xi) = u/(u - \xi) \tag{6.5}$$

with the normalization condition

$$\pi^{-1/2} \int_{-\infty}^{\infty} \frac{u e^{-\xi^2}}{u - \xi}\, d\xi = s + 1 \tag{6.6}$$

If, on the contrary, u is a real number, $g_u(\xi)$ must be treated as a generalized function, and Eq. (6.4) gives

$$g_u(\xi) = P\frac{u}{u - \xi} + p(u; s)\, \delta(u - \xi) \tag{6.7}$$

where $p(u; s)$, fixed by Eq. (6.6), is given by

$$p(u; s) = \pi^{1/2} e^{u^2} s + p(u) \tag{6.8}$$

with $p(u)$ the function introduced in Section 3. Equation (6.7) gives the generalized eigensolutions corresponding to the continuous spectrum $(-\infty < u < \infty)$. The essential point now is to study the possible values of

u which satisfy Eq. (6.6) and therefore constitute the discrete spectrum. Such values coincide, according to Eq. (6.6), with the zeros of the following function of the complex variable z:

$$M(z;s) = 1 - \pi^{-1/2}(s+1)^{-1} \int_{-\infty}^{\infty} \frac{ze^{-t^2}}{z-t}\,dt \tag{6.9}$$

This function is analytic in the complex z plane with a cut along the real axis where $M(z;s)$ has a discontinuity. In fact, Eqs. (3.18) and (3.19) of Chapter I give the following result for the limiting values of $M^{\pm}(u;s) = \lim_{\epsilon\to 0} M(u \pm i\epsilon; s)$ (u real, $\epsilon > 0$):

$$M^{\pm}(u;s) = 1 - \pi^{-1/2}(s+1)^{-1}\left[P\int_{-\infty}^{\infty} \frac{ue^{-t^2}}{u-t}\,dt + \pi i u e^{u^2}\right] \tag{6.10}$$

Equation (6.10) can also be written as follows:

$$M^{\pm}(u;s) = e^{-u^2}(s+1)^{-1}\pi^{-1/2}[p(u;s) \pm \pi i u] \tag{6.11}$$

In the limiting case of s such that

$$p(u;s) \pm \pi i u = 0 \qquad (\text{real } u) \tag{6.12}$$

the discrete spectrum merges into the continuous one. Equation (6.12) is satisfied on a closed heart-shaped curve of the complex s plane (Fig. 5). We have the following parametric representation for such a curve (to be called γ):

$$\begin{aligned} \operatorname{Re} s &= -\pi^{-1/2}e^{-u^2}p(u) \\ \operatorname{Im} s &= -\pi^{1/2}ue^{-u^2} \end{aligned} \qquad (-\infty < u < \infty) \tag{6.13}$$

These equations are obtained from Eqs. (6.12) and (6.8), taking into account that $p(u)$ is real. The equation $M(z;s) = 0$ defines a mapping from the z plane to the s plane: this equation gives unambiguously a point in the s plane once a point z off the real axis has been fixed. When z goes to a real value u it is seen that $M(z;s) = 0$, because of Eq. (6.11), becomes Eq. (6.12); the double sign is connected with the fact that one can go from above or from below. Therefore when u ranges through the real axis, s describes the curve counterclockwise if we think of the real axis as the boundary of the upper half-plane, clockwise if we think of the real axis as the boundary of the lower half-plane. In both cases Eqs. (6.13) establish a one-to-one correspondence between the curve γ of the s plane and the real axis of the z plane. From this fact and the argument principle it follows that both the lower and the upper half-planes of the z plane are mapped into the region

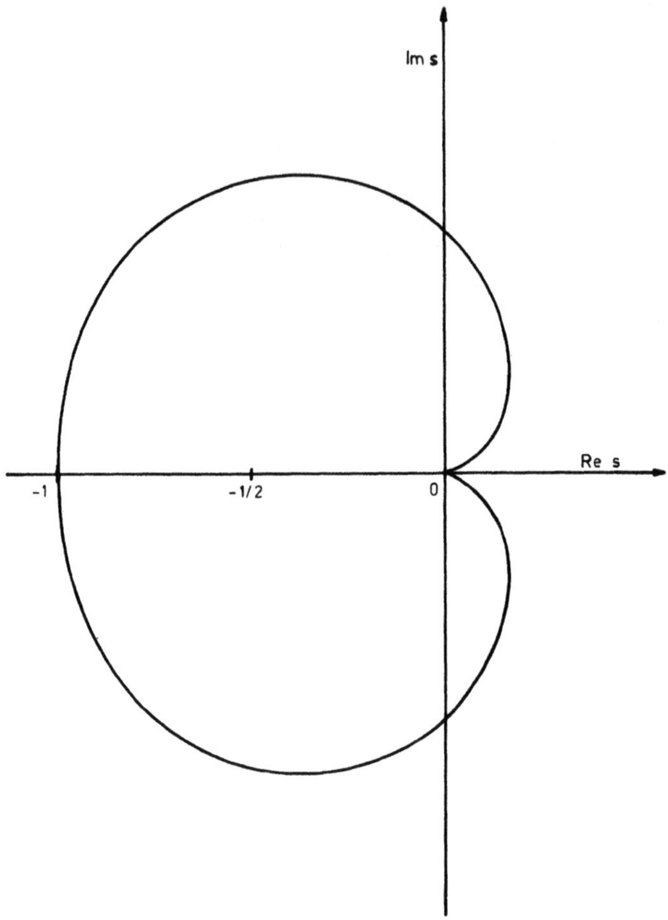

Fig. 5. The γ curve separating in the s plane the region with no discrete spectrum of u's (outer region) from the region with two discrete eigenvalues.

inside γ by the mapping $M(z; s) = 0$, and for each half-plane the mapping is one-to-one. It follows that for any s in the region inside γ there are two complex values of u which satisfy Eq. (6.6), while there are none outside γ. It is easily seen from Eq. (6.6) that these values are negative of each other. We shall denote them by $\pm u_0(s)$.

It is now possible to extend the results of the steady case ($s = 0$) and, in particular, Theorems I and II of Section 3. The completeness results remain true, and there are only slight changes in the equation. Thus in the case of the full range ($-\infty < \xi < \infty$) we can expand any function $g(\xi)$ satisfying the assumptions of Theorem I as follows:

$$g(\xi) = A_+ g_+(\xi) + A_- g_-(\xi) + \int_{-\infty}^{\infty} A(u) g_u(\xi)\, du \tag{6.14}$$

where $g_u(\xi)$ is given by Eq. (6.7) and $g_\pm$ by Eq. (6.5) with $u = \pm u_0$. The coefficients A_+ and A_- are zero outside γ (g_+ and g_- do not exist there), while they are given by

$$A_\pm = \pi^{-1/2}[s(1 - 2u_0^2) + 1]^{-1} \int_{-\infty}^{\infty} \frac{\xi e^{-\xi^2}}{\xi \mp u_0} g(\xi)\, d\xi \tag{6.15}$$

for s inside γ. For any s, $A(v)$ is given by Eq. (3.11) provided that $p(v)$ is replaced by $p(v; s)$ throughout. It is also obvious that $A_\pm$ and $A(v)$, and possibly $g(\xi)$, depend upon s, although this dependence has not been exhibited in the equations. In the case of the half range $0 < \xi < \infty$ a function $g(\xi)$ can be expanded as follows:

$$g(\xi) = A_+ g_+(\xi) + \int_0^{\infty} A(u) g_u(\xi)\, du \tag{6.16}$$

where A_+ is zero for s outside γ, and is given by

$$A_+ = [2u_0^2 s \pi^{1/2} P(u_0)]^{-1} \int_0^{\infty} \xi g(\xi) g_+(\xi) P(\xi) e^{-\xi^2}\, d\xi \tag{6.17}$$

for s inside γ. Here

$$P(u; s) = \frac{u}{u_0 + u} \exp\left[-\frac{1}{\pi} \int_0^{\infty} \tan^{-1} \frac{\pi t}{p(t, s)} \frac{dt}{t + u} \right] \tag{6.18}$$

For any s, $A(u)$ is given by Eq. (3.15), where $P(u)$ is given by Eq. (6.18) when s is inside γ, and is given by the same equation with $u_0 = 0$ when s is outside γ. The function $P(u; s)$ again satisfies certain identities which make its manipulation simpler than would be expected.

7. Analytical Solutions of Specific Problems

The theory sketched in Section 6 can be used to solve analytically problems of shear flows when the region filled by gas is the whole space or a half-space. One can, e.g., solve the following problem: let two half-spaces be separated by the plane $x = 0$, and assume that initially the gas has the same density ρ_0 and temperature T_0 in both regions, while the gas in the region $x > 0$ flows uniformly in the z direction with velocity U and the gas in the region $x < 0$ flows uniformly in the same direction with velocity $-U$; we want to find the evolution of the gas, i.e., the smoothing out and diffusion of the velocity discontinuity. The problem can be solved (Cercignani and Tambi, ref. 11) by using the theorem of full-range completeness to construct the Laplace transform of the solution. One can even obtain an analytic inversion of the Laplace transform and write the solution for the mass velocity as follows:

$$v(x,t) = U \operatorname{sgn} x\Bigl[1 - 2\pi^{-1/2}\int_{-\infty}^{\infty} H\left(\frac{u}{x} - \frac{1}{t}\right)\exp\left[-\frac{y}{\theta u} + q(u)\frac{(x/u) - t}{\theta}\right] \times \left\{e^{-u^2}\cos\left[\frac{\pi^{1/2}u}{\theta}\left(t - \frac{x}{u}\right)e^{-u^2}\right] + \frac{1 - q(u)}{u\pi^{1/2}}\sin\left[\frac{\pi^{1/2}u}{\theta}\left(t - \frac{y}{u}\right)e^{-u^2}\right]\right\}du \tag{7.1}$$

where H is the Heaviside step function and

$$q(u) = \pi^{-1/2}e^{-u^2}p(u) \tag{7.2}$$

We can use the exact solution to obtain asymptotic expansions for both short and long times and numerical tabulation of the space-time behavior of the gas. The solution shows that the velocity profile becomes more and more flattened as time goes on, but a disagreement of about 10% from the Navier–Stokes equations is still present after 12 collision times.

Half-space problems are more difficult to solve, since they require using the half-range completeness theorem, and, consequently, equations involving $P(u)$. However, the solution can always be reduced to a double quadrature for initial value problems and a single quadrature for problems of steady oscillations. As an example of the latter, we consider the propagation of Rayleigh waves in a half-space: let a half-space be filled with a gas of density ρ_0 and temperature T_0 and bounded by an infinite plane wall which is oscillating in its own plane with frequency ω. We shall consider the system in a steady state, when the transients have disappeared. Therefore if the velocity of the wall is the real part of $Ue^{i\omega t}$ (U being a constant), the solution of the linearized problem will be the real part of a function h having time dependence $e^{i\omega t}$ and satisfying

$$i\omega h + \xi_1\, \partial h/\partial x = Lh \tag{7.3}$$

The linearized boundary condition at the wall is

$$h(0, \boldsymbol{\xi}, t) = 2Ue^{i\omega t}\xi_3 \qquad (\xi_1 > 0) \tag{7.4}$$

We also require that the solution be bounded at infinity. If the BGK model is assumed to describe collisions, we can write

$$h(x, \boldsymbol{\xi}, t) = 2Ue^{i\omega t}\xi_3 \tilde{Y}(x, \xi_1) \tag{7.5}$$

where $\tilde{Y}(x, \xi)$ satisfies Eq. (6.2) (with $s = i\omega$) and the boundary condition

$$\tilde{Y}(0, \xi) = 1 \qquad (\xi > 0) \tag{7.6}$$

We shall write s in place of $i\omega$ because the following results are valid for any complex s ($\mathrm{Re}\, s \geq -1$) and the more general form will be useful later. Using the boundary condition, Eq. (7.6) and half-range completeness (see Section 6) together with boundedness at infinity in space, we find

$$\tilde{Y}(x, \xi) = -\pi^{1/2}s \int_0^\infty \frac{g_u(\xi)[P(u;s)]^{-1}}{[p(u;s)]^2 + \pi^2 u^2} \exp\left[-(s+1)\frac{x}{u} + u^2\right] du \tag{7.7}$$

when s is outside γ [then $P(u;s)$ is given by Eq. (6.18) with $u_0 = 0$]. Analogously, when s is inside γ we find

$$\tilde{Y}(x, \xi) = -2g_+(\xi) \exp\left[-(s+1)\frac{x}{u_0}\right][P(u_0)]^{-1}$$

$$-\pi^{1/2}s \int_0^\infty \frac{[P(u)]^{-1} g_u(\xi) \exp[-(s+1)x/u]}{[p(u;s)]^2 + \pi^2 u^2} du \tag{7.8}$$

where $u_0 = u_0(s)$ is selected between the two possible values in such a manner that

$$\mathrm{Re}\left[\frac{s+1}{u_0(s)}\right] > 0 \tag{7.9}$$

$P(u)$ in Eq. (7.8) is given by Eq. (6.18).

Equations (7.7) and (7.8) can be used to obtain, by integration, the mass velocity and the shearing stress. Particularly simple expressions can be found at $x = 0$ (Cercignani and Sernagiotto, ref. 12; Cercignani, ref. 28, Chapter VI).

Let us now briefly discuss the solution. First of all we note that there is a limiting frequency ω_0 ($i\omega_0 \in \gamma$) such that for $\omega > \omega_0$ we have only the eigensolutions of the continuous spectrum. It seems, therefore, that for $\omega > \omega_0$ no plane shear wave exists. However, we are able to exhibit a discrete term for $\omega > \omega_0$; as a matter of fact, we can extend the integration in Eq. (7.8) to the half-straight-line $\mathrm{Re}\, u + m\, \mathrm{Im}\, u = 0$ ($\mathrm{Re}\, u > 0$), where

$m = \max(\omega, 1)$, provided that we add the contribution from any poles of the integrand between this half-straight-line and the real semiaxis. Now it is easily seen that, at least for frequencies larger than, but still close to ω_0, there is one such pole, u_0, which satisfies

$$p(u_0; s) - \pi i u_0 = 0 \tag{7.10}$$

where $p(u; s)$ is given by Eq. (6.8) and the last expression in Eq. (3.6), which has a meaning for any complex u. Equation (7.10) is the analytical continuation of Eq. (6.6) for $\omega > \omega_0$. From this point of view ω_0 loses its character of being a critical frequency. Another feature of our results is that for any fixed frequency if we go sufficiently far from the wall the contribution from the continuous spectrum dominates the discrete term, since the former is less than exponentially damped. This feature is strictly related to the fact that the spectrum of values of v extends to infinity and would not be present if the collision frequency increased at least linearly with molecular velocity for large values of the latter; the experimental verification of these results seems to be outside the available techniques.

Another solvable problem is the following: Let a half-space be filled with a gas of density ρ_0 and temperature T_0 and bounded by a plane wall; the gas is initially in absolute equilibrium and the wall is at rest, and then the plate is set impulsively into motion in its own plane with uniform velocity U. The propagation into the gas of the disturbance produced by the motion of the plate is to be studied. This problem is known as Rayleigh's problem; we want to solve it analytically by using the linearized BGK model.

The perturbed distribution function satisfies the linearized Boltzmann equation and the following initial and boundary conditions:

$$h(x, 0, \boldsymbol{\xi}) = 0 \tag{7.11}$$

$$h(0, t, \boldsymbol{\xi}) = 2U\xi_3 \qquad (\xi_1 > 0) \tag{7.12}$$

Furthermore, when $x \to \infty$, $h(x, t, \boldsymbol{\xi})$ must be bounded for any fixed t and ξ. If we use the BGK model, we have

$$h(x, t, \boldsymbol{\xi}) = 2U\xi_3 Y(x, t, \xi_1) \tag{7.13}$$

where $Y(x, t, \xi)$ satisfies Eq. (6.1) and the following initial and boundary conditions:

$$Y(x, 0, \xi) = 0 \tag{7.14}$$

$$Y(0, t, \xi) = 1 \qquad (\xi > 0) \tag{7.15}$$

By introducing the Laplace transform of Y, $\tilde{Y}(x, s, \xi)$, we reduce Eq. (6.1) to Eq. (6.2), while the boundary condition at the wall becomes

$$\tilde{Y}(0, s, \xi) = 1/s \qquad (\xi > 0) \tag{7.16}$$

therefore $\tilde{Y}$ is obtained from the equations for the oscillating wall by multiplying the right-hand side of the equations by $1/s$; the same is also true for the mass velocity and the stress. It is important to note that $\tilde{Y}$ defined by Eqs. (7.7) and (7.8) for s outside and inside γ is an analytic function of s not only outside and inside γ, but also through this curve; in other words, Eq. (7.8) is the analytic continuation of Eq. (7.7) inside γ. This follows from the fact that the functions and integrals appearing in Eq. (7.7) undergo a discontinuity when s crosses γ, and these discontinuities contribute to a discrete term equal to the limiting value of the discrete term in Eq. (7.8). This circumstance allows the integration path in inverting the Laplace transform to be deformed through γ, provided that in every region the appropriate expression is used. On the other hand, the segment $(-1, 0)$ of the real axis is easily seen to be a discontinuity line because of the choice of u_0 [Eq. (7.9)]. According to well-known theorems on the Laplace transform, $Y(x, t, \xi)$ is given by

$$Y(x, t, \xi) = \frac{1}{2\pi i} \int_{c-i\infty}^{c+i\infty} \frac{e^{st}}{s} \tilde{Y}(x, s, \xi)\, ds \tag{7.17}$$

where $\tilde{Y}$ is given by Eq. (7.7) and the path of integration is a vertical straight line to the right of γ. Owing to the analyticity properties of $\tilde{Y}$, this integration path in the s plane can be deformed to a path indented on the segment $(-1, 0)$ of the real axis and along the vertical line $\mathrm{Re}(s + 1) = 0$. The resulting integrals can be put into a completely real form; in this way the problem is solved in terms of quadratures, which can, in principle, be performed with any desired accuracy. But we can also use our results for obtaining interesting information by analytical manipulations. We can, e.g., expand our results for short and long times (Cercignani and Sernagiotto, ref. 12). It is particularly interesting to quote the result for the mass velocity valid for long times:

$$v_3(x, t)/U \approx 1 - (\pi \nu t)^{-1/2} w(x) \qquad (t \to \infty) \tag{7.18}$$

where ν is the kinematic viscosity ($\theta = 2\nu$), and $w(x)$ the mass velocity corresponding to a unitary gradient at infinity in the Kramers problem [Eqs. (4.11), (4.8), and (4.9) with $k = 1$]. Therefore the flow shows the same structure for the kinetic boundary layer both in the steady and in the time-dependent flows (for large values of the time); this is not surprising, because we know that the equations which describe the kinetic layers (Chapter V) do not depend upon the particular problem. In particular, in this time-dependent flow the slip coefficient retains the same value as in steady flows.

Other applications of the analytical solution consist of evaluating the velocity and the stress at the plate. Rather simple expressions are found (ref. 12); as an example, we quote

$$v_3(0,t)/U = 1 - (1/\pi)\int_0^1 (v^{-1} - 1)^{1/2} e^{-vt}\,dv$$

$$= \tfrac{1}{2} + \tfrac{1}{2}\int_0^{t/2} e^{-v} v^{-1} I_1(v)\,dv$$

where I_1 denotes, as usual, the modified Bessel function of the first kind and order one. These expressions show that $v_3(0,t)$ slowly increases from an initial value $U/2$ to the final value U (Fig. 6).

8. More General Models

After the detailed treatment of both steady and timedependent shear flows it seems natural to consider Eqs. (2.10) and (2.11), which describe steady heat transfer processes according to the BGK model. However, these equations are coupled. By adopting matrix notation we can repeat the familiar

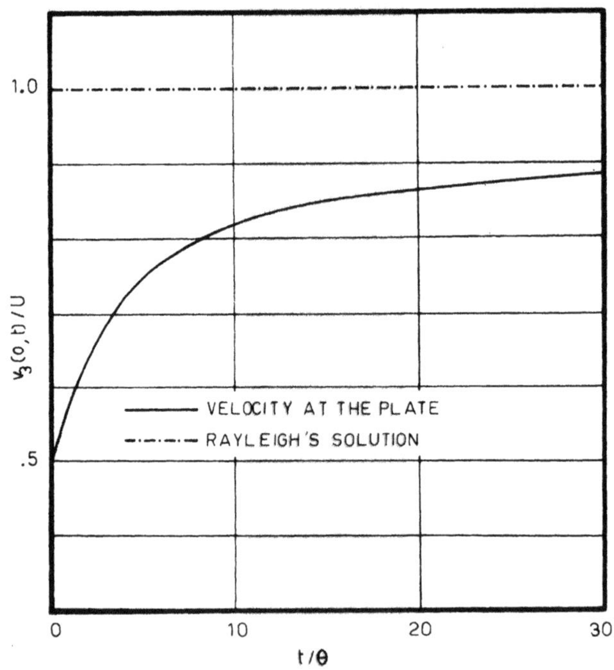

Fig. 6. Time evolution of the mass velocity at the plate in the linearized Rayleigh problem according to the BGK model.

procedures of previous sections as far as linear operations are concerned. In such a way, e.g., the completeness of the elementary solutions in the full range can easily be shown, and explicit formulas for the coefficients of the expansion can be given. Unfortunately, the matter is not so easy for the half-range expansion, where strongly nonlinear operations were required to construct $P(\xi)$ from $p(\xi)$ [Eq. (3.16)]. Therefore a different procedure must be used, but we lose a part of the explicit information, which has to be recovered by approximate procedures. Since the same methods can be applied to more general models with the only difference being the use of $n \times n$ instead of 2×2 matrices, there is no point in considering the BGK model separate from higher-order models when heat transfer problems or unsteady problems with normal stresses are considered. Accordingly, in this section we shall consider the more general models described in Chapter IV, Section 2. The more general model equation with constant collision frequency can be written as follows (one-dimensional problems):

$$\frac{\partial h}{\partial t} + \xi_1 \frac{\partial h}{\partial t} = \sum_{k,j=1}^{M} \alpha_{jk}(h, \psi_j)\psi_k - \nu h \tag{8.1}$$

where the ψ_j are the eigenfunctions of the Maxwell collision operator.

First of all we eliminate the transverse components of the molecular velocity by decomposing $h(x, t, \boldsymbol{\xi})$ according to

$$h = \sum_{j=1}^{n} h_j(x, t, \xi_1)g_j(\xi_2, \xi_3) + h_R(x, t, \boldsymbol{\xi}) \tag{8.2}$$

where $\{g_j\}$ denotes the set of polynomials orthogonal according to the inner product (2.8) and h_R is unambiguously defined as being orthogonal to the g_j ($j = 1, 2, \ldots, n$); n is an integer ($\leq M$) such that every ψ_k ($k = 1, \ldots, M$) can be expressed in terms of the g_j ($j = 1, \ldots, n$) with coefficients which are polynomials in ξ_1. Then Eq. (8.1) can be transformed into the following system:

$$\frac{\partial h_j}{\partial t} + \xi \frac{\partial h_j}{\partial x} + h_j(x, t, \xi) = \sum_{r=1}^{n} \sum_{k=0}^{m} (\varphi_k, h_r)_1 X_{jkr}(\xi) \qquad (j = 1, \ldots, n) \tag{8.3}$$

$$\frac{\partial h_R}{\partial t} + \xi_1 \frac{\partial h_R}{\partial x} + h_R(x, t, \boldsymbol{\xi}) = 0 \tag{8.4}$$

where $\{\varphi_k(\xi)\}$ is the set of polynomials which are orthogonal with respect to the scalar product

$$(h, g)_1 = \pi^{-1/2} \int_{-\infty}^{\infty} h^*(\xi) g(\xi) e^{-\xi^2} \, d\xi \tag{8.5}$$

the asterisk denoting complex conjugation; in (8.3) m is an integer ($\leq M$) which is given by the maximum degree of the coefficients of the expansions of the ψ_k ($k = 1, \ldots, M$) in terms of the g_j ($j = 1, \ldots, n$). The X_{jkr} are suitable polynomials (having maximum degree m). Here v has been assumed to be unity.

Equation (8.4) is immediately solved, and since it is not coupled with the system given by Eq. (8.3), will not be considered further. Separating the variables in the usual way in Eq. (8.3), we obtain

$$(s + 1)\left(1 - \frac{\xi}{u}\right)h_j(\xi; u, s) = \sum_{r=1}^{n} \sum_{k=0}^{m} (\varphi_k, h_r)_1 X_{jkr}(\xi) \tag{8.6}$$

From Eq. (8.6) we obtain the following moment equations:

$$u(s + 1)(\varphi_r, h_j)_1 - (\xi\varphi_r, h_j)_1 = u \sum_{k=0}^{m} \sum_{s=1}^{n} (\varphi_k, h_s)_1(\psi_r, X_{jks})_1 \tag{8.7}$$

$$r = 0, 1, \ldots, m - 1; \quad j = 1, \ldots, n$$

Since $\xi\varphi_r$ can be expressed as a combination of φ_{r-1} and φ_{r+1}, the latter system of $m \times n$ equations can be solved immediately with respect to the $m \times n$ quantities (φ_r, h_j) ($r = 1, 2, \ldots, m; j = 1, \ldots, n$), which are accordingly expressed in terms of $(\psi_0, h_j)_1$. The coefficients of the combination are rational functions of u and s (polynomials for Maxwell molecules). Then Eq. (8.6) can be written as follows:

$$\left(1 - \frac{\xi}{u}\right)h_j(\xi; u, s) = \sum_{k=0}^{m} T_{jk}(\xi; u; s)A_k(u, s) \qquad (j = 0, \ldots, m) \tag{8.8}$$

where the T_{jk} are polynomials in ξ, u, and s and

$$A_k(u; s) = (1, h_k)[D(u, s)(s + 1)]^{-1} \tag{8.9}$$

with $D(u, s)$ the determinant of Eq. (8.7) [essentially, $D(u, s) = 1$ for Maxwell molecules, but is a polynomial in u and s for more general models]. Equation (8.10) can be transcribed in matrix notation:

$$\left(1 - \frac{\xi}{u}\right)\mathbf{h}(\xi; u; s) = \mathbf{T}(\xi; u, s)\mathbf{A}(u; s) \tag{8.10}$$

where $\mathbf{h}$ and $\mathbf{A}$ are vectors and $\mathbf{T}$ a matrix. In correspondence to the usual continuous spectrum, we now have the following eigenfunctions:

$$\mathbf{h}(\xi; u, s) = \mathbf{T}(\xi; u, s)\mathbf{A}(u, s)P\frac{u}{u - \xi} + \mathbf{p}(u; s)\mathbf{A}(u, s)\,\delta(u - \xi) \tag{8.11}$$

where, because of Eq. (8.9),

$$\mathbf{p}(u; s) = e^{v^2}\left[\pi^{1/2}D(u; s)(s + 1)\mathbf{I} - P\int_{-\infty}^{\infty}\frac{\mathbf{T}(\xi, u, s)ue^{-\xi^2}}{u - \xi}\,d\xi\right] \quad (8.12)$$

where $\mathbf{I}$ is the identity matrix. The integrals appearing in Eq. (8.12) can easily be expressed in terms of the function $p(u)$ introduced in Section 3 and therefore in terms of tabulated functions. Let us introduce the following matrix, which has functions of the complex variable z as elements:

$$\mathbf{M}(z; s) = \pi^{1/2}D(u; s)\mathbf{I} + \frac{1}{s + 1}\int_{-\infty}^{\infty}\frac{\mathbf{T}(\xi; z; s)ze^{-\xi^2}}{\xi - z}\,d\xi \quad (8.13)$$

then it is easily seen that the discrete spectrum (if any) is given by the complex values of u solving

$$\text{Det } \mathbf{M}(u; s) = 0 \quad (8.14)$$

It is relatively easy to prove the completeness of eigensolutions in the full range ($-\infty < \xi < \infty$) and construct the coefficients of the functions in terms of quadratures. The expansion for a general function $\mathbf{g}(\xi)$ takes on the form

$$\mathbf{g}(\xi) = \sum_{i=1}^{N}\mathbf{g}_i(\xi; s)\mathbf{A}_i(s) + \int_{-\infty}^{\infty}\mathbf{g}(\xi; u, s)\mathbf{A}(u; s)\,du \quad (8.15)$$

where

$$\mathbf{g}(\xi; u, s) = \mathbf{T}(\xi; u, s)P\frac{u}{u - \xi} + \mathbf{p}(u; s)\,\delta(u - \xi) \quad (8.16)$$

and

$$\mathbf{g}_i(\xi; s) = \mathbf{T}(\xi; u_i, s)P\frac{u_i}{u_i - \xi} \quad (8.17)$$

u_i ($i = 1, \ldots, N$) being the possible solutions of Eq. (8.14). The coefficients $\mathbf{A}_i(s)$ and $\mathbf{A}(u; s)$ can be expressed explicitly in terms of quadratures involving the given function $\mathbf{g}(\xi)$. A key role is played by $\mathbf{M}(z; s)$ and its obvious relations to $\mathbf{p}(u; s)$, which allow us to solve the necessary integral equations by algebraic operations on matrices (ref. 28, Chapter VI). We note also that in the steady case ($s = 0$) we have no complex u and therefore no discrete spectrum. However, in this case we have to add the solutions arising from the collision invariants as well as from the particular solution (linear in x)

which is obtained when specializing Eq. (8.1) of Chapter VI to one-dimensional problems.

It is an easy guess to anticipate that the elementary solutions have partial-range completeness properties. However, the straightforward procedure of demonstration which proves useful for the case of one equation cannot be extended to the case of a system, since a closed form cannot be found for a certain matrix [the analog of the function $P(u)$ in the scalar case] which plays a fundamental role in the analytical process of solution. However, for the case of the half-range ($0 < \xi < \infty$), which is the most important for applications, a demonstration that the completeness property holds can actually be set up (ref. 28, Chapter VI). Constructive proofs have been found for some special cases only (see the next section and the appendix). However, one can devise simple approximation procedures which should prove very good: one can either approximate the matrix analog of $P(u)$ (which is a very smooth function for $u > 0$) (ref. 28, Chapter VI) or make a reasonable guess (with adjustable constants) for the distribution function of the arriving molecules, which is also very smooth and nearly a polynomial (see Section 4) (Buckner and Ferziger, ref. 13).

9. Some Special Cases

The results of Section 8 imply that shear flow problems with the BGK model are very special, in the sense that analytical tools can be used to a larger extent than for more general problems and models. One can, however, consider some other cases which allow the analytical machinery to be developed to the same extent. A first class of cases arises from simplifications of the BGK model: one can drop conservation of energy for the purpose of studying the so-called isothermal waves (Ostrowski and Kleitman, ref. 14; Mason, ref. 15), one can conserve energy but allow only one-dimensional collisions (Weitzner, ref. 16), or, finally, one can decouple one of the three degrees of freedom of the molecules from the remaining two (Cercignani, ref. 28, Chapter VI). All these modifications allow a simplification of the equations in such a way that we can reduce the problem to one of solving a single equation instead of a system.

Another interesting case is offered by the ES model [Eq. (2.10) of Chapter IV with $N = 9$]. If we consider shear flow problems in one dimension, the only difference from the BGK model is the appearance of an integral proportional to the shearing stress, but using conservation of momentum one can eliminate it in favor of the integral proportional to mass velocity and obtain an equation very similar to the BGK model. The situation is very simple in steady problems (Cercignani and Tironi, ref. 9, Chapter IV), and we can express the solution of shear problems with the ES model in terms of the analogous solution with the BGK model. In particular, one can show that the slip coefficient according to the ES model has exactly the same value as the BGK one.

Another special case is offered by a model with velocity-dependent collision frequency proportional to speed, $\nu(\xi) = \sigma\xi$ (σ = const). In this case the heat transfer equations, analogous to Eqs. (2.10) and (2.11), can be reduced to equations involving a single moment in the right-hand side, as was shown by Cassel and Williams (ref. 17) and hence can be solved by the techniques used in the previous sections. On the other hand, the solution of the heat transfer equations for the BGK model, Eqs. (2.10) and (2.11), requires considerable more effort. The first solution was published by the author in 1977 (ref. 18) and was worked out in more detail by Siewert and Kelley in 1980 (ref. 19). In both papers use is made of an idea of Darrozès (ref. 20), who proposed to diagonalize the relevant Riemann–Hilbert problem. New singularities (branch cuts) then arise, which must be canceled. Darrozès was not able to master this new problem, which was brought to completion in 1982 (ref. 21), when a procedure to compute analytically the partial indices of the Riemann–Hilbert problem was proposed. The method, which will be sketched in the appendix to this chapter, was later extended to time-dependent problems (refs. 22 and 23) and used to treat the half-range completeness problem for sound propagation (see Section 13) in a closed form.

Another interesting case is offered by the model with velocity-dependent collision frequency, given by Eq. (3.3) of Chapter IV. This model is not included in the treatment given in Section 8, but one can again split the equation in such a way as to separate shear flows from changes in density, normal velocity, and temperature (Cercignani, ref. 10, Chapter IV). The relevant equation for the shear flow problem is

$$\frac{\partial Y}{\partial t} + \xi\mu\frac{\partial Y}{\partial x} + \nu(\xi)Y = \frac{\nu(\xi)}{\bar{\nu}}\int_0^\infty d\xi_1 \int_{-1}^{+1} d\mu_1 \nu(\xi_1)\xi_1^4 \exp(-\xi_1^2)(1-\mu_1^2) \times Y(x, t, \xi_1, \mu_1) \tag{9.1}$$

where

$$\bar{\nu} = \tfrac{4}{3}\int_0^\infty \nu(\xi)\xi^4 e^{-\xi^2}\, d\xi \tag{9.2}$$

$\xi\mu$ is the x component of $\boldsymbol{\xi}$ (so that μ is the cosine of the angle between $\boldsymbol{\xi}$ and the x axis), and Y is related to $h(x, t, \boldsymbol{\xi})$ by

$$h = 2\xi_3 Y(x, t, \xi, \mu) \tag{9.3}$$

In the steady case the treatment of Eq. (9.1) is rather easy. If we set

$$w = \xi\mu/\nu(\xi) \tag{9.4}$$

Eq. (9.1) (with $\partial Y/\partial t = 0$) can be written as follows:

$$w\frac{\partial Y}{\partial x} + Y = \frac{1}{\bar{\nu}}\int_{-k}^{k} dw_1\, Z(x, w_1) \tag{9.5}$$

where k is defined by

$$k = \lim_{\xi\to\infty}[\xi/\nu(\xi)] \tag{9.6}$$

and

$$Z(x, w) = \int_{\xi(w)}^{\infty} d\xi_1\, \xi_1\{\xi_1^2 - w^2[\nu(\xi_1)]^2\}\exp(-\xi_1^2)[\nu(\xi_1)]^2 Y(x, w, \xi_1) \tag{9.7}$$

Here we have introduced the function $\xi(w)$ which inverts the relation

$$\xi/\nu(\xi) = |w| \tag{9.8}$$

Multiplying Eq. (9.5) by $e^{-\xi^2}[\nu(\xi)]^2\{\xi^2 - w^2[\nu(\xi)]^2\}$ and integrating with respect to ξ from $\xi(w)$ to ∞, we have

$$w\frac{\partial Z}{\partial x} + Z = \Phi(w)\int_{-k}^{k} dw_1\, Z(x, w_1) \tag{9.9}$$

where

$$\Phi(w) = \bar{\nu}^{-1}\int_{\xi(w)}^{\infty} d\xi_1\, \xi_1\{\xi_1^2 - w^2[\nu(\xi_1)]^2\}[\nu(\xi_1)]^2\exp(-\xi_1^2) \tag{9.10}$$

It is useful to note that $\Phi(w) \geq 0$, $w\Phi'(w) \leq 0$, and

$$\int_{-k}^{k} \Phi(w)\, dw = 1 \tag{9.11}$$

The latter result is easily obtained by a partial integration. Equation (9.10) is very similar to the equation for the BGK model, and can be treated by a completely analogous procedure. The general solution is

$$Z(x, w) = \Phi(w)\left\{A_0 + A_1(x - w) + \int_{-k}^{k} A(u)e^{-x/u}g_u(w)\,du\right\} \tag{9.12}$$

where

$$g_u(w) = P\frac{u}{u - w} + p(u)\,\delta(u - w) \tag{9.13}$$

$$p(u) = [\Phi(u)]^{-1}P\int_{-k}^{k} \frac{w\Phi(w)}{w - u}\,du \tag{9.14}$$

The general solution of Eq. (9.5) can now be easily written, since it is a matter of solving an ordinary first-order differential equation with constant coefficients and given source term. We obtain

$$Y(x, w, \xi) = \bar{v}^{-1}\left\{A_0 + A_1(x - w) + \int_{-k}^{k} A(u)e^{-x/u}f_u(w)\,du + B(w, \xi)e^{-x/w}\right\} \tag{9.15}$$

where $B(w, \xi)$ is an arbitrary function such that

$$\int_{\xi(w)}^{\infty} d\xi_1\,\xi_1\{\xi_1^2 - w^2[\nu(\xi_1)]^2\}\exp(-\xi_1^2)[\nu(\xi_1)]^2 B(w, \xi_1) = 0 \tag{9.16}$$

The mass velocity $v_3(x)$ is given by

$$v_3(x) = \bar{v}^{-1}\left\{A_0 + A_1 x + \int_{-k}^{k} A(u)n(u)e^{-x/u}\,du + \int_{-k}^{k} B(u)\,e^{-x/u}\,du\right\} \tag{9.17}$$

where

$$n(u) = \int_{-k}^{k} dw f_u(w)\Psi(w) \tag{9.18}$$

$$\psi(w) = 2\pi^{-1/2}\int_{\xi(w)}^{\infty} d\xi_1\xi_1\{\xi_1^2 - w^2[\nu(\xi_1)]^2\}\nu(\xi_1)\exp(-\xi_1^2) \tag{9.19}$$

Completeness in the full range and in the half range is easily proved. In this case the function $P(u)$ which occurs in the weight of the half-range orthogonality relations is given by

$$P(u) = u \exp\left\{-\frac{1}{\pi}\int_0^k \tan^{-1}\left[\frac{\pi t}{p(t)}\right]\frac{dt}{t+u}\right\} \tag{9.20}$$

and satisfies identities and relations analogous to those valid in the case of the BGK model.

As a consequence of these results, all the particular problems which can be solved with the BGK model can be solved with the more general model we are considering. In particular, one can solve exactly the Kramers problem and find the slip coefficient (Cercignani, ref. 10, Chapter IV; Cercignani and Sernagiotto, ref. 24):

$$\zeta = -\pi^{-1}\int_0^k \tan^{-1}[\pi t/p(t)]\, dt \tag{9.21}$$

This expression has been used to evaluate the dependence of ζ upon the form of $\nu(\xi)$ (Cercignani, Foresti, Sernagiotto, ref. 25). It has been found that for $\nu(\xi)$ increasing linearly when $\xi \to \infty$ the value of ζ is somewhat lower (3%) than for the BGK model (for a fixed value of the viscosity coefficient).

Knudsen layers for kinetic models have been considered by many authors. Part of this work was stimulated by the experimental procedures developed (refs. 26 and 27) to measure velocity profiles in the Knudsen layer on a flat wall. In particular, Reynolds *et al.* reported that in the Knudsen layer the deviation of the velocity defect (i.e., the deviation of the actual velocity from the continuum velocity profile) shows a behavior quantitatively different from the results obtained by the BGK model and described in Section 4. Loyalka (ref. 28) pointed out that such a discrepancy could be due to a basic deficiency of the BGK model, in that it does not allow for the velocity dependence of the collision frequency. With this in view, he carried out a detailed numerical study of the model with a velocity-dependent collision frequency, which has been just discussed. The solution of ref. 10 in Chapter IV was considered by Loyalka not to be useful for the purpose of numerical evaluation, for which he preferred to use a direct numerical technique. His results show that the velocity dependence of the collision frequency appropriate to rigid spheres does indeed have an important effect on the velocity defect, and in fact the numerical solution practically coincides with the upper boundary of the region containig 80% of the experimental data reported by Reynolds *et al.* Another well-known deficiency of the BGK model is that it yields a Prandtl number Pr not appropriate to a monatomic gas (Pr = 1 instead of Pr = 2/3). This aspect motivated the above-mentioned work with the ES model by Cercignani and

Tironi (ref. 9, Chapter IV). Unfortunately the solution for the Knudsen layer in the latter paper contained a trivial mistake: a numerator and a denominator were erroneously interchanged. When the mistake is corrected the results are in reasonably good agreement with the experiments, in fact, almost indistinguishable from the results presented by Loyalka. This fact was pointed out in ref. 29. At about the same time, Abe and Oguchi (ref. 30) examined the Knudsen layers with a model equation of the Gross and Jackson hierarchy, involving 13 models in the collision term. The model contains two nondimensional parameters, the Prandtl number and a further parameter taking the value unity when the model reduces to the linearized ES model. They conclude that the solution indicates a "weak dependence" on this parameter (which should take the value 4/9, according to Gross and Jackson's prescription).

Gorelov and Kogan (ref. 31) reported the results of Monte Carlo calculations which appear to fall on the lower boundary of the aforementioned region containing 80% of the data. These Monte Carlo calculations were confirmed by Bird (ref. 32). Then it was pointed out (ref. 33) that one can concoct a new model having the desirable properties of both a correct Prandtl number and a variable collision frequency, while amenable to a simple solution. The velocity profile computed by means of this model turns out to be in an exceptionally good agreement with the experiments of Reynolds *et al.*

The results just described show that no essentially new situations arise with respect to the BGK model as long as we restrict ourselves to steady problems. The same thing cannot be said about problems depending on time as well as on a space coordinate. This circumstance is related to the appearance of a spectrum filling a two-dimensional region, which was pointed out in Chapter VI, Section 7 and will be investigated in some detail in the next section.

Completely different problems arise in connection with evaporating and condensing flows. They have been briefly discussed in Section 6 of the previous chapter.

10. Unsteady Solutions of Kinetic Models with Velocity-Dependent Collision Frequency

If we take the Laplace transform of Eq. (9.1) and disregard, as usual, a possible inhomogeneous term related to the initial datum, we have

$$[\nu(\xi) + s]\tilde{Y} + \xi\mu\frac{\partial \tilde{Y}}{\partial x} = \frac{\nu(\xi)}{\bar{\nu}}\int_0^\infty d\xi_1 \int_{-1}^{+1} d\mu_1\, \nu(\xi_1)\xi_1^4 \exp(-\xi_1^2)(1-\mu_1^2)\tilde{Y}(x, s, \xi_1, \mu_1) \tag{10.1}$$

where $\tilde{Y}$ is the Laplace transform of Y. The same equation (with $s = i\omega$) governs the state of a gas forced to undergo steady transverse oscillations with frequency ω. If s is real, then the treatment is very similar to the steady case; it is sufficient to introduce a variable w related to ξ and μ by Eq. (9.4) with $\nu(\xi) + s$ in place of $\nu(\xi)$. If s is complex, however, w turns out to be complex:

$$w = \alpha + i\beta = \xi\mu/[\nu(\xi) + s] \tag{10.2}$$

For any given complex s, when ξ ranges from 0 to $+\infty$ and μ from -1 to $+1$, w covers a region $G(s)$ of the plane (α, β). This region reduces to a segment of the real axis for real values of s, but is a two-dimensional set when $\operatorname{Im} s \neq 0$. It is useful to introduce

$$Z(x, w) = \frac{\nu(\xi) + s}{\nu(\xi)}\tilde{Y} \tag{10.3}$$

where the dependence of Z on s is not explicitly indicated in order to simplify the notation. It is also useful to remark that the notation $Z(x, w)$ does not mean that Z is an analytic function of the complex variable w; it is only a shorthand notation for $Z(x, \alpha, \beta)$. In terms of the new quantities Eq. (10.1) can be written as follows:

$$Z + w\frac{\partial Z}{\partial x} = \iint_{G(s)} d\alpha_1\, d\beta_1\, \Phi(w_1)Z(x, w_1) \tag{10.4}$$

Here

$$\Phi(w) = \frac{1}{\bar{\nu}}\frac{[\nu(\xi)]^2\xi^4}{\nu(\xi) + s - \xi\nu'(\xi)}e^{-\xi^2}\left(\frac{1 - \mu^2}{|\mu|}\right)|\nu(\xi) + s|^2 \tag{10.5}$$

where ξ and μ are to be replaced by their expressions in terms of α and β as obtained from Eq. (10.2). Separating the variables in Eq. (10.4) gives

$$Z_u(x, w) = g_u(w)e^{-x/u} \tag{10.6}$$

where $u = \gamma + i\delta$ is the separation parameter and $g_u(w)$ satisfies

$$[1 - (w/u)]g_u(w) = \iint_{G(s)} d\alpha_1\, d\beta_1\, \Phi(w_1)g_u(w_1) \tag{10.7}$$

The right-hand side does not depend on w and can be normalized to unity. Accordingly, we are led to a typical division problem, in complete analogy with the previous cases. If the factor $(u - w)/u$ cannot be zero, $f_u(w)$ is an ordinary function given by

$$f_u(w) = u/(u - w) \tag{10.8}$$

with the normalization condition

$$\iint\limits_{G(s)} d\alpha\, d\beta\, \Phi(w)\frac{u}{u - w} = 1 \tag{10.9}$$

If, on the contrary, the factor $(u - w)/u$ is zero for some w, $g_u(w)$ must be allowed to be a generalized function. This happens if $u \in G(s)$. We must then allow for a deltalike term which becomes zero when multiplied by $(u - w)$. If we write $\delta(u - w)$ for $\delta(\alpha - \gamma)\,\delta(\beta - \delta)$, we have

$$g_u(w) = \frac{u}{u - w} + p(u, s)\,\delta(u - w) \tag{10.10}$$

where $p(u, s)$ is given by

$$p(u, s) = [\Phi(u)]^{-1}\left[1 - \iint\limits_{G(s)} d\alpha\, d\beta\, \Phi(w)\frac{u}{u - w}\right] \tag{10.11}$$

Equation (10.10) gives the eigensolutions corresponding to the continuous spectrum. We note that in this case $u/(u - w)$ does not need to be interpreted through the "principal part" concept, since double integrals with a first-order pole at a point of the integration domain exist in the ordinary sense.

Concerning the discrete spectrum, a discussion of Eq. (10.9) (Cercignani, ref. 30, Chapter VI) shows that the situation is qualitatively the same as for the BGK model: i.e., a curve γ exists such that there are two points $\pm u_0$ of the discrete spectrum when s is inside γ and none when γ is outside. One can obtain an implicit parameter representation of γ and show that γ is symmetric with respect to the real axis. The curve γ has two points in common with the real axis: the abscissa of one of them is $-\nu(0)$, while the other one has a positive abscissa (0 in the limiting case $k = \infty$).

The general solution of Eq. (10.4) can now be written as follows:

$$Z(x, w) = \iint\limits_{G(s)} \frac{A(u, s)u}{u - w} e^{-x/u}\, d\gamma\, d\delta + p(w, s)A(w, s)e^{-x/w}$$
$$+ A_+\frac{u_0}{u_0 - w}\exp(-x/u_0) + A_-\frac{u_0}{u_0 + w}\exp(x/u_0) \tag{10.12}$$

where $\pm u_0(s)$ are the possible eigenvalues of the discrete spectrum, $A_\pm$ are arbitrary coefficients ($=0$ if no discrete spectrum exists), and $A(u, s)$ is an "arbitrary function."

The main difference between Eq. (10.12) and the general solutions of the models considered in previous cases is that here we have a double integral (which exists in the ordinary sense) in place of a simple integral (of the Cauchy type). A disadvantage of the present situation is that no standard theory exists for equations having the complex Cauchy kernel $(u - w)^{-1}$ and involving two-dimensional integrations; such a theory is needed to prove the theorems of completeness and orthogonality in a constructive fashion. It is possible, however, to construct such a theory (Cercignani, ref. 30, Chapter VI) by using some results from the theory of generalized analytic functions (Vekua, ref. 34). In general, if w is a function of a complex variable, $w = \alpha + i\beta$, a generalized analytic function $f = \varphi + i\psi$ is a complex function which satisfies

$$\frac{\partial f}{\partial \bar{w}} + g_1(w)f(w) + g_2(w)\bar{f}(w) = h(w) \qquad (w \in G) \tag{10.13}$$

where the bar denotes complex conjugation, g_1, g_2, and h are given functions, and

$$\frac{\partial f}{\partial \bar{w}} = \frac{1}{2}\left(\frac{\partial f}{\partial \alpha} + i\frac{\partial f}{\partial \beta}\right) \tag{10.14}$$

The latter definition is meaningful if f is differentiable with respect to α and β: otherwise $\partial/\partial\bar{w}$ is to be understood as the Sobolev generalized derivative or the Pompeju areolar derivative (Vekua, ref. 34). It is obvious that any function of α and β can be made to satisfy an equation of the type (10.13) by a suitable choice of g_1, g_2, h, but this approach is useless; generalized analytic functions are useful when we have a whole class of functions which satisfy Eq. (10.13) with fixed coefficients g_1, g_2 (h can vary). The name "generalized analytic functions" obviously comes from the fact that when $g_1 = g_2 = h = 0$ we obtain the Cauchy–Riemann equations for the analytic function $f(w) = \varphi(\alpha, \beta) + i\psi(\alpha, \beta)$. If we have any integrable function of α and β ($\alpha, \beta \in \bar{G}$, $\bar{G}$ being the closure of G), $h(w)$, we can immediately construct a generalized analytic function in G in the following way:

$$f = T_G h = -\frac{1}{\pi}\iint_G \frac{h(u)}{u - w}\, d\gamma\, d\delta \qquad (w \in G) \tag{10.15}$$

where $u = \gamma + i\delta$ is an integration variable. In order to see that f is a generalized analytic function, we observe that

$$\begin{aligned}\frac{\partial}{\partial \bar{w}}\frac{1}{u - w} &= -\frac{\partial}{\partial \bar{w}}\frac{\partial}{\partial w}\log(u - w) \\ &= -\frac{1}{4}\Delta\{\log[\alpha - \gamma) + i(\beta - \delta)]\} - \frac{\pi}{2}\delta(u - w)\end{aligned} \tag{10.16}$$

where Δ is the Laplace operator in the (α, β) plane and the last term arises from the fact that one cannot commute the differentiations with respect to α and β at $\alpha = \beta = 0$. If we also recall that

$$\Delta\{\log[(\alpha - \gamma) + i(\beta - \delta)]\} = 2\pi\,\delta(\alpha - \gamma)\,\delta(\beta - \delta) = 2\pi\,\delta(u - w) \quad (10.17)$$

Equation (10.16) gives

$$\frac{\partial}{\partial \bar{w}}\left(\frac{1}{u - w}\right) = -\pi\,\delta(u - w) \quad (10.18)$$

Using this result in Eq. (10.15), we have (for a less formal derivation, see Vekua, ref. 34)

$$\frac{\partial f}{\partial \bar{w}} = \iint h(u)\,\delta(u - w)\,d\gamma\,d\delta = h(w) \qquad (w \in G) \quad (10.19)$$

i.e., Eq. (10.13) with $g_1 = g_2 = 0$. We note that when $w \notin \bar{G}$, then f is an analytic function of w which goes to zero when $w \to \infty$. This proves that for any integrable h, $T_G h$ is analytic outside G, and goes to zero at infinity, and

$$\frac{\partial}{\partial \bar{w}} T_G h = h \qquad (w \in G) \quad (10.20)$$

If, vice versa, $\partial f/\partial \bar{w} = h$ in G, and h is analytic outside g and goes to zero when $w \to \infty$, then $f = T_G h$. This follows because $f - T_G h$ would be analytic everywhere and zero at infinity, which implies its vanishing according to Liouville's theorem. These results allow us to solve Eq. (10.13) by quadratures for the case $g_2(w) = 0$ (Cercignani, ref. 30, Chapter VI). On the other hand, if we want to prove full-range or partial-range completeness, we have to solve equations of the following kind:

$$p(w)A(w) - \pi T_G[wA(w)] = h(w) \qquad (w \in G) \quad (10.21)$$

where T_G is the operator defined by Eq. (10.15). The domain G can be either the whole region $G(s)$ or a subset (typically one half of G, corresponding to $\operatorname{Re} w \geq 0$). It is clear that if we put

$$f(w) = T_G[wA(w)] \quad (10.22)$$

we obtain

$$p(w)\frac{\partial f}{\partial \bar{w}} - \pi w f = wh(w) \quad (10.23)$$

i.e., an equation of the type (10.13) with $g_2(w) = 0$. It is obvious now that this equation can be solved analytically and hence $A(w)$ can be found by means of the relation

$$wA(w) = \partial f/\partial \bar{w} \tag{10.24}$$

It follows that all the problems which can be solved by means of the BGK model can now be solved with the above model, because one has an algorithm completely equivalent to the one available for the BGK model (for more detail, see Cercignani, ref. 30, Chapter VI, and Cercignani and Sernagiotto, ref. 24). We note also that it can be convenient, once we have established the basic formulas, to return to the original variables ξ and μ; in fact, the complex variable w is very useful in order to establish the connection between the present theory and the basic theorems about the generalized analytic functions, but can be cumbersome in dealing with specific problems.

11. Two-Dimensional and Three-Dimensional Problems

As already noted (Chapter VI, Section 7), the method of separating the variables can be used not only for one-dimensional problems, but in general. The main difficulty is not in separating the variables and discussing the possible eigensolutions, but in singling out a complete set and proving its completeness. The situation is very simple if, when looking for plane wave solutions, one assumes a real unit vector $\mathbf{e}$ for the wave vector $\mathbf{k} = k\mathbf{e}$, as we did in Section 6. In fact, if this assumption is made, one can repeat the one-dimensional treatment, provided one substitutes $\boldsymbol{\xi} \cdot \mathbf{e}$ for ξ_1 and $(\mathbf{e} \times \boldsymbol{\xi}) \times \mathbf{e}$ for $\xi_2\mathbf{j} + \xi_3\mathbf{k}$ ($\mathbf{j}$ and $\mathbf{k}$ being unit vectors along the y and z axes). The only essential difference is that now the coefficients of the expansions depend arbitrarily upon $\mathbf{e}$, and an integration with respect to all the possible orientations of $\mathbf{e}$ must be introduced. However, the solution obtained in this way is by no means general, because, in general, $\mathbf{e}$ also must be complex (i.e., must be specified by complex angles). This is already apparent in the collisionless case [Eq. (7.3) of Chapter VI]. In fact, a strange feature of the general solution, Eq. (7.5) of Chapter VI, may already have drawn attention: the coefficient $A(\boldsymbol{\xi}, \mathbf{k}, \omega)$ depends upon one of the variables, $\boldsymbol{\xi}$ (as was explicitly noted), while it should depend only on the separation parameters, $\mathbf{k}$ and ω. On the other hand, if we drop the $\boldsymbol{\xi}$ dependence, we do not obtain the general solution of Eq. (7.8); it can be verified, however, that if we let $\mathbf{e}$ be complex (so that $\mathbf{k}$ is determined by three complex numbers in general) and integrate over $\operatorname{Im} \mathbf{e}$ as well, then we can drop the dependence on $\boldsymbol{\xi}$ and yet obtain the general solution. This circumstance becomes extremely important when we have a collision term, because in this case the trick of letting A depend upon $\boldsymbol{\xi}$ obviously does not work (the operators acting upon the $\boldsymbol{\xi}$ dependence are not simply multiplication operators) and we must let $\mathbf{e}$ be complex.

If $\mathbf{e}$ is complex, then the analysis is no longer one-dimensional, because $\boldsymbol{\xi} \cdot \mathbf{e}$ is not real, and we cannot decompose the real vector $\boldsymbol{\xi}$ into $\boldsymbol{\xi} \cdot \mathbf{e}$ and $(\mathbf{e} \times \boldsymbol{\xi}) \times \mathbf{e}$ while using properties of real vectors. The analysis can be always made two-dimensional by writing $\mathbf{k} = \operatorname{Re} \mathbf{k} + i \operatorname{Im} \mathbf{k}$ and decomposing $\boldsymbol{\xi}$ into a part in the plane of the real vectors $\operatorname{Re} \mathbf{k}$ and $\operatorname{Im} \mathbf{k}$ and a part orthogonal to this plane. Once this decomposition has been made, however, it seems more convenient to use the complex vector $\mathbf{k} = k\mathbf{e}$ with k and $\mathbf{e}$ complex, instead of the two real vectors $\operatorname{Re} \mathbf{k}$ and $\operatorname{Im} \mathbf{k}$. In fact, if we regard $\mathbf{e}$ as given and look for the admissible k, the combination $\boldsymbol{\xi} \cdot \mathbf{e}$ defines a complex variable which can be treated by methods analogous to those employed in Section 10 (generalized analytic functions). Accordingly, although the matter treated in this section has not been investigated in detail, it seems that an investigation is worthwhile undertaking, and that the basic tools should be analogous to those employed for time-dependent problems when the collision frequency depends upon the molecular speed.

12. Connections with the Chapman–Enskog Method

The method of elementary solutions has relations with the Chapman–Enskog method from at least two points of view. First, we see that the separation of the solution into a discrete part and a continuum reflects (at least in the simplest models) the separation between a Chapman–Enskog solution (valid far from solid boundaries and a certain initial stage) and a transient part which describes the kinetic layers. Second, the elementary solutions seem particularly useful for solving the connection problems for the Hilbert and Chapman–Enskog methods (especially the problem of matching boundary conditions). This has been shown by evaluating the slip coefficient with the BGK model. For more general models and matching problems one cannot in general obtain an analytical solution. However, one can always obtain a fairly accurate description of the solution by estimating the coefficients of the expansions or the corrections to lowest-order models. In particular, the temperature-jump coefficient can be obtained in terms of quadratures by artificially decoupling the normal and transverse degrees of freedom or evaluated exactly by means of the more complicated method mentioned in Section 9 and discussed in the appendix.

Another relation between the methods of elementary solutions and Chapman–Enskog theory is offered by two-dimensional flows; in fact, conditions at infinity cannot be satisfied, in general, by a linearized treatment (Chapter VI, Section 6), and therefore one has to obtain the outer solution from continuum plus Chapman–Enskog, while the inner solution can be constructed in terms of elementary solutions.

13. Sound Propagation and Light Scattering in Monatomic Gases

One of the problems for which the theory expounded in Section 8 is useful is the problem of sound propagation. A plate oscillates in the direction of its normal with frequency ω; a periodic disturbance propagates through the gas which fills the region at one side of the plate. If the frequency ω is very high, the continuum theory is not good even at ordinary densities, because $1/\omega$ can be of the same order as the mean free time. One can measure experimentally the phase speed and the attenuation of disturbance by assuming that the latter is locally a plane wave (it is not in general because there is the continuum contribution). One of the difficulties is that there is the receiver, which in principle does not allow us to treat the problem as a half-space problem, especially since the receiver is usually kept very close to the plate (at a distance of a mean free path or less). If we disregard the disturbance produced by the receiver, then we can treat the problem as a half-space problem (Buckner and Ferziger, ref. 13) by means of the method of elementary solutions. The results show that the discrete spectrum determines the situation for $\omega\theta \lesssim 1$ (θ being a suitable mean free time). For $\omega\theta \gtrsim 1$ the continuum modes are important; yet the discrete contribution (or its analytical continuation—see Section 7) gives a qualitatively correct picture for moderate values of $\omega\theta$. The latter remark explains why the results by Sirovich and Thurber (ref. 35), based only on the discrete contribution and its analytical continuation, show a reasonable agreement with experimental data. The two methods mentioned so far give a good fit with the experiments (for a more detailed discussion see ref. 36 and ref. 15, Chapter I). This is to be particularly stressed because other methods, based on expanding the solution of the linearized Boltzmann equation into a series of orthogonal polynomials, have failed. A first method, used by Wang Chang and Uhlenbeck (ref. 13, Chapter III) and Pekeris *et al.* (ref. 37), was to expand the solution into the eigenfunctions of the Maxwell operator; the results for the attenuation rate are in complete disagreement with the experiments. Since Pekeris *et al.* used 483 moments (!), one can conclude that their expansion, if convergent, does not converge to the correct solution for large values of ω. Another approach is due to Maidanik *et al.* (ref. 38) and Kahn and Mintzer (ref. 39); in their solution the unknown of the linearized Boltzmann equation is expanded into a series of orthogonal polynomials whose weight function is based on a free-molecular solution rather than a Maxwellian. Unexpectedly, their results turned out to possess the correct continuum limit. Because of this circumstance, the method of Kahn and Mintzer attracted considerable attention and favorable comments (refs. 13 and 40–43). Subsequent work showed, however, that this unexpected accuracy was due to some mistakes made by Kahn and Mintzer; in fact Toba (ref. 44) pointed out an error in the boundary conditions and Hanson and Morse (ref. 45), a mistake in the asymptotic evaluation of certain integrals. Hanson and Morse found, in fact, that the behavior at low frequencies was completely wrong and even physically nonsensical (growing rather than damped modes). The agreement for very high frequencies is reasonably good, as was to be expected.

Thomas and Siewert (ref. 46) applied an accurate numerical method for the half-range analysis of the BGK model developed by Siewert and Burniston (ref. 47) to improve upon the solution of Buckner and Ferziger (ref. 13). There is a problem when comparing theoretical results with experimental data, because for high frequencies the solution remains wave-like, but is no longer a classical plane wave, with the consequence that the phase speed and the attenuation parameter are no longer clearly defined. The solutions of refs. 13 and 46 produce nearly linear plots for the real and imaginary parts of the logarithm of the pressure, but the average slopes are determined by the range of the space coordinates within which the calculations are performed. Buckner and Ferziger correctly adopted the range chosen in the experiments to produce the results already mentioned. This work seems not to have been done, so far, for the results of Thomas and Siewert.

The influence of the spectrum of the collision operator can be felt in a critical way only far from the plate, where no available receiver can detect it; therefore the detection of the behavior of the collision frequency at high speeds cannot be achieved by an experiment on sound propagation, at least not one feasible at present.

An interesting procedure for experimentally testing the details of the collision term should be the scattering of light from density fluctuations of a gas (Nelkin and Yip, ref. 48). When light of incident wavelength λ is scattered by a monatomic gas the light scattered through an angle θ will have a frequency distribution proportional to $S(K, \omega)$, where $K = (4\pi/\lambda)\sin(\theta/2)$, $\omega/2\pi$ is the frequency shift due to scattering, and $S(K, \omega)$ is the double Fourier transform of the density correlation function $G(r, t)$. The latter is given by

$$G(r, t) = \int d\boldsymbol{\xi} f_0(\boldsymbol{\xi}) h(\mathbf{x}, \boldsymbol{\xi}, t) \qquad (r = |\mathbf{x}|) \tag{13.1}$$

where $f_0(\boldsymbol{\xi})$ is the equilibrium Maxwellian corresponding to the density and temperature of the gas and h satisfies the linearized Boltzmann equation in infinite space, with the initial datum

$$h(\mathbf{x}, \boldsymbol{\xi}, 0) = \delta(\mathbf{x}) \tag{13.2}$$

This problem can be solved numerically for Maxwell's molecules and exactly for all the models considered so far (only full-range completeness is needed). Some of these solutions have been computed by Ranganathan and Yip (ref. 49).

14. Nonlinear Problems

The solutions we have considered so far can be easily criticized because they are obtained by solving linearized models. Those who do not accept approximations will in fact claim that in kinetic theory one should only consider solutions of the nonlinear, exact Boltzmann equation. Unfortunately, we are looking for solutions of boundary value problems and it is by no means easy to produce such solutions for the nonlinear Boltzmann equation. In fact, the only known solutions are either space homogeneous or depend on space coordinates in such a simple way that they cannot describe the Knudsen layers. Yet, we shall give a short description of these exact solutions of the nonlinear Boltzmann equation in this section.

In 1956, C. Truesdell (ref. 50) and V. S. Galkin (ref. 51) independently investigated the steady homoenergetic flows of a gas of Maxwellian molecules according to the infinite system of moment equations associated with the Boltzmann equation. Later Galkin (refs. 52–54) extended his analysis to some typical unsteady homoenergetic affine flows. The book by Truesdell and Muncaster (ref. 1 of Chapter III) gives a unified discussion of all these works, and provides further calculations linking these flows to more current research in kinetic theory.

Let us recall the basic ideas about homoenergetic affine flows. The defining properties are the following:

1. The body force (per unit mass) $\mathbf{X}$ acting on the molecules is constant:

$$\mathbf{X} = \text{const} \tag{14.1}$$

2. The density ρ, the internal energy per unit mass e, the stress tensor P and the heat flux $\mathbf{q}$ may be functions of time but not of the space coordinates.
3. The bulk velocity $\mathbf{v}$ is an affine function of position $\mathbf{x}$:

$$\mathbf{v} = \mathsf{K}(t)\mathbf{x} + \mathbf{v}_0(t) \tag{14.2}$$

This definition holds for a general material; for a gas described by the kinetic theory, a natural extension of property (2) is immediate:

2′. The moments

$$M_{i_1 i_2 i_3 \ldots i_n} = \int c_{i_1} c_{i_2} c_{i_3} \cdots c_{i_n} f \, d\boldsymbol{\xi} \tag{14.3}$$

formed with the peculiar velocity $\mathbf{c} = \boldsymbol{\xi} - \mathbf{v}$ may be functions of time but do not depend upon space coordinates. Here $\boldsymbol{\xi}$ is the molecular velocity with respect to an inertial frame.

The condition (2′) holds for the solutions obtained by Truesdell (ref. 50) and Galkin (refs. 51–54). For analyses relating directly to the distribution function f, this condition is transformed into

2″. The variable $\mathbf{x}$ appears in f only through $\mathbf{v}$, given by Eq. (14.2), i.e.:

$$f = f(\mathbf{c}, t) \tag{14.4}$$

While the analyses in refs. 50–54 have the great advantage of leading to explicit solutions, which lend themselves to a detailed discussion of their properties and a comparison with the corresponding Chapman–Enskog and Hilbert expansions, they suffer from the drawback of providing solutions to the system of equations for moments, but no corresponding solution of the Boltzmann equation itself. In ref. 52 of the previous chapter a proof was given of the existence of corresponding solutions for the Boltzmann equation itself (even for non-Maxwellian molecules).

The solutions in refs. 50–54 lead to an explicit relation between the stress tensor $\mathbf{P}$ and the velocity gradient $\mathbf{K}$, which is an explicit example of what a normal solution should be able to yield. One can also investigate the series expansion of this relation into powers of $\mathbf{K}$ to find that it converges for small enough values of $\|\mathbf{K}\| = \max_{i,j} K_{ij}$ and diverges for higher values. Thus one can say that in this case a Chapman–Enskog expansion converges or diverges, rather than being simply asymptotic. This does not mean anything, however, for more complex solutions, having moments with nonpolynomial dependence upon $\mathbf{x}$ (see Section 8 of the previous chapter).

We remark that the relation between $\mathbf{P}$ and $\mathbf{K}$ can be found by using just a few moment equations, which happen to be identically the same for the BGK model and the Boltzmann equation for Maxwell molecules. Thus there is no difference, in this case, between the true equation and the model. This was first remarked by Zwanzig (ref. 55) for the case of shear flow. Zwanzig also determined the full distribution function satisfying the BGK equation (with constant collision frequency) for this case. He also indicated that in the case of a collision frequency proportional to the square root of the temperature (as appropriate for a BGK model trying to mimic hard sphere molecules) one can write a second-order differential equation that determines the relation between $\mathbf{P}$ and $\mathbf{K}$. Recently Gomez Ordonez *et al.* (ref. 56) compared these solutions with the Monte Carlo solutions for a gas of hard spheres and found a fairly good agreement over a wide range of values of $\mathbf{K}/\nu$, where ν is the collision frequency.

A subject which has developed in the last ten years deals with exact solutions of the space-homogeneous Boltzmann equation. Although Boltzmann's H theorem guarantees that any assigned initial space-homogeneous distribution function will decay to a Maxwellian, and this result can even be proved in all rigor, the details of this decay are not known explicitly. An exception is offered by Maxwell molecules, for which Maxwell derived a set of exact equations satisfied by the moments. However, he did not write out all the terms; these were first published by Grad (ref. 57) and thoroughly discussed by Ikenberry and Truesdell (ref. 58) and Truesdell (ref. 59). In the latter paper, the general solution for the moments of the distribution function is given by the following expression:

$$M_n = \sum_{i=1}^{N_n} A_n^{(i)}(t) \exp(-k_n^{(i)} t) + M_n^{(0)} \tag{14.5}$$

where M_n is any moment and n a suffix to identify it; $M_n^{(0)}$ is the equilibrium value of M_n; $A_n^{(i)}(t)$ is a polynomial in t; $k_n^{(i)}$ is a positive constant and N_n a positive integer; and $A_n^{(i)}(t)$, $k_n^{(i)}$, and $M_n^{(0)}$ depend on the constant values of density, velocity, and temperature. It was not proved, but only conjectured, by Truesdell that the $A_n^{(i)}(t)$ are in fact constants, i.e., zeroth-degree polynomials. The explicit expressions for $A_n^{(i)}(t)$ can be found by recursion; easy, albeit tedious, computations seem to confirm the conjecture.

It is to be remarked that, although a knowledge of all the moments is theoretically equivalent to a knowledge of the distribution function, no simple expression of the latter is obtained from Eq. (14.5). This explains the interest which surrounded the publication (refs. 60 and 61) of an exact solution of the nonlinear Boltzmann equation for Maxwell molecules. The solution corresponds to a special initial datum, i.e.:

$$f = e^{-a\xi^2}(b + c\xi^2) \tag{14.6}$$

where a, b, and c are positive constants. The basic property of a distribution function of the form (14.6) is that it evolves preserving its shape, the only change being in the coefficients a, b, and c, which vary in time until, ultimately, c tends to zero, a and b to positive constants, and f to a Maxwellian. It is to be noted that the same solution was independently published by Bobylev (ref. 62) and had appeared in an unpublished master's thesis as early as 1967 (ref. 63). Also, if one is aware of the fact that the Boltzmann equation for Maxwell molecules has a solution of the form indicated in Eq. (14.6), it is not difficult to check this and compute explicitly a, b, and c as functions of time t (ref. 64).

The solution published by Bobylev (ref. 62) and Krook and Wu (refs. 60 and 61) (frequently called the BKW mode in the literature) approaches an equilibrium distribution when $t \to \infty$ in a nonuniform fashion; this is due to the high-speed tail of the distribution and indicates that linearization does not hold for high speeds even if we are close to a Maxwellian in some sense. This was already known from other facts (loss of positivity of certain linearized solutions), but is immediately obvious here.

The importance of the BKW solution was overestimated initially, because of the following conjecture formulated by Krook and Wu: An arbitrary initial state tends toward a BKW mode; then a relaxation toward equilibrium takes place according to that solution. This conjecture can be rephrased in more mathematical terms, but this is not necessary since analytical and numerical evidence against this conjecture has been found by many authors. An enormous literature on this subject and related solutions of the Boltzmann equation for molecules other than Maxwell molecules is available. For more details, the reader is referred to the papers by Ernst (refs. 65 and 66).

We also remark that Nikol'skii (refs. 67 and 68) gave a formula to generate space-inhomogeneous solutions from any given spatially homogeneous solution. The solutions which are thus generated belong to the class of homoenergetic affine solutions discussed above. In particular, when applied to the BKW solution, this transformation produces a solution first published by Muncaster (ref. 69).

Appendix

Proving the completeness of the singular eigenfunctions obtained by the method of elementary solutions means solving a singular integral equation with a Cauchy kernel, where the function to be expanded is given and the coefficient of the expansion, $A(u)$, is the unknown. The coefficients of the discrete spectrum, is any, are constants available for satisfying auxiliary requirements (see below). The proof then rests on the techniques for solving the above-mentioned singular integral equations. These techniques are presented in ref. 4, to which we refer for further details and proofs. Here we collect some of the relevant results needed in the completeness proof.

We start by recalling some properties of the analytic functions regular in the complex plane with a cut along a line L.

Let $f(z)$ be an analytic function which tends to zero when $z \to \infty$ and is regular in the complex plane with a cut along the oriented open line L; let the limits $f^+(t)$ and $f^-(t)$ of $f(z)$ when z tends to a point $t \in$ L from the left or from the right exist; also, if c denotes either endpoint of L, let $|f(z)| < A(z - c)^{-\gamma}$ for some A and $\gamma < 1$ when $z \to c$. Let

$$\Delta(t) = f^+(t) - f^-(t) \tag{A.1}$$

denote the jump of $f(z)$ when going through the cut. Then $\Delta(t)$ is a Hölder continuous function on L (i.e., $|\Delta(t_1) - \Delta(t_2)| < A|t_1 - t_2|^{\alpha}$), for $t_1, t_2 \in$ L, $0 \leq \alpha \leq 1$) and

$$f(z) = \frac{1}{2\pi i} \int_{L} \frac{\Delta(t)}{t - z}\, dt \tag{A.2}$$

Also

$$f^{+}(t) + f^{-}(t) = \frac{1}{\pi i} P \int_{L} \frac{\Delta(t')}{t' - t}\, dt' \tag{A.3}$$

where P means that we consider the Cauchy principal value of the integral.

If, on the contrary, $\Delta(t)$ is a function of the Hölder type on the open line L and $|\Delta(t)| < A(t - c)^{-\gamma}$ at both endpoints c for some A and $\gamma < 1$ when $z \to c$, let us define $f(z)$ through Eq. (A.2). Then $f(z)$ is regular in the complex plane with a cut along L and tends to zero when $z \to \infty$, and its limits when approaching L from the left and from the right exist, are finite, and are related to $\Delta(t)$ by Eqs. (A.1) and (A.3). These equations, which are usually referred to as Plemelj's formulas, can be written in the equivalent form:

$$f^{\pm}(t) = \frac{1}{2\pi i} P \int_{L} \frac{\Delta(t')}{t' - t}\, dt' \pm \tfrac{1}{2}\Delta(t) \tag{A.4}$$

The exact behavior of $f(z)$ when $z \to c$ is given by

$$f(z) = \pm \frac{\Delta(c)}{2\pi i} \log(z - c) + O(1) \tag{A.5}$$

if $\Delta(c) \neq \infty$; the sign is positive for the upper limit of the integral A(2) and negative for the lower one.

Let us consider now a singular integral equation with a Cauchy kernel:

$$A(t)y(t) + B(t)P \int_{L} \frac{y(t')}{t' - t}\, dt' = C(t) \tag{A.6}$$

where $A(t)$, $B(t)$, and $C(t)$ are given functions and L is a given open line in the complex plane. A procedure for solving Eq. (A.6) in a closed form is as follows. Let us introduce the following function of the complex variable z:

$$N(z) = \frac{1}{2\pi i} P \int_{L} \frac{y(t)}{t - z}\, dt \tag{A.7}$$

Then Eq. (A.6) can be rewritten as follows:

$$[A(t) + \pi i B(t)]N^{+}(t) - [A(t) - \pi i B(t)]N^{-}(t) = C(t) \tag{A.8}$$

where Eqs. (A.1) and (A.3) (with $\Delta = y$, $f = N$) have been used. Given the analiticity of $N(z)$, we must determine $N(z)$ by means of Eq. (A.8); this problem is known as the Riemann-Hilbert problem. We shall assume that $A \pm iB \neq 0$ on L. Once $N(z)$ is known, $y(t)$ is obtained by means of Plemelj's formulas.

In order to solve the Hilbert problem we remark that if we were to know a function $X(z)$ that was analytic and nonzero in the complex plane with a cut along L and such that

$$[A(t) + \pi i B(t)]X^{-}(t) = [A(t) - \pi i B(t)]X^{+}(t) \tag{A.9}$$

then Eq. (A.8) could be written as follows:

$$X^{+}(t)N^{+}(t) - X^{-}(t)N^{-}(t) = \frac{X^{-}(t)C(t)}{A(t) - \pi i B(t)} \tag{A.10}$$

and because of Plemelj's formulas applied to $X(z)N(z)$:

$$N(z) = \frac{1}{X(z)} \frac{1}{2\pi i} P \int_{L} \frac{X^{-}(t)C(t)}{A(t) - \pi i B(t)} \frac{dt}{t - z} \tag{A.11}$$

provided $X(z)/N(z)$ tends to zero for $z \to \infty$. Thus we can solve the inhomogeneous Riemann-Hilbert problem related to Eq. (A.6) provided we know a solution of the homogeneous Hilbert problem (A.9). But (A.9) can be written as follows:

$$\log X^{+}(t) - \log X^{-}(t) = \log \frac{A(t) + \pi i B(t)}{A(t) - \pi i B(t)} \tag{A.12}$$

and Plemelj's formulas applied to $\log X(z)$ give

$$X(z) = (z - a)^{n}(z - b)^{m} \exp\left\{\frac{1}{2\pi i} P \int_{L} \log \frac{A(t) + \pi i B(t)}{A(t) - \pi i B(t)} \frac{dt}{t - z}\right\} \tag{A.13}$$

where a and b are the endpoints of L; n and m are two integers which must be chosen in such a way that $|X(z)| \sim k(z - c)^{\sigma}$ $(-1 < \sigma < 1)$ at both endpoints. This is required for Eq. (A.11) to be valid. Accordingly, it can happen that $N(z)$ is not completely determined if more than one pair (m, n) can be chosen or, on the contrary, a restriction on the function $C(t)$ must be imposed since the behavior at infinity of $X(z)$ can be such that $N(z)$, as given by Eq. (A.11), does not tend to zero when $z \to \infty$ for a general $C(t)$. The latter case occurs for the equations considered in the main text whenever the eigensolutions associated with the continuous spectrum must be complemented by discrete terms; the corresponding restriction on $C(t)$ determines the coefficients of the discrete spectrum terms. A particular but important case is when one of the endpoints is at infinity. We consider only the case where $[A(t) + \pi i B(t)]/[A(t) - \pi i B(t)]$ tends to 1 when $t \to \infty$; then the branch of the logarithm in Eq. (A.13) is to be chosen in such a way that the logarithm tends to zero when $t \to \infty$. Also, $m = 0$ if b is the endpoint at infinity.

Theorems I and II of Section 3 can be obtained by applying the method described above to Eqs. (3.8) and (3.13). In particular, $P(u)$ is simply $[X(-u)]^{-1}$, given by Eq. (A.13) with $A(t) = p(t)$, $B(t) = t$, $a = 0$, $n = -1$, and $m = 0$. This fact can be exploited to prove Eqs. (3.21)-(3.23) by applying Plemelj's formulas to $X(z)$ (refs. 25 and 28, Chapter VI).

It is clear that the above treatment cannot be extended to the case of a system of singular integral equations. In fact there is a general theory (ref. 4) for this case, but, generally speaking, no explicit form of the solution. Yet one can obtain explicit solutions in particular cases, among which is the case of the BGK model, which leads to a system of two equations (refs. 18-23).

The idea behind the method can be explained as follows. When trying to prove half-range completeness we are led to solving the following Riemann-Hilbert problem [analogous to (A.9)]:

$$\mathbf{X}^+ = \mathbf{G}\mathbf{X}^- \tag{A.14}$$

where $\mathbf{G}$ is the 2×2 matrix

$$\mathbf{G} = (\mathbf{A}^+)^T(\mathbf{A}^-)^{-T} \tag{A.15}$$

where, if $\mathbf{I}$ is the 2×2 identity matrix and $P(z)$ a function analytic in the complex plane cut along the real axis, $\mathbf{A}$ is (or can be trivially reduced to) a 2×2 matrix of the form

$$\mathbf{A} = \mathbf{T} + P(z)\mathbf{I} \tag{A.16}$$

where $\mathbf{T}$ is (or is proportional to) a matrix with polynomial entries. Here and in what follows the superscripts T and $-T$ are used to denote the transposed and inverse-transposed matrix, respectively, while the superscripts + and − indicate the limiting values as z approaches the real axis from above and below, respectively.

It is clear that **T** (and hence **A**) can be diagonalized by solving a second-degree algebraic equation; i.e., a 2×2 matrix **G** exists such that

$$\mathbf{G}\mathbf{A}\mathbf{G}^{-1} = \mathbf{A}^D \tag{A.17}$$

where $\mathbf{A}^D$ is diagonal and **G** is analytic everywhere with the exception of branch cuts Γ_k ($k = 0, 1, \ldots, N-1$) joining two subsequent zeroes (in an arbitrary order) of the discriminant $D(z)$ of the above-mentioned second degree equation. In order to avoid unnecessary complications we shall assume that D is of even degree, $2N$ (a condition satisfied in all the particular cases considered so far). The solution $\mathbf{X}^D$ of the diagonalized Riemann problem is trivial, in principle. However, when transforming back from $\mathbf{X}^D$ to **X**, one must be careful because the singularities along the branch cuts Γ_k are extraneous to the problem and should not appear in **X**. In order to circumvent this difficulty, one can remark that the two nonzero entries of the diagonal matrix $\mathbf{X}^D$ are the two branches of the same (two-valued) analytic function and, as a consequence of the fact that **X** and $\mathbf{X}^D$ have the same trace and the same determinant, the product of these entries and the logarithm of their ratio multiplied by $R(z) = \sqrt{D(z)}$ are analytic functions. This remark leads to solving the problem except for a small but important detail: If the degree of $D(z)$ is higher than 2 (a situation occurring in all the applications considered so far) the solutions have an essential singularity at infinity of the form $\exp(bz^{N-1})$, where b is a nonzero constant. This was essentially the difficulty that Darrozès (ref. 20) was not able to master.

In order to avoid the difficulty one can use the fact that a change of $2\pi i$ in a logarithm of a function does not have any influence of the function itself. This can be used to cancel the singularity at infinity. Easy considerations lead to the following problem: Find integers n_k, m_k ($k = 1, \ldots, N-1$) and complex numbers z_k ($k = 1, \ldots, N-1$) such that

$$\sum_{k=1}^{N-1} \int_{\Gamma_k} \frac{z^{r-1}}{R^+(z)}\,dz + \sum_{k=1}^{N-1} m_k \int_{\Delta_k} \frac{z^{r-1}}{R(z)}\,dz + \sum_{k=1}^{N-1} \int_{c_k} \delta_k(z)\frac{z^{r-1}}{R(z)}$$

$$= -\frac{1}{4\pi i}\int_0^\infty \frac{\mu^{r-1}}{R(\mu)} L(\mu)\,d\mu \qquad (r = 1, \ldots, N-1) \tag{A.18}$$

Here $L(\mu)$ is a known function, while Δ_k ($k = 1, \ldots, N-1$) is a system of cuts joining two roots that were endpoints of different cuts of the Γ type. c_k denote $N-1$ arbitrary paths from the origin to the points z_k ($k = 1, \ldots, N-1$). $\delta_k(z)$ is a step function defined as follows: $\delta_k(z) = 1$, if c_k

does not intersect Γ_0; if c_k, on the contrary, intersects Γ_0 at p_k, then $\delta_k(z) = 1$ from 0 to p_k and $\delta_k(z) = -1$ from p_k to z_k. Eq. (A.18) is equivalent to Jacobi's inversion problem (refs. 70 and 71), which can be solved analytically, according to Riemann's ideas (ref. 70).

To this end, we introduce the Riemann surface $\mathcal{R}$ defined by

$$v^2 = D(z) \tag{A.19}$$

and represent $\mathcal{R}$ as a two-sheeted covering of the z-plane cut along Γ_k ($k = 0, 1, \ldots, N-1$). We denote the branch of v which has the expansion $\pm(z^N + \ldots)$ as $z \to \infty$ by $\pm R(z)$ and the points on the "upper" ("lower") sheet by $[z, R(z)]$ ($[z, -R(z)]$). The points on $\mathcal{R}$ are expressed by $[z, v]$ or simply by a single letter such as q. We choose a system of canonical oriented cross sections (ref. 71) $\mathbf{a}_k$ and $\mathbf{b}_k$ ($k = 1, \ldots, N-1$) of $\mathcal{R}$ and denote by $\hat{\mathcal{R}}$ the surface $\mathcal{R}$ cut along all the cross sections $\mathbf{a}_k$ and $\mathbf{b}_k$. We use the symbols $\mathbf{a}_k^{-1}$, $\mathbf{b}_k^{-1}$ when the curves $\mathbf{a}_k$ and $\mathbf{b}_k$ are described in the negative sense. Thus if we describe successively the curves $\mathbf{a}_1\mathbf{b}_1\mathbf{a}_1^{-1}\mathbf{b}_1^{-1} \ldots \mathbf{a}_{N-1}\mathbf{b}_{N-1}\mathbf{a}_{N-1}^{-1}\mathbf{b}_{N-1}^{-1}$, which form the boundary $\partial\hat{\mathcal{R}}$ of $\hat{\mathcal{R}}$, then $\hat{\mathcal{R}}$ remains throughout on the left.

It is known (ref. 71) that the differentials

$$df_1 = \frac{dz}{v}, \quad df_2 = \frac{z\,dz}{v}, \ldots, \quad df_{N-1} = \frac{z^{N-2}\,dz}{v}, \qquad [z, v] \in \mathcal{R} \tag{A.20}$$

form a basis of Abelian differentials of the first kind on $\mathcal{R}$, i.e., three linearly independent differentials everywhere analytic on $\mathcal{R}$. Now we introduce the following notation:

$$a_{ik} = \int_{\mathbf{a}_k} df_i, \qquad b_{ik} = \int_{\mathbf{b}_k} df_i \qquad (i, k = 0, 1, \ldots, N-1) \tag{A.21}$$

and define the differentials

$$df_i = \sum_{j=1}^{N-1} a_{ij}\,du_j$$

$$du_i = \sum_{j=1}^{N-1} \alpha_{ij}\,df_j \qquad ([\alpha_{ij}] = [a_{ij}]^{-1}) \tag{A.22}$$

Then we find that the A periods A_{ij} of du_i are the elements of the $(N-1) \times (N-1)$ unit matrix, i.e.,

$$A_{ij} = \int_{\mathbf{a}_j} du_i = \delta_{ij} \qquad (i, j = 1, \ldots, N-1) \tag{A.23}$$

Thus du_i $(i = 1, \ldots, N-1)$ are a normalized basis of the Abelian differentials of the first kind on $\mathscr{R}$. It is known that the B periods of du_i

$$B_{ij} = \int_{\mathbf{b}_j} du_i \qquad (i, j = 1, \ldots, N-1) \tag{A.24}$$

form a symmetric matrix, whose imaginary part is positive definite (ref. 70).

In terms of the functions u_k $(k = 1, \ldots, N-1)$ (the integrals of the differentials du_k), Jacobi's inversion problem, Eq. (A.18), can be rewritten as follows:

$$\sum_{j=1}^{N-1} u_j(q_k) + n_k' + \sum_{j=1}^{N-1} B_{jk} m_j' = r_k \qquad (k = 1, \ldots, N-1) \tag{A.25}$$

where r_k are given functions [simply related to the integrals appearing in the right-hand side of Eq. (A.18)] and n_k', m_k' integers simply related to n_k, m_k. Solving Eq. (A.25) means finding these integers and the points $q_k = [z_k, v_k]$ $(k = 1, \ldots, N-1)$ on $\mathscr{R}$.

In order to solve the inversion problem, we introduce the Riemann θ-function associated with the problem:

$$\begin{aligned} F(q) &= \theta(u_k(q) - e_k) \\ &= \sum_{l_k=-\infty}^{\infty} \exp\left[\pi i \sum_{j=1}^{N-1} \sum_{s=1}^{N-1} B_{js} l_j l_s + 2\pi i \sum_{s=1}^{N-1} l_s(u_s(q) - e_s)\right] \end{aligned} \tag{A.26}$$

where

$$e_s = r_s + k_s, \qquad k_s = -\tfrac{1}{2} - \tfrac{1}{2}B_{ss} - \sum_{\substack{j=1 \\ j \neq s}}^{N-1} \int_{\mathbf{a}_j} u_s^- \, du_j \tag{A.27}$$

The θ-function is a single-valued analytic function of $q \in \hat{\mathscr{R}}$ and, because of the definition of u_k, it has the property

$$F^+(q) = F^-(q) \exp[\pi i B_{jj} + 2\pi i (u_j^-(q) - e_j)], \qquad q \in \mathbf{a}_j \tag{A.28}$$

$$F^+(q) = F^-(q), \qquad q \in \mathbf{b}_j. \tag{A.29}$$

The following theorem holds:

Theorem. If the Riemann θ-function $F(q)$ is not identically zero, it has $N-1$ zeroes q_j on $\mathscr{R}$, which give a solution of Jacobi's inversion problem (A.25). (The integers m_k' and n_k' are then determined by the equation itself.)

We note that the nontriviality of the Riemann θ-function $F(q)$ can be tested by evaluating it at N distinct points; if by accident it turns out to be trivial, the theory holds for the first nonvanishing partial derivative of θ. The zeroes can be found by solving an $(N - 1)$th-degree equation, with coefficients expressible in terms of the limiting values of the partial derivatives of the Riemann θ-function when its arguments go to infinity and of integrals involving $R(z)$. This brings the analytical solution to completion.

References

Section 1——A general reference for this Chapter is ref. 28 of Chapter VI.

Sections 2-4——These sections follow refs. 25 and 28 of Chapter VI and

1. C. Cercignani, *J. Math. Anal. Appl.* **10**, 93 (1965).

Functions related to $p(u)$ are tabulated in

2. V. N. Faddeyeva and N. M. Terent'ev, *Tables of the Values of the Function*

$$w(z) = e^{-z^2}\left(1 + \frac{2i}{\sqrt{\pi}}\int_0^z e^{t^2}\,dt\right),$$

Pergamon Press, London and New York (1961);

3. B. D. Fried and S. D. Conte, *The Plasma Dispersion Function: The Hilbert Transform of the Gaussian,* Academic Press, New York (1961).

The complete theory of singular integral equations can be found in

4. N. I. Muskhelishvili, *Singular Integral Equations,* Noordhoff, Groningen (1953).

The numerical evaluation of the slip coefficient was presented in

5. S. Albertoni, C. Cercignani, and L. Gotusso, *Phys. Fluids* **6**, 993 (1963).

Section 5——The application of the method of elementary solutions to Couette flow was proposed in

6. C. Cercignani, *J. Math. Anal. Appl.* **11**, 93 (1965).

The analogous treatment for Poiseuille flow summarized in this section was given in

7. C. Cercignani, *J. Math. Anal. Appl.* **12**, 234 (1965).

The case of a temperature gradient was considered in

8. C. Cercignani, "Flows of Rarefied Gases Supported by Density and Temperature Gradients," University of California Report No. AS-64-18 (1964).

A theoretical proof of the existence of the minimum in the flow rate of plane Poiseuille flow was given by

9. C. Cercignani, in: *Rarefied Gas Dynamics* (J. A. Laurmann, ed.), Vol. II, p. 92, Academic Press, New York (1963),

but its experimental existence (for cylindrical tubes) has been known since

10. M. Knudsen, *Ann. Physik* **28**, 75 (1909).

Sections 6, 7——These sections follow refs. 26 and 28 of Chapter VI. The diffusion of a velocity discontinuity is treated in detail in

11. C. Cercignani and R. Tambi, *Meccanica* **2**, 1 (1967),

and the Rayleigh problem in

12. C. Cercignani and F. Sernagiotto, in: *Rarefied Gas Dynamics* (J. H. de Leeuw, ed.) Vol. 1, p. 332, Academic Press, New York (1965).

Section 8——This section follows ref. 28 of Chapter VI, where the procedures for general models were first investigated. Analogous results with some formal changes are given in

13. J. K. Buckner and J. H. Ferziger, *Phys. Fluids* **9**, 2315 (1966),

where an approximating procedure for half-range problems was suggested and applied to the problem of sound propagation.

Section 9——The various simplifications of the BGK model were suggested and used by

14. H. S. Ostrowski and D. J. Kleitman, *Nuovo Cimento* **XLIV** B, 49 (1966);
15. R. J. Mason, in: *Rarefied Gas Dynamics* (C. L. Brundin, ed.), Vol. I, p. 395, Academic Press, New York (1967);
16. H. Weitzner, in: *Rarefied Gas Dynamics* (J. H. de Leeuw, ed.), Vol. I, p. 1, Academic Press, New York (1965),

and in ref. 28 of Chapter VI. The treatment of the shear flows with the ES model was given in ref. 9 of Chapter IV. The special model with collision frequency proportional to molecular speed was treated by

17. J. S. Cassell and M. M. Williams, *Transport Theory and Statistical Physics* **2**, 81 (1972).

The closed-form solution for the heat transport equation arising from the BGK model was discussed in the following papers:

18. C. Cercignani, *Transport Theory and Statistical Physics* **6**, 29 (1977);
19. C. E. Siewert and C. T. Kelley, *Z. Angew. Math. Phys.* **31**, 344 (1980);
20. J. S. Darrozès, *La Recherche Aérospatiale* **119**, 13 (1967);
21. C. Cercignani and C. E. Siewert, *Z. Angew. Math. Phys.* **33**, 297 (1982),

and the same technique was used to discuss sound propagation in

22. K. Aoki and C. Cercignani, *Z. Angew. Math. Phys.* **35**, 127 (1984);
23. K. Aoki and C. Cercignani, *Z. Angew. Math. Phys.* **35**, 345 (1984).

Shear flows with the velocity-dependent collision frequency were treated in ref. 10 of Chapter IV and in

24. C. Cercignani and F. Sernagiotto, in: *Rarefied Gas Dynamics* (C. L. Brundin, ed.), Vol. I, p. 381, Academic Press, New York (1967).

The dependence of the slip coefficient on the form of the collision frequency is treated in

25. C. Cercignani, P. Foresti, and F. Sernagiotto, *Nuovo Cimento* **X**, 57B, 297 (1968);
26. M. A. Reynolds, J. J. Smolderen, and J. F. Wendt, in: *Rarefied Gas Dynamics* (M. Becker and M. Fiebig, eds.), Vol. I, p. A21, DFVLR Press, Porz-Wahn (1974);
27. W. Rixen and F. Adomeit, in: *Rarefied Gas Dynamics* (M. Becker and M. Fiebig, eds.), Vol. I, p. B18, DFVLR Press, Porz-Wahn (1974);
28. S. K. Loyalka, *Phys. Fluids* **18**, 1666 (1975);
29. C. Cercignani, in: *Rarefied Gas Dynamics* (J. L. Potter, ed.), Vol. II, p. 795, AIAA, New York (1977);
30. T. Abe and H. Oguchi, ISAS Report No. 553, **42**(8), Tokyo (1977);
31. S. L. Gorelov and M. N. Kogan, *Fluid Dynamics* **3**, 96 (1968);
32. G. A. Bird, in: *Rarefied Gas Dynamics* (J. L. Potter, ed.), Vol. I, p. 323, AIAA, New York (1977);
33. C. Cercignani, in: *Recent Developments in Theoretical and Experimental Fluid Mechanics* (U. Müller, K. G. Roesner, and B. Schmidt, eds.), p. 187, Springer, Berlin (1979).

Section 10——The results in this section are taken from ref. 30 of Chapter VI. The theory of generalized analytic functions is expounded in

34. I. N. Vekua, *Generalized Analytic Functions*, Pergamon Press, Oxford (1962).

For some details on the solution of particular problems, as well as the use of the original variables ξ and μ in place of w see ref. 24.

Section 12——The temperature-jump coefficient for a simplified model is evaluated in ref. 28 of Chapter VI.

Section 13——The problem of sound propagation has been treated in refs. 13–16 and in

35. L. Sirovich and J. K. Thurber, *J. Acoust. Soc. Am.* **37**, 329 (1965);
36. J. K. Thurber, in: *The Boltzmann Equation* (F. A. Grünbaum, ed.), p. 211, New York University Press, New York (1972);
37. C. L. Pekeris, Z. Alterman, L. Finkelstein, and K. Frankowski, *Phys. Fluids* **5**, 1608 (1962);
38. G. Maidanik, H. L. Fox, and M. Hekl, *Phys. Fluids* **8**, 259 (1965);
39. D. Kahn and D. Mintzer, *Phys. Fluids* **8**, 1090 (1965);
40. S. S. Abarbanel, in: *Rarefied Gas Dynamics* (C. L. Brundin, ed.), Vol. I, p. 369, Academic Press, New York (1967);
41. G. Sessler, *J. Acoust. Soc. Am.* **38**, 974 (1965);
42. L. H. Holway, *Phys. Fluids* **10**, 35 (1967);
43. L. Lees, *SIAM J. Appl. Math.* **13**, 278 (1965);
44. K. Toba, *Phys. Fluids* **11**, 2495 (1968);
45. F. B. Hanson and T. F. Morse, *Phys. Fluids* **12**, 1564 (1969);
46. J. R. Thomas, Jr., and C. E. Siewert, *Transport Theory Statist. Phys.* **8**, 219 (1979);
47. C. E. Siewert and E. E. Burniston, *J. Math. Phys.* **18**, 376 (1973).

The use of scattering of light from gases was proposed as a test of the Boltzmann equation by

48. M. Nelkin and S. Yip, *Phys. Fluids* **9**, 380 (1966),

and related calculations are presented in

49. S. Ranganathan and S. Yip, *Phys. Fluids* **9**, 372 (1966);

Section 14——Analytic solutions of nonlinear problems are discussed in

50. C. Truesdell, *Journal Rat. Mech. Anal.* **5**, 55 (1956);
51. V. S. Galkin, *PMM* **20**, 445 (1956) (in Russian);
52. V. S. Galkin, *PMM* **22**, 532 (1958);
53. V. S. Galkin, *PMM* **28**, 226 (1964);
54. V. S. Galkin, *Fluid Dynamics* **1**, 29 (1966);
55. R. Zwanzig, *J. Chem. Phys.* **71**, 4416 (1979);
56. J. Gomez Ordonez, J. J. Brey, and A. Santos, *Phys. Rev.* **A39**, 3038 (1989);
57. H. Grad, *Commun. Pure Appl. Math.* **2**, 331 (1949);
58. E. Ikenberry and C. Truesdell, *J. Rat. Mech. Anal.* **5**, 1 (1956);
59. C. Truesdell, *J. Rat. Mech. Anal.* **5**, 55 (1956);
60. M. Krook and T. T. Wu, *Phys. Rev. Lett.* **36**, 1107 (1976);
61. M. Krook and T. T. Wu, *Phys. Fluids* **20**, 1589 (1977);
62. A. V. Bobylev, *Sov. Phys. Dokl.* **20**, 820 and 822 (1976) and **21**, 632 (1977);
63. R. Krupp, Masters Thesis, Massachusetts Institute of Technology, Cambridge, Massachusetts (1967);
64. C. Cercignani, in: *Rarefied Gas Dynamics* (R. Campargue, ed.), Vol. I, p. 141, CEA, Paris (1981);
65. M. H. Ernst, *Phys. Rep.* **78**, 1 (1981);

66. M. H. Ernst, in: *Nonequilibrium Phenomena I: The Boltzmann Equation* (J. L. Lebowitz and E. W. Montroll, eds.), p. 51, North-Holland, Amsterdam (1983);
67. A. A. Nikol'skii, *Soviet Physics—Doklady* **8**, 633 (1964);
68. A. A. Nikol'skii, *Soviet Physics—Doklady* **8**, 639 (1964);
69. R. G. Muncaster, *Arch. Rat. Mech. Anal.* **70**, 79 (1979);

Appendix——The mathematical tools used in this appendix are described in

70. B. Riemann, *Collected Works* (H. Weber, ed.), p. 88, Dover, New York (1953);
71. G. Springer, *Introduction to Riemann Surfaces*, Addison-Wesley, Reading, Massachusetts (1957).

Chapter VIII

OTHER METHODS OF SOLUTION

1. Introduction

In the previous chapters we reviewed some methods of solution for the Boltzmann equation based on perturbation expansions, i.e., the Hilbert and Chapman–Enskog expansions and the linearization of the Boltzmann equation. The latter procedure has usually been coupled with the use of kinetic models. These models, however, have been shown to be capable of arbitrarily approximating not only the linearized Boltzmann equation, but also its solutions (Chapter VI); hence the procedures presented in Chapter VII can be considered to be exact, as long as the use of the linearized Boltzmann equation is justified. The complexity of the results obtained in Chapter VII even for relatively simple problems suggests that for more complicated problems, of linear or nonlinear nature, one should look for less sophisticated procedures yielding approximate but essentially correct results. Such procedures can be easily constructed for linearized problems or in the limit of either large or small Knudsen numbers; the intermediate range of Knudsen numbers (transition region) in nonlinear situations is at present a matter of interpolation procedures of more or less sophisticated nature. In addition, it is to be noted that a good procedure does not necessarily mean a good method of solution, since in many cases the procedure consists in deducing a system of nonlinear partial differential equations. The latter have to be solved in correspondence with particular problems, and in general they are tougher than the Navier–Stokes equations for the same problem. As a consequence one has to resort to numerical procedures to solve them. The approximation procedures can be grouped under two general headings: moment methods and integral equation methods. In the former case one constructs certain partial differential equations as mentioned above, and in the latter one tries to obtain either expansions valid for large Knudsen numbers or numerical solutions. In connection with both methods one can simplify the calculations by the use of models (sometimes in an essential manner), but one has to remember that the accuracy of kinetic models in nonlinear problems is less obvious than in the linearized ones. Finally, in connection with both methods one can apply variational procedures; again, the latter are more significant, and probably much more accurate for linearized problems.

In this chapter we shall briefly review this matter, with a certain emphasis on problems which have been solved accurately and methods which seem to provide a systematic approach to the transition regime of rarefaction.

2. Moment Methods

If one multiplies both sides of the Boltzmann equation by functions $\varphi_i(\boldsymbol{\xi})$ $(i = 1, \ldots, N, \ldots)$ forming a complete set, and integrates over the molecular velocity, one obtains infinitely many relations to be satisfied by the distribution function

$$\frac{\partial}{\partial t}\int \varphi_i(\boldsymbol{\xi})f(\mathbf{x}, \boldsymbol{\xi}, t)\, d\boldsymbol{\xi} + \frac{\partial}{\partial \mathbf{x}}\cdot\int \varphi_i(\boldsymbol{\xi})f(\mathbf{x}, \boldsymbol{\xi}, t)\boldsymbol{\xi}\, d\boldsymbol{\xi}$$

$$= \int \varphi_i(\boldsymbol{\xi})Q(f, f)\, d\boldsymbol{\xi} \qquad i = 1, \ldots, N, \ldots \tag{2.1}$$

This system of infinitely many relations (Maxwell's transfer equations) is equivalent to the Boltzmann equation because of the completeness of the set $\{\varphi_i\}$. The common idea behind the so-called moment methods is to satisfy only a finite number of transfer equations or moment equations.

This leaves the distribution function f largely undetermined, since only the infinite set (2.1) (with proper initial and boundary conditions) can determine f. This means that we can choose, to a certain extent, f arbitrarily and then let the moment equations determine the details which we have not specified. The different "moment methods" differ in the choice of the set φ_i and the arbitrary input for f. Their common feature is to choose f in such a way that f is a given function of $\boldsymbol{\xi}$ containing N undetermined parameters depending upon $\mathbf{x}$ and t, M_i $(i = 1, \ldots, N)$; this means that if we take N moment equations, we obtain N partial differential equations for the unknowns $M_i(\mathbf{x}, t)$. In spite of the large amount of arbitrariness, it is hoped that any systematic procedure yields, for sufficiently large N, results essentially independent of the arbitrary choices. On practical grounds another hope is that for sufficiently small N and a judicious choice of the arbitrary elements one can obtain accurate results. We shall not attempt to describe all the possible choices, and not even all those which have actually been proposed, but shall limit ourselves to the simplest and most often used ones.

The simplest choice (Grad, ref. 1) is to assume f to be a Maxwellian f_0 times a polynomial:

$$f = f_0 \sum_{k=0}^{N-1} Q_k(\mathbf{x}, t)H_k(\boldsymbol{\xi}) \tag{2.2}$$

where for convenience the polynomial is expressed in terms of the three-dimensional Hermite polynomials $H_k(\xi)$ orthogonal with respect to the weight f_0. It is also convenient to choose f_0 to be the local Maxwellian (this implies that $Q_0 = 1$, $Q_1 = Q_2 = Q_3 = Q_4 = 0$). There are N arbitrary quantities which can be identified with the basic moments ($\rho, v_i, T, p_{ij}, q_i$ and higher-order moments) and can be determined by taking $\varphi_i = H_i$ in Eq. (2.1) ($i = 0, \ldots N - 1$). A reasonable choice is $N = 13$; in such a case the unknowns are $\rho, v_i, T, p_{ij} - p\,\delta_{ij}$, and q_i and the equations are known as Grad's 13-moment equations.

There is an obvious disadvantage in Grad's choice: the distribution function is assumed to be continuous in the velocity variables, and this is not true at the boundaries. Furthermore, even the question of the convergence of polynomial approximations can be given a definite negative answer in certain nonlinear flows (Holway, ref. 2).

In order to avoid the first of the above-mentioned consequences, one can take for f a distribution function which already takes into account the discontinuities in velocity space, e.g., a function which piecewise reduces to Eq. (2.2). Then one can choose between piecewise continuous functions (Gross, Jackson, and Ziering, ref. 14, Chapter IV) or continuous functions (Lees, ref. 3) for the set $\{\varphi_i\}$. The second method seems to be more satisfactory and easier to handle; in its crudest version it reduces to assuming that the distribution function is piecewise Maxwellian, the discontinuities being located exactly where predicted by the free-molecular solution.

In order to avoid the absence of convergence in nonlinear flows, especially those involving shock waves, one can use linear combinations of Maxwellians (Mott-Smith, ref. 4) and take suitable moments.

It is obvious that these procedures can be used to obtain reasonably good results by judicious choice of the arbitrary elements, but it is to be remembered that they eventually become very complicated and generate equations which can be solved only numerically.

A particular feature is offered by Maxwell's molecules; in fact, if we choose for $\{\varphi_i\}$ a set of polynomials, the integrals in the right-hand side of Eq. (2.1) can be evaluated explicitly in terms of the full-range moments $\int \varphi_i f \, d\xi$, and only moments of degree inferior to the degree of φ_i appear in the right-hand sides of the ith equation. This means that in the space-homogeneous case one can evaluate exactly the time evolution of subsequent moments by solving ordinary differential equations (essentially linear, even in the nonlinear case) (Truesdell, ref. 59, Chapter VII).

3. The Integral Equation Approach

In Chapter VI, Section 3 it was shown that one can transform the linearized Boltzmann equation from an integrodifferential to a purely integral form. This can be repeated for the nonlinear equation, since the differential part is always linear. As in the linear case the transformation can be achieved in many ways (now the arbitrary function $\mu(\boldsymbol{\xi})$ is also, in general, a functional of f). The net result is

$$f = \bar{f}_0 + N(f) \tag{3.1}$$

where $\bar{f}_0$ is a "source term" (in general depending on f) and N a nonlinear operator acting upon f; the separation between $\bar{f}_0$ and $N(f)$ is purely formal, as always happens in nonlinear problems. One can use Eq. (3.1) as a tool for an iteration method: we start with a guess $f^{(0)}$ for f and then evaluate $f^{(1)}$, $f^{(2)}, f^{(3)}, \ldots$, etc. by means of the iteration scheme

$$f^{(n)} = \bar{f}_0^{(n-1)} + N(f^{(n-1)}) \tag{3.2}$$

Since there are different integral equations, there will be different iteration methods. Formally, the simplest one corresponds to the integral equation obtained by a simple integration along the characteristics. This iteration procedure is called the Knudsen iteration, but, as mentioned in Chapter VI, it presents serious difficulties, especially in one-dimensional plane geometries.

Better results are obtained by taking $\mu(\xi) > 0$ in the general procedure sketched in Chapter VI, Section 3. In this case the iteration method (integral iteration or successive collisions method—Willis, ref. 5) gives satisfactory results for sufficiently large Knudsen numbers.

A more far-reaching advantage of the integral equation approach can be obtained in connection with the use of models. In fact, all the collision models (linear or not) can be split into two parts, one of which (say, $\nu\Phi$) depends only on a finite number of moments of the distribution function (five, if Φ is the local Maxwellian, as is the case for the BGK model), while the second one (say, νf) can be written as the distribution function times a smooth functional of f (a constant in the linearized case). If we choose $\mu = \nu$ in building the integral equation, Eq. (3.1), then $N(f)$ and $\bar{f}_0$ depend only on a finite number of moments (plus the initial and boundary conditions). This means that f is known when these moments are known. Since, however, these moments are defined in terms of f, one can use Eq. (3.1) to construct integral equations for the basic moments. These equations can be very complicated, but have the essential advantage that the independent variables are only four $(\mathbf{x}, t)$ instead of seven $(\mathbf{x}, \boldsymbol{\xi}, t)$, a very important feature for numerical computations. In fact, in the case of one-dimensional steady problems one is reduced to a system of a few integral equations with one independent variable, something feasible on a personal computer. In the linearized case one can even achieve reasonably accurate results with limited amounts of computation time. The solutions obtained by these accurate numerical procedures are very important for two reasons: (1) By comparison with experimental data they can show the essential accuracy of certain models in describing certain experimental situations, and (2) they can be used to test the accuracy of any approximate method of solution.

Most of the problems treated so far have been solved with the BGK and ES problems. The typical problems were first solved in the following papers:

1. The shock wave structure (Liepmann *et al.*, ref. 6).
2. Heat transfer between parallel plates (Willis, ref. 7, with the BGK model; Cercignani and Tironi, ref. 5, Chapter IV, with the ES model).
3. Heat transfer between concentric cylinders (Anderson, ref. 8).

In the linearized case we have:

1. Plane Couette flow (Willis, ref. 9, with the BGK model; Cercignani and Tironi, ref. 9, Chapter IV, with the ES model).
2. Plane Poiseuille flow (Cercignani and Daneri, ref. 10, with the BGK model; Cercignani and Tironi, ref. 9, Chapter IV, with the ES model).
3. Cylindrical Poiseuille flow (Cercignani and Sernagiotto, ref. 11, with the BGK model; Cercignani and Tironi, ref. 16, with the ES model).
4. Poiseuille flow in annular tubes (Bassanini *et al.*, ref. 12).
5. Rotating cylinder (Bassanini *et al.*, ref. 13).
6. Cylindrical Couette flow (Cercignani and Sernagiotto, ref. 14).
7. Heat transfer between parallel plates (Bassanini *et al.*, ref. 15).

4. The Variational Principle

As we have seen, the most efficient procedures for solving the Boltzmann equation in its full or linearized form are based on moment methods or on the use of kinetic models. When going from the Boltzmann equation to moment methods or kinetic models one gives up any intent of accurately investigating the distribution function, and restricts oneself to the study of the space variation of some moments of outstanding physical significance, such as density, mass velocity, temperature, and heat flow. However, it is to be noted that even such restricted knowledge is not required for the purpose of making comparisons with experimental results. As a matter of fact, the typical output of an experimental investigation of Poiseuille flow is a plot of the flow rate versus the Knudsen number, while the actual velocity profile is usually the object of purely theoretical considerations. Analogously, the outstanding quantity is the stress constant in Couette flow, the heat flux constant in heat transfer problems, the drag on the body in the flow past a body. From the point of view of evaluating these overall quantities any computation of the flow fields appears to be a waste of time. Of course, the knowledge of flow fields is always interesting and illuminating, but frequently it happens that we have such a clear qualitative insight of the space behavior of the unknowns that we can imagine simple approximated analytical expressions for them containing a small number of adjustable constants. It therefore appears that a method which would succeed in giving both a precise rule for determining the above-mentioned constants and a highly accurate evaluation of the overall quantities of outstanding interest should turn out to be most useful. The features which we have just mentioned are typical of variational procedures; it is therefore in this direction that one has to look for the desired method. The procedure is well established, and useful only in the case of the linearized Boltzmann equation; accordingly, we shall restrict ourselves to this case.

A nontrivial variational procedure for a linear equation is usually based on the self-adjointness of the linear operator $\mathscr{L}$ which appears in the equation

$$\mathscr{L}h = S \tag{4.1}$$

where S is a source term. Once $\mathscr{L}$ has been shown to be self-adjoint with respect to a certain scalar product $((\ ,\ ))$ then the functional

$$J(\tilde{h}) = ((\tilde{h}, \mathscr{L}\tilde{h})) - 2((S, \tilde{h})) \tag{4.2}$$

is easily shown to satisfy

$$\delta J = 0 \tag{4.3}$$

for infinitesimal departures $\delta h = \tilde{h} - h$ from the solution of Eq. (4.1). If additional conditions are satisfied, then $J(\tilde{h})$ has a relative or absolute maximum for $\tilde{h} = h$. In the case of the steady linearized Boltzmann equation, S is a surface term (related, in general, to the inhomogeneous part of the boundary conditions), while

$$\mathscr{L} = L - \boldsymbol{\xi} \cdot \partial/\partial \mathbf{x} \tag{4.4}$$

with homogeneous boundary conditions of the type discussed in Chapter VI, Section 1. Although L is self-adjoint in $\mathscr{H}$ and hence also in the space of square summable functions of $\boldsymbol{\xi}$ and $\mathbf{x}$, $\boldsymbol{\xi} \cdot \partial/\partial \mathbf{x}$ is not self-adjoint in such a space or in any space with positive-definite metric. Let us consider, however, the complex linear space with indefinite Hermitian metric based on the following scalar product:

$$((h, g)) = \iint \bar{h}(-\boldsymbol{\xi}, \mathbf{x}) g(\boldsymbol{\xi}, \mathbf{x})\, d\boldsymbol{\xi}\, d\mathbf{x} \tag{4.5}$$

where the bar denotes complex conjugation and we have eliminated the Maxwellian weighting the integration in velocity space by adopting the notation of Chapter VI. The scalar product defined by Eq. (4.5) has all the usual properties, except for the fact that it produces a "norm" which is a real but not necessarily a positive number, as is immediately verified by splitting h into its odd and even part with respect to $\boldsymbol{\xi}$ and evaluating $((h, h))$. Now, a simple calculation which takes into account the properties of the homogeneous boundary conditions shows that $\mathscr{L}$ is formally self-adjoint with respect to the considered pseudometric, i.e.,

$$((h, \mathscr{L}g)) = ((\mathscr{L}h, g)) \tag{4.6}$$

This property can be used to discuss the eigenvalues of $\mathscr{L}$; if $\mathscr{L}$ were self-adjoint in an ordinary Hilbert space, its eigenvalues would be real, but now we can only show that if there are complex eigenvalues, the corresponding eigenfunctions belong to the isotropic cone in the pseudo-Hilbert space, i.e., have zero norm. The indefinite character of the metric, however, does not destroy the possibility of obtaining a variational principle, because for Eq. (4.3) to hold only Eq. (4.6) is needed. The functions $\tilde{h}$ for which Eq. (4.3) holds must satisfy the same boundary conditions as the solution h of the linearized Boltzmann equation. Accordingly, we can make a guess containing a certain number of parameters for $\tilde{h}$ and then determine them, according to the variational principle, by making $J(\tilde{h})$ stationary. If the boundary conditions have an inhomogeneous term of the form $2\mathbf{U}\cdot\boldsymbol{\xi}$, where $\mathbf{U}$ is the boundary velocity, or $\Delta T(\xi^2 - \frac{5}{2})$, where ΔT is the deviation of the boundary from the average temperature, then $-((S, h))$, i.e. the value of $J(\tilde{h})$ for $\tilde{h} = h$, takes on the form of a surface integral of $p_{ji}n_iU_j$, which gives the drag, or q_in_i, which gives the heat transfer through the body surface. This means that if by means of the variational principle one approximates h with an average error of 10%, one can approximate the drag or the heat transfer with an error of order 1% because terms in the δh cancel according to Eq. (4.3) and the error is of order $|\delta h|^2$. In the case of Poiseuille flow, Eq. (5.6) of Chapter VII, the boundary conditions are homogeneous and the source term is $S = k\xi_3$; this means that $((S, h))$ is proportional to the volume flow rate, which, as a consequence, can be evaluated accurately. The usefulness of the variational principle as explained above is somewhat limited by two circumstances: we have to make guesses about the distribution function, and these guesses are restricted to satisfy the boundary conditions. In such a situation either we make poor guesses or we obtain very complicated trial functions. The advantages of the variational approach are much enhanced if we use it in connection with two other tools: kinetic models and the integral equation approach. In fact, if we use kinetic models in the integral form, a guess about a finite number of moments implies a guess about the distribution function which automatically satisfies the boundary conditions (since the latter are built in the equations). If we denote the basic moments by ρ_j $(j = 1, \ldots, N)$ and $\boldsymbol{\rho}$ is a column vector which summarizes the ρ_j, then $\boldsymbol{\rho}$ satisfies (Cercignani and Pagani, ref. 18):

$$\boldsymbol{\rho} = \boldsymbol{\rho}^0 + \mathbf{A}\boldsymbol{\rho} \tag{4.7}$$

where $\mathbf{A}$ is a matrix integral operator. In the case of the models defined by Eq. (8.1) of Chapter VII if $\mathbf{a} = \|\alpha_{jk}\|$ one can show that

$$\mathbf{A}^+\mathbf{as} = \mathbf{saA} \tag{4.8}$$

where $\mathbf{A}^+$ is the transpose of the operator $\mathbf{A}$ with respect to the scalar product

$$(\boldsymbol{\rho}, \boldsymbol{\psi})_V = \int_V \boldsymbol{\rho}(\mathbf{x}) \cdot \boldsymbol{\psi}(\mathbf{x})\, d\mathbf{x} \tag{4.9}$$

and $\mathbf{s}$ is a diagonal matrix with components $(-1)^p \delta_{jk}$, $(-1)^p = \pm 1$ depending on the parity of ρ_j. Then the functional

$$J(\tilde{\boldsymbol{\rho}}) = (\mathbf{as}\tilde{\boldsymbol{\rho}}, \tilde{\boldsymbol{\rho}} - A\tilde{\boldsymbol{\rho}} - 2\boldsymbol{\rho}^{(0)})_V \tag{4.10}$$

is stationary for $\tilde{\boldsymbol{\rho}} = \boldsymbol{\rho}$ if $\boldsymbol{\rho}$ satisfies Eq. (4.7). Also, if the symmetry of the problem is such that only those ρ_j with a definite parity are different from zero, then $\mathbf{s}$ coincides with either $+\mathbf{I}$ or $-\mathbf{I}$, $\mathbf{I}$ being the unit matrix, and one can easily show that the stationary principle becomes a minimum or a maximum principle. Finally, one can relate the value attained by J for $\tilde{\boldsymbol{\rho}} = \boldsymbol{\rho}$ to the global quantities of basic interest (drag, flow rate, heat flux) and hence retain the feature of having a very precise procedure for evaluating these quantities.

The ideas described before in connection with the pseudo-scalar product appearing in Eq. (4.5) were exploited in a paper by the author (ref. 19), who obtained a variational principle for the integrodifferential form of the Boltzmann equation. The functional to be varied was generalized in order to include a contribution from the boundary. In this way one can avoid the inconvenient constraint that the trial functions must satisfy the boundary conditions exactly. Again, in most cases the value attained by the functional when $\tilde{h} = h$ is related to a quantity of physical interest.

It was remarked by Lang and Loyalka (refs. 20 and 21) that it is possible to generalize the variational principle under consideration by introducing, along with h, the solution h^* of a linearized Boltzmann equation with a different source term S^*; in this case we then introduce a functional $J(\tilde{h}, \tilde{h}^*)$ which is stationary for $\tilde{h} = h$ and $\tilde{h}^* = h^*$ and we can choose the quantity related to $J(\tilde{h}, \tilde{h}^*)$ by suitably choosing S^*.

5. Examples of Specific Problems

In this section we shall consider very simple problems as examples of application of both the integral equation approach and the variational procedure. The first example is the Kramers problem with the BGK model (Chapter VII, Section 4); as we know, this problem can be solved exactly, but we shall ignore this at present. If we follow the procedure indicated in Section 3, we can express the perturbation of the distribution function in terms of a single moment, the z component of the mass velocity $v_3(x)$; in fact, density, temperature, and the remaining components of $\mathbf{v}$ remain unperturbed in a linearized treatment (as we know from the splitting procedure used in Chapter VII). Accordingly, we can construct an integral equation for v_3 by taking the corresponding moment of the expression of h in terms of v_3. If we put $\varphi(x) = [v_3(x) - kx]/(k\theta)$, i.e., if we subtract the asymptotic behavior and make the unknown nondimensional, we obtain the following integral equation:

$$\pi^{1/2}\varphi(x) = T_1(x) + \int_0^\infty T_{-1}(|x-y|)\varphi(y)\,dy \qquad (x \geq 0) \tag{5.1}$$

where we have put

$$T_n(x) = \int_0^\infty t^n \exp[-t^2 - (x/t)]\,dt \tag{5.2}$$

θ has been taken to be unity, while the velocity unit is, as usual, $(2RT)^{1/2}$. Equation (5.2) defines a family of transcendental functions, which one always runs into when using the integral equation approach for solving models with constant collision frequency. The T_n functions have some notable properties:

$$dT_n/dx = -T_{n-1} \tag{5.3}$$

$$T_n(x) = \tfrac{1}{2}(n-1)T_{n-2} + \tfrac{1}{2}xT_{n-3} \tag{5.4}$$

$$T_n(0) = \frac{1}{2}\Gamma\left(\frac{n+1}{2}\right) \tag{5.5}$$

One can derive expansions valid for small x and expansions for large x and compute the T_n to any desired accuracy. Note that many authors use the notation J_n in place of T_n.

Let us put

$$\mu = \lim_{x\to\infty} \varphi(x) = \pi^{1/2}\zeta/2l \tag{5.6}$$

where ζ is the slip coefficient and l the mean free path; then if we define

$$\psi(x) = \varphi(x) - \mu \tag{5.7}$$

we have

$$\pi^{1/2}\psi(x) - \int_0^\infty T_{-1}(|x-y|)\psi(y)\,dy = T_1(x) - \mu T_0(x) \qquad (x \geq 0) \tag{5.8}$$

where Eq. (5.3) has been used.

Since $\psi(x)$ is integrable on $(0, \infty)$, we can integrate Eq. (5.8) from x to ∞ twice and obtain

$$\int_0^\infty T_1(|x-y|)\psi(y)\,dy = \mu T_2(x) - T_3(x) \qquad (x \geq 0) \tag{5.9}$$

where Eqs. (5.3) and (5.5) have been used.

$$J(\tilde{\varphi}) = \pi^{1/2}\int_0^\infty dx\left\{[\tilde{\varphi}(x)]^2 - \int_0^\infty T_{-1}(|x-y|)\tilde{\varphi}(x)\tilde{\varphi}(y)\,dx - 2\tilde{\varphi}(x)T_1(x)\right\} \tag{5.10}$$

This functional attains a minimum when $\tilde{\varphi} = \varphi$, φ being the solution of Eq. (5.1). Equation (5.9) with $x = 0$ gives

$$\int_0^\infty T_1(y)\psi(y)\,dy = \tfrac{1}{4}\mu\pi^{1/2} - \tfrac{1}{2} \tag{5.11}$$

and using Eq. (5.6) we have

$$\zeta = l\left[(2/\pi) + (4/\pi)\int_0^\infty \varphi(y)T_1(y)\,dy\right] \tag{5.12}$$

We note now that for $\tilde{\varphi} = \varphi$ we have

$$J(\varphi) = \min[J(\tilde{\varphi})] = -\int_0^\infty \varphi(y)T_1(y)\,dy \tag{5.13}$$

and Eq. (5.12) becomes

$$\zeta = l\{(2/\pi) - (4/\pi)\min[J(\tilde{\varphi})]\} \tag{5.14}$$

In this way we have found a direct connection between the slip coefficient and the minimum value attained by the functional $J(\tilde{\varphi})$; this means that even a poor estimate for $\tilde{\varphi}$ can give an accurate value for ζ. Accordingly, we make the simplest choice for the trial function: $\tilde{\phi} = c$ (constant)! We find by straightforward computation:

$$J(c) = \tfrac{1}{2}c^2 - \tfrac{1}{2}\pi^{1/2}c \tag{5.15}$$

The minimum value is attained when

$$c = \tfrac{1}{2}\pi^{1/2} \tag{5.16}$$

corresponding to

$$J\{\tfrac{1}{2}\pi^{1/2}\} = \tfrac{1}{8}\pi \tag{5.17}$$

Eq. (5.14) then gives for ζ

$$\zeta = [\tfrac{1}{2} + (2/\pi)]l = (1.1366)l \tag{5.18}$$

which is to be compared with the exact value $\zeta = (1.1466)l$ given in Chapter VII. We see that even a very simple choice for $\tilde{\phi}$, a choice which we know to be very inadequate to describe the kinetic boundary layer, yields a rather accurate estimate of the slip coefficient ζ.

As the next example we consider Couette flow between two parallel plates (described in Chapter VII, Section 5). If $\varphi(x)$ now denotes the ratio of the mass velocity to $U/2$ ($\pm U/2$ being the velocities of the plates), we obtain the following integral equation (Willis, ref. 9):

$$\pi^{1/2}\varphi(x) = T_0(\tfrac{1}{2}\delta - x) - T_0(\tfrac{1}{2}\delta + x) + \int_{-\delta/2}^{\delta/2} T_{-1}(|x - y|)\varphi(y)\,dy \tag{5.19}$$

where δ is the distance between the plates in θ units.

Analogously, we can find

$$\pi_{xz} = T_1(\tfrac{1}{2}\delta - x) + T_1(\tfrac{1}{2}\delta + x) - \int_{-\delta/2}^{\delta/2} \operatorname{sgn}(x - y)T_0(|x - y|)\varphi(y)\,dy \tag{5.20}$$

where π_{xz} is the ratio of the stress tensor p_{xz} to its free-molecular value $(-\rho U\pi^{-1/2}/2)$. Note that π_{xz} is constant, in spite of the fact that Eq. (5.20) shows an x-dependence; the constancy of π_{xz} is obvious from conservation of momentum, and can be recovered from Eq. (5.20) by differentiating and comparing with Eq. (5.19). Therefore we can take Eq. (5.20) at any point for evaluating π_{xz}; the simplest choice is to take the arithmetical mean of the expressions given by Eq. (5.20) for $x = \pm\delta/2$. We obtain

$$\pi_{xz} = \tfrac{1}{2} + T_1(\delta) + \tfrac{1}{2}\int_{-\delta/2}^{\delta/2} [T_0(\tfrac{1}{2}\delta + y) - T_0(\tfrac{1}{2}\delta - y)]\varphi(y)\,dy \qquad (5.21)$$

Equation (5.19) was solved numerically by Willis (ref. 9) for δ ranging from 0 to 20. If we want to use the variational procedure, we construct the functional

$$J(\tilde{\varphi}) = \int_{-\delta/2}^{\delta/2} [\tilde{\varphi}(x)]^2\,dx - \pi^{-1/2}\iint_{-\delta/2}^{\delta/2} T_{-1}(|x - y|)\tilde{\varphi}(x)\tilde{\varphi}(y)\,dx\,dy$$

$$+ 2\pi^{-1/2}\int_{-\delta/2}^{\delta/2} [T_0(\tfrac{1}{2}\delta + x) - T_0(\tfrac{1}{2}\delta - x)]\tilde{\varphi}(x)\,dx \qquad (5.22)$$

which attains its minimum value

$$J(\varphi) = \min J(\tilde{\varphi}) = \pi^{-1/2}\int_{-\delta/2}^{\delta/2} [T_0(\tfrac{1}{2}\delta + x) - T_0(\tfrac{1}{2}\delta - x)]\varphi(x)\,dx \qquad (5.23)$$

when $\tilde{\varphi} = \varphi(x)$ solves Eq. (5.19). Comparing Eq. (5.21) and (5.23), we deduce

$$\pi_{xz} = \tfrac{1}{2} + T_1(\delta) + \tfrac{1}{2}\pi^{1/2}\min J(\tilde{\varphi}) \qquad (5.24)$$

Thus the stress has a direct connection with the minimum value attained by $J(\tilde{\varphi})$. As a trial function we can take the continuum solution, i.e.,

$$\tilde{\varphi}(x) = Ax \qquad (5.25)$$

where A is an indeterminate constant. After some easy manipulations based on the properties of the T_n functions we find (Cercignani and Pagani, ref. 18) an expression for $J(Ax)$ in terms of $T_n(\delta)$ $(1 \le n \le 3)$. The minimum condition is easily found, the corresponding $A = A_0(\delta)$ and $J(A_0x)$ again being expressed in terms of the $T_n(\delta)$. Inserting $J(A_0x)$ into Eq. (5.24), we obtain π_{xz}; the resulting values are tabulated *versus* δ in Table I (third column). The second column of this table gives the results obtained by Willis by means of a numerical solution of Eq. (5.19). It will be noted that the agreement is excellent. But one can see something more, that the results obtained by the variational procedure are more accurate than those of Willis. As a matter of fact, the variational procedure gives for π_{xz} a value approximated from above, and it is noted that values in the third column are never larger

Table I
Stress versus Inverse Knudsen Number for Couette Flow*

δ	Willis's result, ref. 9	Linear trial function	Cubic trial function
0.01	0.9913	0.9914	0.9914
0.10	0.9258	0.9258	0.9258
1.00	0.6008	0.6008	0.6008
1.25	0.5517	0.5512	0.5511
1.50	0.5099	0.5097	0.5096
1.75	0.4745	0.4743	0.4742
2.00	0.4440	0.4438	0.4437
2.50	0.3938	0.3935	0.3933
3.00	0.3539	0.3537	0.3535
4.00	0.2946	0.2945	0.2943
5.00	0.2526	0.2524	0.2523
7.00	0.1964	0.1964	0.1963
10.00	0.1474	0.1474	0.1473
20.00	0.0807	0.0805	0.0805

* From Cercignani and Pagani, ref. 18.

than the corresponding ones in the second column. The fourth column gives the results obtained by using a cubic trial function:

$$\tilde{\varphi}(x) = Ax + Bx^3 \tag{5.26}$$

In this case the algebra is more formidable, but still straightforward. It is to be noted that the resulting values for π_{xz} are only slightly different from the previous one; this is another test of the accuracy of the method.

Another example is offered by Poiseuille flow. The starting point is Eq. (5.6) of Chapter VII. If we adopt the BGK model, we can obtain an integral equation for the mass velocity $v_3(\mathbf{x})$, where $\mathbf{x}$ is the two-dimensional vector describing the cross section of the channel. If we put $v_3(\mathbf{x}) = \frac{1}{2}k\theta[1 - \varphi(\mathbf{x})]$, where $k = p^{-1}\,\partial p/\partial z$, the integral equation to be solved can be written as follows:

$$\varphi(\mathbf{x}) = 1 + \pi^{-1} \iint_{\Sigma(\mathbf{x})} T_0(|\mathbf{x} - \mathbf{y}|)|\mathbf{x} - \mathbf{y}|^{-1}\varphi(\mathbf{y})\,d\mathbf{y} \tag{5.27}$$

where $\mathbf{x}$ is measured in θ units and $\Sigma(\mathbf{x})$ is the part of the cross section whose points can be reached from $\mathbf{x}$ by straight lines without intersecting boundaries [$\Sigma(\mathbf{x})$ is the whole cross section Σ when the boundary curvature has constant sign]. The functional to be considered is now

$$J(\tilde{\varphi}) = \int_{\Sigma} [\tilde{\varphi}(\mathbf{x})]^2 \, d\mathbf{x} - \int_{\Sigma} \tilde{\varphi}(\mathbf{x}) \int_{\Sigma(\mathbf{x})} T_0(|\mathbf{x} - \mathbf{y}|)|\mathbf{x} - \mathbf{y}|^{-1} \tilde{\varphi}(\mathbf{y}) \, d\mathbf{y}$$

$$- 2 \int_{\Sigma} \tilde{\varphi}(\mathbf{x}) \, d\mathbf{x} \tag{5.28}$$

The value of this functional attains a minimum when $\tilde{\varphi} = \varphi$, where φ satisfies Eq. (5.27), and the minimum value is

$$J(\tilde{\varphi}) = -\int_{\Sigma} \tilde{\varphi}(\mathbf{x}) \, d\mathbf{x} \tag{5.29}$$

a quantity obviously related to the flow rate. Actual calculations have been performed for the case of a slab (Cercignani and Pagani, ref. 18) and a cylinder (Cercignani and Pagani, ref. 22) by assuming a parabolic profile (i.e., a continuum-like solution). In the plane case one can obtain the flow rate in terms of T_n functions, while in the cylindrical case one also has to perform numerical quadratures involving these functions.

In both cases one can evaluate the flow rate with great accuracy and compare the results with the numerical solutions by Cercignani and Daneri (ref. 10) and Cercignani and Sernagiotto (ref. 11). As shown by Tables II and III, the agreement is very good. The results also compare very well with experimental data (Figs. 7 and 8), and, in particular, the flow rate exhibits à minimum for $\delta \approx 1.1$ in the plane case (δ is the distance of the plates in θ units) and for $\delta \approx 0.3$ in the cylindrical case (δ is the radius of the cylinder in θ units). The location of the minimum is in excellent agreement with experimental data, while the values of the flow rate show deviations of order 3% from experimental data in the cylindrical case. It is to be noted that the same problem can be treated with the ES model (Cercignani and Tironi, ref. 16) to show that once one knows the BGK solution (Pr = 1) one can obtain the ES for any Prandtl number; in particular, the dimensionless flow rate in the cylindrical case is given by

$$Q(\delta, \mathrm{Pr}) = Q(\delta \mathrm{Pr}, 1) + (1 - \mathrm{Pr})(\delta/4) \tag{5.30}$$

If the value $\mathrm{Pr} = \frac{2}{3}$ is chosen, as is appropriate for a monatomic gas, we obtain results deviating from experimental data by only 1% or 2% (Cercignani, ref. 17; see Fig. 8).

Other problems which have been treated with the variational technique are heat transfer between parallel plates (Bassanini, Cercignani,and Pagani, (ref. 15) and from a sphere (Cercignani and Pagani, ref. 22); flow past a solid

Table II
Flow Rate versus the Inverse Knudsen Number for Plane Poiseuille Flow*

δ	Results of Cercignani and Daneri, ref. 10 From above	From below	Variational method
0.01	3.0499	—	3.0489
0.10	2.0331	2.0326	2.0314
0.50	1.6025	1.6010	1.6017
0.70	1.5599	1.5578	1.5591
0.90	1.5427	1.5367	1.5416
1.10	1.5391	1.5352	1.5379
1.30	1.5441	1.5390	1.5427
1.50	1.5546	1.5484	1.5530
2.00	1.5963	1.5862	1.5942
2.50	1.6497	1.6418	1.6480
3.00	1.7117	1.7091	1.7092
4.00	1.8468	1.8432	1.8440
5.00	1.9928	1.9863	1.9883
7.00	2.2957	2.2851	2.2914
10.00	2.7669	2.7447	2.7638

* From Cercignani and Pagani, ref. 18.

Table III
Flow Rate versus the Inverse Knudsen Number for Cylindrical Poiseuille Flow*

δ	Numerical solution†	Variational method
0.01	1.4768	1.4801
0.10	1.4043	1.4039
0.20	1.3820	1.3815
0.40	1.3796	1.3788
0.60	1.3982	1.3971
0.80	1.4261	1.4247
1.00	1.4594	1.4576
1.20	1.4959	1.4937
1.40	1.5348	1.5321
1.60	1.5753	1.5722
2.00	1.6608	1.6559
3.00	1.8850	1.8772
4.00	2.1188	2.1079
5.00	2.3578	2.3438
7.00	2.8440	2.8245
10.00	3.5821	3.5573

* Cercignani and Pagani, ref. 22.
† Cercignani and Sernagiotto, ref. 11.

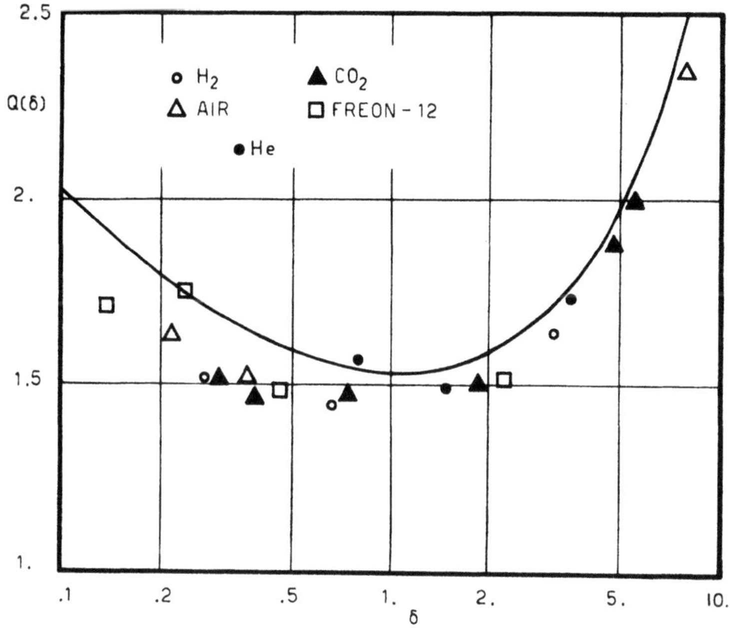

Fig. 7. Comparison between the BGK solution for plane Poiseuille flow and the experimental data of Dong, ref. 28. Here δ is the ratio of the distance between the plates to $\theta(2RT)^{1/2}$, and Q is the ratio of the mass flow rate to $-d^2(2RT)^{-1/2}(dp/dz)$. (As usual, θ is the mean free time, T the temperature, dp/dz the pressure gradient, d the distance between the plates, and R the gas constant.)

body (Cercignani and Pagani, ref. 23; Cercignani, Pagani, and Bassanini, ref. 24); and heat transfer between concentric cylinders (Bassanini, Cercignani, and Pagani, ref. 25). Figures 9 and 10 show comparisons between the experimental data and the theoretical results for heat transfer between parallel plates and from a sphere. In the plane case curves for different values of the accommodation coefficient are shown; the theoretical calculations are based on the boundary conditions given by Eq. (4.10) of Chapter IV.

Figure 11 shows a comparison between the variational results for the drag coefficient C_D of a sphere and the semiempirical formula proposed by Millikan, ref. 33, to interpolate his experimental data. In the same plot Stokes's classical result and Sherman's interpolating formula for the drag are also reported. The latter formula is as follows:

$$C_D(R)/C_{DFM} = [1 + (0.685)R]^{-1} \tag{5.31}$$

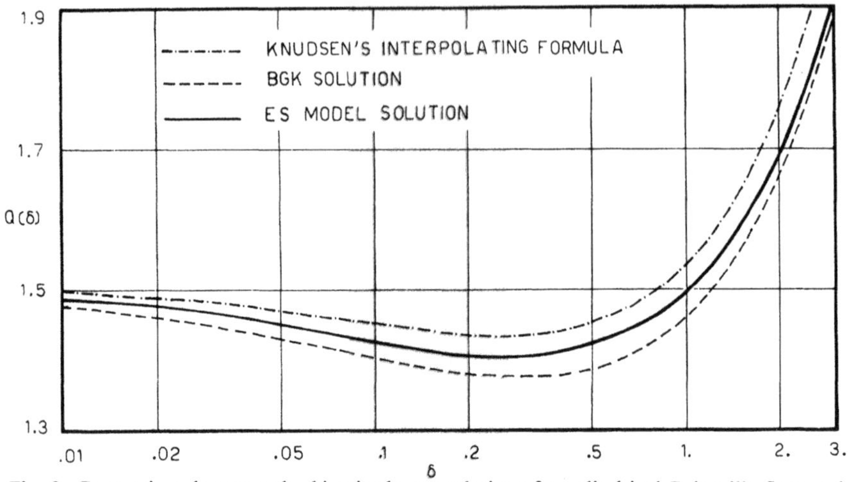

Fig. 8. Comparison between the kinetic theory solutions for cylindrical Poiseuille flow and the semiempirical formula proposed by Knudsen to interpolate his experimental data (ref. 29). Here δ is the ratio of the radius a of the cylinder to $\theta(2RT)^{1/2}$, and Q is the ratio of the mass flow to $-\pi a^3(2RT)^{-1/2}(dp/dz)$. The existence and location of the minimum constitute an interesting check of the theory with experiments.

where R is the radius of the sphere in $\theta(2RT)^{1/2}$ units and $C_{DFM} = C_D(0)$ is the drag coefficient in free-molecular flow. Equation (5.31) is a particular case of a universal formula (Sherman, ref. 27) relating a quantity $F(\mathrm{Kn})$, as a function of the Knudsen number Kn (here $\mathrm{Kn} = 1/R$), to its free-molecular value F_{FM} and its value according to continuum theory, $F_C(\mathrm{Kn})$, provided $F_C(\mathrm{Kn})/F_{FM} \to 0$ as $\mathrm{Kn} \to 0$. Sherman's general formula is as follows:

$$F/F_{FM} = (1 + F_{FM}/F_C)^{-1} \tag{5.32}$$

and is in reasonably good agreement with experimental data for most subsonic experiments. A notable exception is offered by Poiseuille flows in long tubes, for which Sherman's formula [applied to $1/Q(\delta)$] does not predict a minimum in the flow rate.

These problems have been treated by many authors with different methods; in particular the variational principle in integrodifferential form received an early application to plane Couette flow (ref. 19) and the problem of heat transfer between parallel plates (ref. 34).

Finally, the variational method has been applied to models with velocity-dependent collision frequency, especially in connection with the

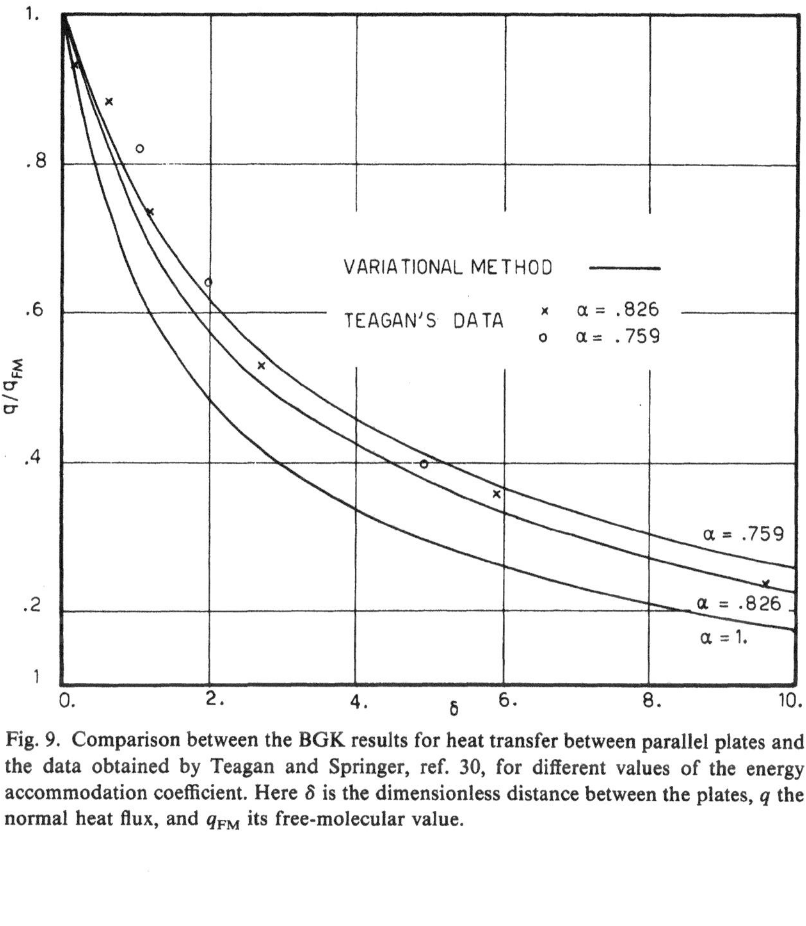

Fig. 9. Comparison between the BGK results for heat transfer between parallel plates and the data obtained by Teagan and Springer, ref. 30, for different values of the energy accommodation coefficient. Here δ is the dimensionless distance between the plates, q the normal heat flux, and q_{FM} its free-molecular value.

evaluation of the slip coefficient (Loyalka and Ferziger, ref. 11, Chapter IV; Cercignani, Foresti, and Sernagiotto, ref. 25, Chapter VII).

6. Concluding Remarks

The survey of methods and problems presented in the previous sections is by no means exhaustive. Too many methods are still to be tested and made systematic, too many problems are yet to be solved. Yet one begins to see the basis for a systematic theory, at least in the linearized case. Furthermore, some conclusions can be drawn, e.g., that the BGK model is essentially correct for describing linearized shear flows; this is a small piece of knowledge, it is true, but by no means an obvious one. By solving more

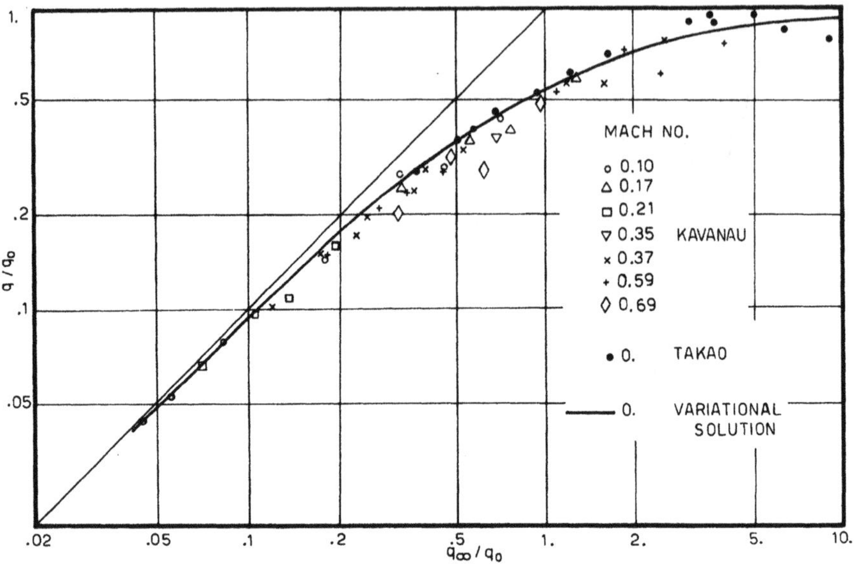

Fig. 10. Comparison between the BGK solution and the experimental data for heat transfer from a sphere. The data are taken from Kavanau, ref. 31, and Takao, ref. 32. Following Sherman, ref. 27, the ratio of the radial heat flux q to its free-molecular value, q_0, is plotted against the ratio of the heat flux according to continuum theory, q_∞, to q_0. Both q and q_∞ are, of course, functions of the Knudsen number. The bisector of the axes clearly represent the continuum solution, while the upper border of the plot represents the free-molecular value.

problems one can obtain more general statements concerning the accuracy of the models; for linearized problems one can probably find out that a certain model (say, a combination of the ES model with the velocity-dependent collision frequency) is essentially exact for the purpose of evaluating the low-order moments. For nonlinear problems, of course, the situation is less clear, and probably will require years of investigation to be elucidated. In this research methods which have not been reviewed in the previous sections because they are essentially numerical can play an important role; we just mention discrete ordinate techniques (refs. 35, 36) and Monte Carlo methods (refs. 37–41), which have been employed by different authors in the last 25 years. It would not be appropriate to mention all of them here. The Proceedings of the Biannual Symposia on Rarefied Gas Dynamics (refs. 42–57) and the book by Bird (ref. 58) should give a fairly good idea of the developments and trends in this area.

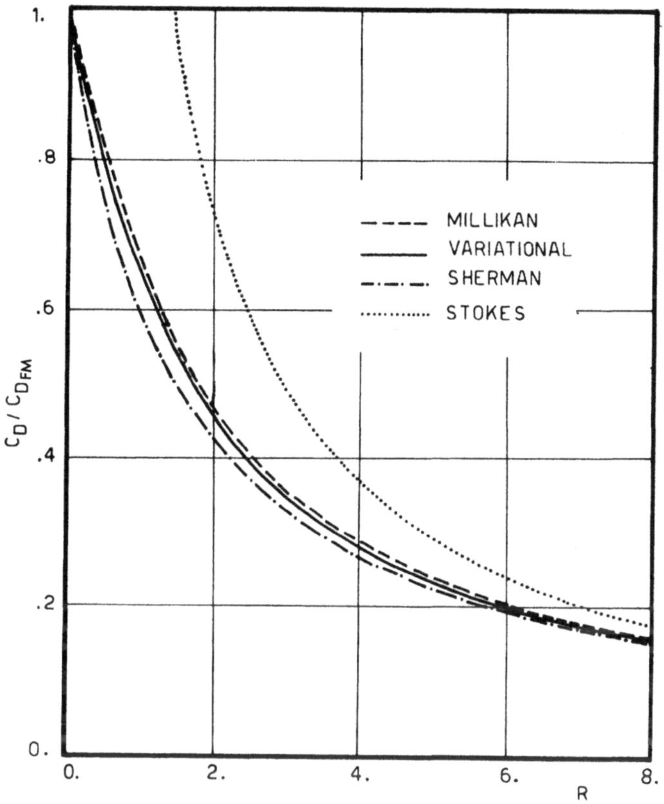

Fig. 11. Comparison between the variational result for the drag of a sphere according to the BGK model and the formula proposed by Millikan, ref. 33, to interpolate his experimental data. Stokes's classical solution and Sherman's formula are also reported. Here R is the radius of the sphere in $\theta(2RT)^{1/2}$ units, C_D the drag coefficient, and C_{DFM} its free-molecular value.

References

Section 2——Grad's 13-moment method was proposed in

1. H. Grad, *Commun. Pure Appl. Math.* **2**, 331 (1949).

The absence of convergence of polynomial expansions in the problem of shock wave structure was indicated by

2. L. W. Holway, Jr., *Phys. Fluids* **7**, 911 (1964).

Half-range approximations have been proposed in ref. 14, Chapter IV and in

3. L. Lees, "A Kinetic Theory Description of Plane Compressible Couette Flow," GALCIT Hypersonic Research Project, Memorandum No. 51 (1959).

The use of "multimodal" procedures based on more than one Maxwellian started with

4. H. M. Mott-Smith, *Phys. Rev.* **82**, 885 (1951).

A detailed treatment of full-range moment equations for Maxwell molecules was given in ref. 1, Chapter III. For more details on moment methods see Kogan's book, ref. 10, Chapter I.

Section 3——The method of successive collisions or integral iteration, although used by many authors for abstract existence theory in the last forty years, was first used for solving problems by

5. D. R. Willis, "A Study of Some Nearly-Free-Molecular Flow Problems," Ph.D. Thesis, Princeton University (1958).

The integral equation formulation has been used to solve nonlinear problems by

6. H. W. Liepmann, R. Narasimha, and M. T. Chahine, *Phys. Fluids* **5**, 1313 (1962);
7. D. R. Willis, in: *Rarefied Gas Dynamics* (J. A. Laurmann, ed.), Vol. I, p. 209, Academic Press, New York (1963),

in ref. 5, Chapter IV, and in

8. D. Anderson, "On the Steady Krook Kinetic Equation—Part 2," *J. Plasma Phys.* **1**, 255 (1968).

The integral equation for linearized problems has been solved numerically in the following papers:

9. D. R. Willis, *Phys. Fluids* **5**, 127 (1962);
10. C. Cercignani and A. Daneri, *J. Appl. Phys.* **34**, 3509 (1963);
11. C. Cercignani and F. Sernagiotto, *Phys. Fluids* **9**, 40 (1966);
12. P. Bassanini, C. Cercignani, and F. Sernagiotto, *Phys. Fluids* **9**, 1174 (1966);
13. P. Bassanini, C. Cercignani, and P. Schwendimann, in: *Rarefied Gas Dynamics* (C. L. Brundin, ed.), Vol. I, p. 505, Academic Press, New York (1967);
14. C. Cercignani and F. Sernagiotto, *Phys. Fluids* **10**, 1200 (1967);
15. P. Bassanini, C. Cercignani, and C. D. Pagani, *Int. J. Heat and Mass Transfer* **10**, 447 (1967).

See also ref. 9, Chapter IV and

16. C. Cercignani and G. Tironi, in: *Proc. of the AIDA-AIR Meeting*, p. 174, AIDA-AIR, Rome (1967);
17. C. Cercignani, *Phys. Fluids* **10**, 1859 (1967).

Section 4——The variational procedure was proposed in

18. C. Cercignani and C. D. Pagani, *Phys. Fluids* **9**, 1167 (1966).

Section 5——Examples of applications of the variational principle can be found in ref. 18; in ref. 11, Chapter IV; in ref. 25, Chapter VII; and in

19. C. Cercignani, *J. Stat. Phys.* **1**, 297 (1969);
20. S. K. Loyalka and H. Lang, in: *Rarefied Gas Dynamics* (D. Dini, ed.), Vol. II, p. 779, Edizioni Tecnico-Scientifiche, Pisa (1971);
21. H. Lang, *Acta Mech.* **5**, 163 (1968);

22. C. Cercignani and C. D. Pagani, in: *Rarefied Gas Dynamics* (C. L. Brundin, ed.), Vol. I, p. 555, Academic Press, New York (1967);
23. C. Cercignani and C. D. Pagani, *Phys. Fluids* **11**, 1395 (1968);
24. C. Cercignani, C. D. Pagani, and P. Bassanini, *Phys. Fluids* **11**, 1399 (1968);
25. P. Bassanini, C. Cercignani, and C. D. Pagani, *Int. J. Heat Mass Transfer* **11**, 1359 (1968).

The influence of the accommodation coefficients was studied in ref. 26 and in

26. C. Cercignani and C. D. Pagani, in: *Rarefied Gas Dynamics* (L. Trilling and H. Wachman, eds.), Vol. I, p. 269, Academic Press, New York (1969).

Sherman's formula was proposed in

27. F. S. Sherman, in: *Rarefied Gas Dynamics* (J. A. Laurmann, ed.), Vol. II, p. 228, Academic Press, New York (1963).

The experimental data reported in the different figures are taken from the following papers:

28. W. Dong, University of California Report UCRL 3353 (1956);
29. M. Knudsen, *Ann. Physik* **28**, 75 (1909);
30. W. P. Teagan and G. S. Springer, *Phys. Fluids* **11**, 497 (1968);
31. L. Kavanau, *Trans. ASME* **77**, 617 (1955);
32. K. Takao, in: *Rarefied Gas Dynamics* (J. A. Laurmann, ed.), Vol. II, p. 102, Academic Press, New York (1963);
33. R. A. Millikan, *Phys. Rev.* **22**, 1 (1923).

The application to a heat transfer problem of the variational principle for the integrodifferential form of the linearized Boltzmann equation was given in

34. C. Cercignani and J. Cipolla, in: *Rarefied Gas Dynamics* (D. Dini, ed.), Vol. II, p. 767, Edizioni Tecnico-Scientifiche, Pisa (1971).

Section 6——Discrete ordinate techniques have been proposed in many papers, among which we cite:

35. A. B. Huang and D. P. Giddens, in: *Rarefied Gas Dynamics* (C. L. Brundin, ed.), Vol. I, p. 481, Academic Press, New York (1967);
36. M. Wachman and B. B. Hamel, in: *Rarefied Gas Dynamics* (C. L. Brundin, ed.), Vol. I, p. 675, Academic Press, New York (1967).

For Monte Carlo methods see

37. J. K. Haviland and M. L. Lavin, *Phys. Fluids* **5**, 1399 (1962);
38. M. Perlmutter, in: *Rarefied Gas Dynamics* (C. L. Brundin, ed.), Vol. I, p. 455, Academic Press, New York (1967);
39. A. Nordsieck and B. L. Hicks, in: *Rarefied Gas Dynamics* (C. L. Brundin, ed.), Vol. I, p. 695, Academic Press, New York (1967);
40. J. O. Ballance, in: *Rarefied Gas Dynamics* (C. L. Brundin, ed.), Vol. I, p. 575, Academic Press, New York (1967);
41. G. A. Bird, in: *Rarefied Gas Dynamics* (J. H. de Leeuw, ed.), Vol. I, p. 216, Academic Press, New York (1965).

Details on these methods can be found in Kogan's book (ref. 10, Chapter I) and in

42. F. M. Devienne, ed., *Rarefied Gas Dynamics*, Pergamon Press, London (1960);
43. L. Talbot, ed., *Rarefied Gas Dynamics*, Academic Press, New York (1961);
44. J. A. Laurmann, ed., *Rarefied Gas Dynamics*, 2 vols., Academic Press, New York (1963);
45. J. H. de Leeuw, ed., *Rarefied Gas Dynamics*, 2 vols., Academic Press, New York (1965);
46. C. L. Brundin, ed., *Rarefied Gas Dynamics*, 2 vols., Academic Press, New York (1967);
47. L. Trilling and H. Wachman, eds., *Rarefied Gas Dynamics*, 2 vols., Academic Press, New York (1969);

48. D. Dini, ed., *Rarefied Gas Dynamics*, 2 vols., Edizioni Tecnico-Scientifiche, Pisa (1971);
49. K. Karamcheti, ed., *Rarefied Gas Dynamics*, 2 vols., Academic Press, New York (1974);
50. M. Becker and M. Fiebig, eds., *Rarefied Gas Dynamics*, 2 vols., DFVLR Press, Porz-Wahn (1974);
51. L. Potter, ed., *Rarefied Gas Dynamics*, 2 vols., AIAA, New York (1977);
52. R. Campargue, ed., *Rarefied Gas Dynamics*, 2 vols., CEA, Paris (1981);
53. S. S. Fisher, ed., *Rarefied Gas Dynamics*, 2 vols., AIAA, New York (1981);
54. O. Belotserkovski, M. N. Kogan, S. S. Kutateladze, and A. K. Rebrov, eds., *Rarefied Gas Dynamics*, 2 vols., Plenum Press, New York (1985);
55. H. Oguchi, ed., *Rarefied Gas Dynamics*, 2 vols., University of Tokyo Press, Tokyo (1984);
56. V. Boffi and C. Cercignani, eds., *Rarefied Gas Dynamics*, 2 vols., Teubner, Stuttgart (1986);
57. D. P. Weaver, E. P. Muntz, and R. Campbell, eds., *Rarefied Gas Dynamics*, 3 vols., AIAA, New York (1989);
58. G. A. Bird, *Molecular Gas Dynamics*, Clarendon Press, Oxford (1976).

AUTHOR INDEX

SUBJECT INDEX

9 780306 434600

MIX
Papier aus verantwortungsvollen Quellen
Paper from responsible sources
FSC® C105338

If you have any concerns about our products,
you can contact us on
ProductSafety@springernature.com

In case Publisher is established outside the EU,
the EU authorized representative is:
Springer Nature Customer Service Center GmbH
Europaplatz 3, 69115 Heidelberg, Germany

Printed by Libri Plureos GmbH
in Hamburg, Germany